Advanced Information and Knowledge Processing

Gregoris Mentzas, Dimitris Apostolou, Andreas Abecker and Ron Young
Knowledge Asset Management
1-85233-583-1

Michalis Vazirgiannis, Maria Halkidi and Dimitrios Gunopulos
Uncertainty Handling and Quality Assessment in Data Mining
1-85233-655-2

Asunción Gómez-Pérez, Mariano Fernández-López, Oscar Corcho
Ontological Engineering
1-85233-551-3

Arno Scharl (Ed.)
Environmental Online Communication
1-85233-783-4

Shichao Zhang, Chengqi Zhang and Xindong Wu
Knowledge Discovery in Multiple Databases
1-85233-703-6

Jason T.L. Wang, Mohammed J. Zaki, Hannu T.T. Toivonen and Dennis Shasha (Eds)
Data Mining in Bioinformatics
1-85233-671-4

C.C. Ko, Ben M. Chen and Jianping Chen
Creating Web-based Laboratories
1-85233-837-7

Manuel Grana, Richard Duro, Alicia d'Anjou and Paul P. Wang (Eds)
Information Processing with Evolutionary Algorithms
1-85233-886-0

Nikhil R. Pal and Lakhmi Jain (Eds)
Advanced Techniques in Knowledge Discovery and Data Mining
1-85233-867-9

Colin Fyfe

Hebbian Learning and Negative Feedback Networks

With 117 Figures

 Springer

Colin Fyfe
Applied Computational Intelligence Research Unit, The University of Paisley, UK

Series Editors
Xindong Wu
Lakhmi Jain

British Library Cataloguing in Publication Data
Fyfe, Colin
  Hebbian learning and negative feedback networks.–
  (Advanced information and knowledge processing)
  1. Neural Networks (Computer science)   2. Data mining
  I. Title
  006.3'2

Library of Congress Cataloging-in-Publication Data

AI&KP ISSN 1610-3947

ISBN 978-1-84996-945-1   e-ISBN 978-1-84628-118-1
Springer is a part of Springer Science+Business Media
springeronline.com

© Springer-Verlag London Limited 2010
Printed in the United States of America

34/3830-543210   Printed on acid-free paper

To my wife, Mary Teresa, for her unswerving help and support.

Contents

Part II Dual Stream Networks

Acronyms

ANN Artificial Neural Network
ASSOM Adaptive Subspace Self-Organising Map
CCA Canonical Correlation Analysis
ECA Exploratory Correlation Analysis
EPP Exploratory Projection Pursuit
FA Factor Analysis
HEPP Hierarchical Exploratory Projection Pursuit
ICA Independent Component Analysis
KCCA Kernel Canonical Correlation Analysis
KPCA Kernel Principal Component Analysis
LDA Linear Discriminant Analysis
MCA Minor Components Analysis
ML Maximum Likelihood
MOB Minimum Overcomplete Basis
MSE Mean Square Error
NLCCA Nonlinear Canonical Correlation Analysis
NLPCA Nonlinear Principal Components Analysis
NLPLS Nonlinear Partial Least Squares
PCA Principal Component Analysis
PFA Principal Factor Analysis
PLS Partial Least Squares
SIM Scale Invariant Map
SOM Self-Organising Map

Preface

This book is the outcome of a decade's research into a specific architecture and associated learning mechanism for an artificial neural network: the architecture involves negative feedback and the learning mechanism is simple Hebbian learning. The research began with my own thesis at the University of Strathclyde, Scotland, under Professor Douglas McGregor which culminated with me being awarded a PhD in 1995 [52], the title of which was "Negative Feedback as an Organising Principle for Artificial Neural Networks".

Naturally enough, having established this theme, when I began to supervise PhD students of my own, we continued to develop this concept and this book owes much to the research and theses of these students at the Applied Computational Intelligence Research Unit in the University of Paisley. Thus we discuss work from

- Dr. Darryl Charles [24] in Chapter 5.
- Dr. Stephen McGlinchey [127] in Chapter 7.
- Dr. Donald MacDonald [121] in Chapters 6 and 8.
- Dr. Emilio Corchado [29] in Chapter 8.

We briefly discuss one simulation from the thesis of Dr. Mark Girolami [58] in Chapter 6 but do not discuss any of the rest of his thesis since it has already appeared in book form [59]. We also must credit Cesar Garcia Osorio, a current PhD student, for the comparative study of the two Exploratory Projection Pursuit networks in Chapter 8. All of Chapters 3 to 8 deal with single stream artificial neural networks.

Chapters 9-14 discuss research into dual stream artificial neural networks at the Applied Computational Intelligence Research Unit in the University of Paisley. This work has resulted in four PhDs [60, 67, 108, 113]. I must therefore acknowledge the work done by:

- Dr. Pei Ling Lai [113] in Chapters 9, 10 and 11.
- Dr. Jos Koetsier [108] in Chapters 12 and 14.
- Dr. Zhenkun Gou [60] in Chapters 9, 11 and 13.

- Dr. Ying Han [67] in Chapters 12, 13 and 14.

Without their assistance this book could not have been written. We must also credit current PhD student, Emilio Corchado (yes, his second PhD) with some of the results in Chapter 14. We must also acknowledge other research students whose work does not form part of this book, but whose overall contribution to the life of the group was invaluable: these include Dr. Shang-Jen Chuang, Dr. Tzai-Der Wang, Dr. Juan Corchado, Dr. Danny Livingstone and Dr. Lina Petrakieva.

This book also has sections which were used for undergraduate teaching at the University of Paisley and I must credit these undergraduates with inspiring me to write more clearly.

Needless to say, we cannot cover all the work in these theses in a single book and so the interested reader is invited to consult the originals for more detailed description of any work which commands his or her interest.

Colin Fyfe
Paisley, 2004.

1

Introduction

We report, in this book, on the last decade's research into a single architecture of artificial neural networks. The research first identified the fact that a negative feedback artificial neural network using simple Hebbian learning had important statistical properties in that it could self-organise in order to identify the principal component filters of a data set. For readers unfamiliar with these terms, we will discuss them in more detail in Chapter 2. By adding bells and whistles to the basic architecture, we have discovered quite a bit about this very powerful architecture and have brought much of our research together in this single book. Since the book brings together the work of various PhD theses, it cannot give all the details which appear in these theses but seeks to emphasise the common theme underlying the research - the negative feedback network.

In order to make the book readable as a single entity, some of the nomenclature has been changed from the original so that a uniform notation is used throughout the book; for example, in Chapter 4, we use inputs $\mathbf{x}$, outputs $\mathbf{y}$ and residuals $\mathbf{e}$, not inputs $\mathbf{x}$, outputs $\mathbf{z}$ and residuals $\mathbf{y}$ which is how the work first appeared in [51] and [52]. However, these are presentation changes which do not affect the substance of the research. More important is the development of concepts which takes place throughout the book and the chapters are deliberately organised in such a way as to reflect our growing understanding of the capabilities of the negative feedback network.

Similarly, we first became aware of negative feedback networks from the PhD thesis of Mark Plumbley [153] whose work is briefly discussed in Appendix A. Plumbley used the biological term "interneuron"; we initially adopted this term and early papers continued to use it for some years, but we now, in this book, consistently use "output neuron" for these neurons. We remain interested in them as models of biological information processing but prefer now to discuss their relevance to engineering problems rather than their biological inspirations.

1.1 Artificial Neural Networks

An artificial neural network is typically a software simulation of the hardware on which we run. It is difficult to equate the simple machines which we have so-far been able to create with the sophisticated machines between our ears, yet, nevertheless, we are making advances in understanding what such machines can do and what are the essential features of such machines. An artificial neuron receives a number of floating point values either from the environment or from other neurons. These floating point values compose a vector of inputs which are modulated by being filtered by a set of weights. The weights model synaptic efficiencies, the synapse being the connection between one neuron and the next. Thus typically in this book, we will have a set of inputs, $\mathbf{x}$, which will reach the ith neuron through a set of weights, $\mathbf{w}_i$ to give an activation, y_i, at the ith neuron.

$$y_i = \mathbf{w}_i.\mathbf{x} = \sum_{j=1}^{N} w_{ij} x_j \tag{1.1}$$

where we have assumed there are N inputs. Even allowing for the fact that we often use nonlinear functions of the activations, we see that a single neuron is a very simple processor. The power of artificial neural networks comes from using very many neurons acting in concert so as to produce powerful information processors.

It should be clear that the weight values, w_{ij} are crucial in that they determine the magnitude of the effect of the signal in the jth input stream reaching the ith output neuron. The most interesting aspect of neural networks is how to find the best values of w_{ij} so that the most relevant information is passed on. In my opinion, the *most* interesting aspect of this most interesting aspect is finding rules so that an artificial neural network can *self-organise* so that the most relevant information can be determined automatically. As will become clear throughout this book, different decisions as to what constitutes relevance leads to different learning rules in artificial neural networks. However throughout this book we will use unsupervised learning - learning which does not use a teacher signal to learn from a data set. We know that such learning is both possible and powerful since each of us uses it every day.

Therefore, most of the models which we will discuss in this book will be models which retain biological plausibility - we will attempt to keep within the sphere of models which could correspond to a model possible *in vivo*. Such models will attempt to use only local information to self-organise; there will be no globally collected information available throughout the network. There will be no very tight dependencies or constraints on time or space in the models. In fact, all models will be based on variations of what is known as Hebbian learning (Chapter 2), which is believed to underpin much of our learning.

Having said that, as noted in the Preface, this work is the outcome of several PhD theses which were undertaken in departments of Computer Science and therefore have to model the concerns of that discipline. Thus, there is an on-going need to make models of artificial neural networks which have valuable engineering applications. Many of the networks we develop are powerful data analysis machines capable of data/text mining, revealing hidden structure, identifying outliers, clustering data sets and finding interesting low dimensional projections of high dimensional data sets.

Often in artificial neural network research, we create artificial data and run our networks on them before attempting experiments with real world data. The advantages of artificial data are

1. We have more control over our data. Real world data is often noisy with outliers and, in addition, we are also relying on the efficacy of our sampling methods.
2. We can more easily understand our results. We will create data with the specific aim of aiding our understanding of the network.
3. We know the answers we hope to get from our network. Our data has been specifically created to help test a particular hypothesis.
4. We can control the magnitude of our data set. We will use a limited number of representative samples.

While we cannot have our simulations prove our theories using artificial data, we can certainly use the artificial data to prove that a theoretical argument is incorrect [154]. Perhaps a more mundane argument is also often used: if we cannot have our networks converge to the correct values on nice, clean artificial data, what chance is there on real, noisy data sets with missing values, outliers, etc. However, clearly there is an inherent danger in this in that each researcher is liable to use data sets in which his/her own algorithm is seen in the best light. Therefore, both real and artificial data sets have their place in data analysis.

1.2 The Organisation of this Book

This work is in two parts: the first part deals with methods which extract information from a single stream of data using negative feedback and Hebbian learning while the second part deals with methods for extracting information which is shared over two data streams simultaneously.

Chapter 2 sets the scene for the remainder of the book. It provides a short review of Hebbian learning, Information Theory and Principal Components Analysis, three of the pillars on which the book is based. We also discuss four artificial neural networks which perform Principal Component Analysis. Finally, we introduce Independent Component Analysis and discuss one seminal artificial neural network which performs ICA.

Chapter 3 introduces the negative feedback network, illustrates its capabilities on some simple data sets and investigates a number of variations of the network which perform a Principal Component Analysis of the data. We also change the learning rules of the network so that it now performs a Minor Component Analysis and use this on two simple problems. We have opted to put the more theoretical aspects of this chapter into a separate section since much of this is also standard fare in many textbooks and we wish to keep the exposition of the basic network clear.

Chapter 4 investigates under what conditions lateral connections which also use Hebbian learning to self-organise can be used to force convergence to the Principal Component filters. Two main methods introduce asymmetry into the network but it is shown that we require specific asymmetries to force convergence to the actual principal components.

Chapter 5 introduces a constraint to the basic network introduced in Chapter 3: the constraint involves either keeping the weights non-negative or keeping the outputs non-negative. We discuss the work in this Chapter both in terms of a constrained Principal Component Analysis and in terms of Factor Analysis. We illustrate the network's convergence on a standard problem involving the extraction of a single cause from a data set composed of a number of interfering causes before using the network on real data sets.

Chapter 6 introduces a variation of the network which performs Exploratory Projection Pursuit, a technique for finding structure in high dimensional data sets. We derive several functions which seek for specific types of structure in data sets and illustrate the network's properties on both artificial and real data.

Chapter 7 introduces several variations on the basic network which perform a topology-preserving quantisation of data sets. We show that these methods find quite different quantisations from Kohonen's Self-organising Map (SOM) [111].

Chapter 8 introduces another Exploratory Projection Pursuit network from a different perspective to that discussed in Chapter 6. The change to the basic network is first discussed in terms of its connection to Principal Component Analysis before we discuss its application to Exploratory Projection Pursuit. We compare the algorithm in this Chapter with that of Chapter 6 and then combine them in an algorithm which seems to get the best of all worlds.

The second part of this book deals with dual stream architectures.

Chapter 9 introduces two artificial neural networks which perform Canonical Correlation Analysis (CCA), a statistical method for extracting information from two data streams. Simulations on real and artificial data sets sets the scene for the remainder of the book. In Chapter 10, we digress slightly to investigate alternative derivations of similar learning rules for artificial neural networks from a variety of perspectives: we show how a probabilistic formulation of the problem enables us to derive a robust form of the learning rules and how a previous researcher's algorithm can be modified so that it too can be seen to be a member of the same family of algorithms.

Yet such networks are perhaps no more than interesting oddities: we have after all several existing methods of performing CCA. Therefore in Chapter 11, we introduce nonlinearities in several ways to the existing networks and show that the resulting networks can do more than simple linear CCA. In particular, we use the resulting networks for the "blind separation of sources". In Chapter 13, we consider the problems associated with multicollinearity which occurs when there exist internal correlations within a data stream. Several algorithms for solving this are formed and the best used to extract meaningful structure from a functional data set.

In Chapter 12, we extend a single stream method which looks for higher order structure in a single data set so that the resulting method, Exploratory Correlation Analysis, can find higher order structure shared between two data streams. We repeat the process with the other exploratory single stream method from Chapter 8. A comparison of the two methods of this chapter gives a revealing insight into a family of algorithms which vary in their responses to twinned information and higher order information.

Readers familiar with the background can skip Chapter 2, however we recommend that all readers at least skim Chapter 3 in Part I and Chapter 9 in Part II. The other Chapters are designed to be mainly self-contained though towards the end of Chapter 8, a knowledge of Chapter 6 is necessary and both of these chapters are precursors to Chapter 12.

In Chapter 14, we discuss the single stream method of Principal Curves. By twinning the Principal Curves method we create matching curves in two data spaces simultaneously each of which has something to say about the information in the other. The curves are used in a forecasting role and various extensions of the original algorithm are given.

Chapter 15 speculates on what the future might hold for this negative feedback architecture and where new research in this area might lead.

We have five appendices:

1. Appendix A discusses other negative feedback models.
2. Appendix B discusses other models which deal with multiple cause data (Chapter 5).
3. Appendix C discusses other models which deal with independent component analysis.
4. Appendix D discusses other models of dual stream approaches.
5. Appendix E collects together the various data sets which have been used throughout the book.

Finally a word on notation. We have used a boldface font with lowercase symbols for vectors e.g. $\mathbf{x}, \mathbf{w}$ etc., and uppercase symbols such as W for matrices. Sometimes we wish to identify a particular column of a matrix and to do this we have used the convention that the ith column of W is called $\mathbf{w}_i$; if we go on to identify particular elements of this column vector, we use w_{ij}.

Part I

Single Stream Networks

Chapters 2 to 8 develop artificial neural networks which self-organise on a stream of data which is dealt with as an holistic entity i.e.even though we may be seeking independent factors (Chapter 5) or independent components Chapters 6 or 8), we treat all the input data at any one time as a single entity. This is in strict contrast to every network of Chapters 9 to 14 in which we specifically consider two or more distinct streams of data at each instant in time. Single stream artificial neural networks are by far the most common in the literature.

We set the scene in Chapter 2 by introducing a number of topics which are used throughout the book, particularly Principal Component Analysis and Artificial Neural Networks which find principal components, Information Theory and Independent Components Analysis. In this Chapter, we discuss feedforward neural networks which have a feedback element in their learning rules while in subsequent chapters, all networks have a specific feedforward stage followed by a feedback stage followed by a phase during which weights are updated using simple Hebbian learning.

We have a very gentle progression of ideas throughout these chapters:

In Chapter 3, we introduce the basic negative feedback network and show analytically and experimentally that it self-organises to find the principal components of input data. Any residual left at the inputs contains the minimum squared error with which it is possible to be left, after using a linear network to compress the data. We show various extensions of the basic network, perhaps the most interesting of which is that having different feedforward weights than feedback weights. Subsequently, in Chapter 4, we use trainable lateral connections to ensure convergence to the actual principal components (rather than to the subspace spanned by the principal components). In Chapters 5 onwards, we tend mostly to ignore the fact that we may have different feedforward and feedback rules to make the exposition simpler; nevertheless, most of the networks described in this part will function very well with such separate trainable weights.

● We use the networks developed in Chapters 5, 6 and 8 to search for independence. The first of these chapters develops a network which we relate to the statistical technique of Factor Analysis. The second and third chapters develop networks which we link to the statistical technique of Exploratory Projection Pursuit. The networks in Chapter 5 find underlying causes in a data stream when such causes have been OR-ed together. One of the most interesting findings in that chapter is the fact that noise may be used to determine the number of underlying factors in a data set in an entirely automatic manner. Since the PhD theses upon which this discussion was based were Computer Science rather than neurology, the relevance to biological information processing which is inherently noisy is not stressed yet it is nevertheless obvious. The Exploratory Projection Pursuit network in Chapter 6 predates that of Chapter 8 and recent experiments have tended to demonstrate that it is somewhat superior. However, it is interesting that these two techniques can be easily merged so that at the end of Chapter 8 a combined algorithm is developed. The networks in Chapters 6 and 8 are both used for Independent Component Analysis.

- In Chapter 7, we develop three negative feedback networks which self-organise in order to preserve some aspects of the topology of the data. The idea of topology preservation is somewhat ubiquitous in early sensory information processing but again, since the algorithms were developed in a Computer Science environment rather than a Neurology department, the connection with biological neural networks is merely mentioned and not stressed.

While we have written this book as though these topics were independent of one another, in practice various researchers have combined two or more of these techniques though, again to make the exposition clearer, we have not discussed many such combinations in this book. However, later PhD theses have inevitably built on those which have predated them and interested readers may consult the later theses for examples of such combinations.

To make clear the distinction between these chapters and those of Part II, in this Part we are dealing with a single stream of input data whereas in Part II, we deal with two sets of inputs simultaneously at each instant of time. We will see later that many of the methods and analyses developed in Part I can also be usefully deployed in the context of dual stream networks but we strictly adhere to the single stream of data at all times in Part I.

2

Background

In this chapter, we introduce

- Hebbian learning and discuss its ability to perform statistical operations;
- Information Theory and the Gaussian distribution very briefly;
- Principal Component Analysis (PCA);
- a brief survey of the major most-popular Artificial Neural Networks(ANNs) which perform a PCA.

We will not, in this chapter, provide proofs of convergence of the various nets discussed since such proofs are very similar to those we use in Chapter 3 to prove convergence of the negative feedback network and are available in any case in many other good textbooks (e.g.[71, 73]). We begin by outlining the simplest possible ANNs and review a very simple unsupervised learning rule.

2.1 Hebbian Learning

The aim of unsupervised learning is to present a neural net with raw data and allow the net to make its own representation of the data - hopefully retaining all information which we humans find important. Unsupervised learning in neural nets is generally realised by using a form of Hebbian learning which is based on a proposal by Donald Hebb [72] who wrote:

When an axon of cell A is near enough to excite a cell B and repeatedly or persistently takes part in firing it, some growth process or metabolic change takes place in one or both cells such that A's efficiency, as one of the cells firing B, is increased.

Neural nets which use Hebbian learning are characterised by making the activation of a unit depend on the sum of the weighted activations which feed into the unit. They use a learning rule for these weights which depends on the strength of the simultaneous activation of the sending and receiving neurons. These conditions are usually written as

$$y_i = \sum_j w_{ij} x_j \tag{2.1}$$

$$\text{and } \Delta w_{ij} = \eta x_j y_i \tag{2.2}$$

the latter being the learning mechanism. Here y_i is the output from neuron i, x_j is the jth input, and w_{ij} is the weight from x_j to y_i. η is known as the learning rate and is usually a small scalar which may change with time. Note that the learning mechanism says that if x_j and y_i fire simultaneously, then the weight of the connection between them will be strengthened in proportion to their strengths of firing. However, we will not, as does Kosko [112], rename the Hebbian learning rule when an activation function is used. i.e.when

$$y_i = g\left(\sum_j w_{ij} x_j \right) \tag{2.3}$$

$$\text{and } \Delta w_{ij} = \eta x_j y_i \tag{2.4}$$

for some function, $g()$, we will still call this Hebbian learning.

Substituting (2.1) into (2.2), we can write the Hebb learning rule as

$$\Delta w_{ij} = \eta x_j \sum_k w_{ik} x_k = \eta \sum_k w_{ik} x_k x_j \tag{2.5}$$

$$\text{which is equivalent to } \frac{d}{dt} W(t) \propto CW(t) \tag{2.6}$$

where C_{ij} is the correlation coefficient calculated over all input patterns between the ith and j^{th} terms of the inputs and $W(t)$ is the matrix of weights at time t. In moving from the stochastic equation (2.5) to the averaged differential equation (2.6), we must place certain constraints on the process, particularly on the learning rate η. These are usually taken to be $\eta_k \geq 0, \sum \eta_k^2 < \infty, \sum \eta_k = \infty$ (see e.g.[129]). The advantage of this formulation is that it emphasises the fact that the resulting weights depend on the second order statistical properties of the input data. A review of the importance of this aspect of the Hebbian learning rule is given in Section 2.3.

Because of these statistics-based properties, Hebbian learning has found applications in a number of early associative-type memories e.g.Steinbuch's Learning Matrix [173], Anderson's linear associative memory [2], Kohonen's Adaptive Associative Memory [109] and the model of Willshaw et al [184].

However, a major difficulty with this learning rule is that unless there is some limit on the growth of the weights, they tend to grow without bound: we have a positive feedback loop - a large weight will produce a large value of y (2.1) which will produce a large increase in the weight (2.2). It is instructive to follow e.g.[73], in examining the Hebb rule's stability:

Recall first that a matrix A has an eigenvector $\mathbf{x}$ with a corresponding eigenvalue λ if

$$Ax = \lambda \mathbf{x}$$

In other words, multiplying the vector $\mathbf{x}$ or any of its multiples by A is equivalent to multiplying the whole vector by a scalar λ. Thus the direction of $\mathbf{x}$ is unchanged — only its magnitude is affected.

Consider a one output neuron network and assume that the Hebbian learning process does cause convergence to a stable direction, $\mathbf{w}^*$; then if w_k is the weight value linking x_k to y,

$$0 = E(\Delta w_i^*) = E(yx_i) = E\left(\sum_j w_j x_j x_i\right) = \sum_j R_{ij} w_j$$

where $E()$ indicate the expected value taken over the distribution and R is the correlation matrix of the distribution. Now this happens for all i, so $R\mathbf{w} = 0$. The correlation matrix, R, is a symmetric, positive semidefinite matrix and so all its eigenvalues are non-negative. But the above formulation shows that $\mathbf{w}^*$ must have eigenvalue 0. Now consider a small disturbance, ϵ, in the weights in a direction with a nonzero (i.e. positive) eigenvalue. Then

$$E(\Delta w*) = R(\mathbf{w}^* + \epsilon) = R\epsilon = \lambda\epsilon > 0$$

i.e. the weights will grow in any direction with nonzero eigenvalue (and such directions must exist). Thus there exists a fixed point at $W = 0$, but this is an unstable fixed point. In fact, it is well known that, in time, the weights of nets which use simple Hebbian learning tend to be dominated by the direction corresponding to the largest eigenvalue of the correlation matrix. We will often implicitly discuss zero mean data in this book (it is simple to make this a valid assumption for any data set) and so we will equate the correlation matrix of a data set with its covariance matrix.

We will later discuss in detail one of the major ways of limiting this growth of weights while using Hebbian learning and review its important side effects. However, we begin with short reviews of three subjects which will be important in the development of this book: Information Theory, Principal Component Analysis and Independent Component Analysis.

2.2 Quantification of Information

Shannon [168] devised a measure of the information content of an event in terms of the probability of the event happening. He wished to parameterise the intuitive concept that the occurrence of an unlikely event tells you more than that of a likely event. He defined the information in an event i, to be $-\log p_i$ where p_i is the probability that the event labelled i occurs.

Using this, we define the entropy (or uncertainty or information content) of a set of N events to be

$$H = -\sum_{i=1}^{N} p_i \log p_i$$

That is, the entropy is the information we would expect to get from one event happening where this expectation is taken over the ensemble of possible outcomes.

For a pair of random variables X and Y, if $p(i,j)$ is the joint probability of X taking on the ith value and Y taking on the jth value, we define the entropy of the joint distribution as

$$H(X,Y) = -\sum_{i,j} p(i,j) \log p(i,j)$$

Similarly, we can define the conditional entropy (or equivocation or remaining uncertainty in X if we are given Y) as:

$$H(X|Y) = -\sum_{i,j} p(i,j) \log p(i|j)$$

Shannon also showed that if X is a transmitted signal and Y is the received signal, then the information which receiving Y gives about X is

$$I(X;Y) = H(X) - H(X|Y) \tag{2.7}$$
$$\text{or } I(X;Y) = H(Y) - H(Y|X) \tag{2.8}$$
$$\text{or } I(X;Y) = H(X) + H(Y) - H(X,Y) \tag{2.9}$$

Because of the symmetry of the above equations, this term is known as the mutual information between X and Y.

The channel capacity is defined to be the maximum value over all possible values of X and Y of this mutual information.

The basic facts in which we will take an interest are the following:

- Because the occurrence of an unlikely event has more information than that of a likely event, it has a higher information content.
- Hence, a data set with high variance is liable to contain more information than one with small variance.
- A channel of maximum capacity is defined by 100% mutual information i.e. $I(X;Y) = H(X)$

2.2.1 Entropy and the Gaussian Distribution

Let us attempt to find the distribution which has the greatest entropy. This task means little in this form since we can merely keep adding points to the distribution to increase the uncertainty/entropy in the distribution. We must constrain the problem in some way before it is soluble.

Haykin [71] puts it this way:

With the differential entropy of a random variable X defined by

$$h(X) = -\int_{-\infty}^{\infty} f(x) \log f(x) dx \qquad (2.10)$$

find the probability density function $f(x)$ for which $h(X)$ is a maximum, subject to the two constraints,

$$\int_{-\infty}^{\infty} f(x) dx = 1 \qquad (2.11)$$

and

$$\int_{-\infty}^{\infty} (x - \mu)^2 f(x) dx = \sigma^2 = \text{a constant.} \qquad (2.12)$$

where μ is the mean of the distribution and σ^2 is its variance.

The first constraint simply ensures that the function $f()$ is a proper probability density function; the second constrains the variance of the distribution. We will show that the distribution with greatest entropy for a given variance is the Gaussian distribution: there is more uncertainty/information in a Gaussian distribution than in any other comparable distribution.

So we have an optimisation problem (maximise the entropy) under certain constraints. We incorporate the constraints into the optimisation problem using Lagrange multipliers so that we wish to find the maximum of

$$-\int_{-\infty}^{\infty} f(x) \log f(x) dx + \lambda_1 \int_{-\infty}^{\infty} f(x) dx + \lambda_2 \int_{-\infty}^{\infty} (x - \mu)^2 f(x) dx$$

$$= -\int_{-\infty}^{\infty} \{f(x) \log f(x) - \lambda_1 f(x) - \lambda_2 (x - \mu)^2 f(x)\} dx$$

where λ_1 and λ_2 are the Lagrange multipliers. This maximum is achieved when the derivative of the integrand with respect to the function $f(x)$ is zero. i.e. when

$$0 = -1 - \log f(x) + \lambda_1 + \lambda_2 (x - \mu)^2$$
$$\log f(x) = -1 + \lambda_1 + \lambda_2 (x - \mu)^2 \qquad (2.13)$$
$$f(x) = \exp(-1 + \lambda_1 + \lambda_2 (x - \mu)^2)$$

Substituting this into (2.11) and (2.12) gives

$$\int_{-\infty}^{\infty} \exp(-1 + \lambda_1 + \lambda_2 (x - \mu)^2) dx = 1$$

$$\int_{-\infty}^{\infty} (x - \mu)^2 \exp(-1 + \lambda_1 + \lambda_2 (x - \mu)^2) dx = \sigma^2$$

which gives us two equations in the two unknowns λ_1 and λ_2 which can be solved to give

$$\lambda_1 = 1 - \log(\sqrt{2\pi\sigma^2})$$
$$\lambda_2 = -\frac{1}{2\sigma^2}$$

which can be inserted in (2.13) to give

$$f(x) = \frac{1}{\sqrt{2\pi}\sigma} \exp\left(-\frac{(x-\mu)^2}{2\sigma^2}\right) \tag{2.14}$$

the probability density function of a Gaussian distribution. When we use this to calculate the entropy of the Gaussian distribution we get

$$h(X) = \frac{1}{2}\{1 + \log(2\pi\sigma^2)\} \tag{2.15}$$

In summary, we have shown the following

1. The Gaussian distribution is the distribution with the greatest entropy for a given variance: if X and Y are both random variables with a given variance σ^2 and if X is a Gaussian random variable, then

$$h(X) \geq h(Y) \tag{2.16}$$

2. The entropy of a Gaussian random variable is totally determined by its variance. We will later see that this is not true for other distributions.

2.3 Principal Component Analysis

Inputs to a neural net generally exhibit high dimensionality i.e.the N input lines can each be viewed as one dimension so that each pattern will be represented as a coordinate in N-dimensional space.

A major problem in analysing data of high dimensionality is identifying patterns which exist across dimensional boundaries. Such patterns may become visible when a change of basis of the space is made, however an *a priori* decision as to which basis will reveal most patterns requires fore-knowledge of the unknown patterns.

A potential solution to this impasse is found in Principal Component Analysis (PCA) which aims to find the orthogonal basis which maximises the variance of the projection of the data onto the basis for a given dimensionality of basis. The usual tactic is to find the filter which accounts for most of the data's variance; this becomes the first basis vector. One then finds the direction which accounts for most of the remaining variance; this is the second basis vector and so on. If one then projects data onto the Principal Component directions, we perform a dimensionality reduction which will be accompanied by the retention of as much variance in the data as possible.

In general, it can be shown [93] that the kth basis vector from this process is the same as the kth eigenvector of the covariance matrix, C, where

$$c_{ij} = E[(x_i - E(x_i))(x_j - E(x_j))]$$

For zero-mean data, the covariance matrix is equivalent to a simple correlation matrix. Of course, it is difficult to show high-dimensional data on these pages but a two-dimensional example is shown in Fig. 2.1.

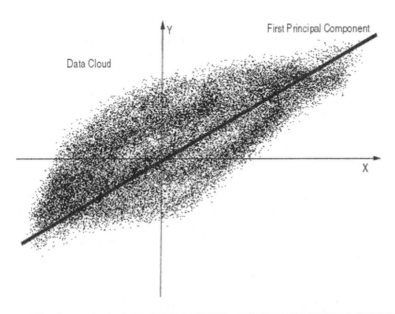

Fig. 2.1. The first principal component direction of a two-dimensional data cloud.

Now, if we have a set of weights which are the eigenvectors of the input data's covariance matrix, C, then these weights will transmit the largest values to the outputs when an item of input data lies on those eigenvectors with the largest eigenvalues. Thus, if we can create a situation in an artificial neural network where one set of weights (into a particular output neuron) converges to the first eigenvector (corresponding to the largest eigenvalue), the next set of weights converges to the second eigenvector, and so on, we will be in a position to maximally recreate at the outputs the projections with the largest variance in the input data.

Note that representing data as coordinates using the basis found by a PCA means that the data will have greatest variance along the first principal component, the next greatest variance along the second, and so on. While it is strictly only true to say that information and variance may be equated in Gaussian distributions, it is a good rule-of-thumb that a direction with more variance contains more information than one with less variance. Thus, PCA provides a means of compressing the data whilst retaining as much information within the data as possible. It can be shown that if a set of Gaussian data has eigenvalues $\{\lambda_1, \lambda_2, ..., \lambda_n\}$ and if we represent the data in coordinates on a basis spanned by the first m eigenvectors, the loss of information due to the compression is

$$E = \sum_{i=m+1}^{n} \lambda_i \qquad (2.17)$$

Artificial neural networks and PCA come together in two ways:

1. There are some networks which use Principal Components as an aid to learning e.g.[82],
2. Some networks have been explicitly designed to (and in fact, do) calculate Principal Components.

It is the latter in which we are most interested.

2.4 Weight Decay in Hebbian Learning

As noted in Section 2.1, if there are no constraints placed on the growth of weights under Hebbian learning, there is a tendency for the weights to grow without bounds. It is possible to renormalise weights after each learning epoch; however, this adds an additional operation to the network's processing.

Another possibility is to allow the weights to grow until each reaches some limit [119] e.g. have an upper limit of w^+ and a lower limit of w^- and clip the weights when they reach either of these limits. Clearly a major disadvantage of this is that if all weights end up at one or other of these limits[1] the amount of information which can be retained in the weights is very limited.

A third possibility is to prune weights which do not seem to have importance for the network's operation. However, this is an operation which must be performed using nonlocal knowledge: typically which weights are of much smaller magnitude than their peers.

Hence, interest has grown in the use of decay terms embedded in the learning rule itself (e.g. [126], Chapter 17). Ideally such a rule should ensure that no single weight should grow too large while keeping the total weights on connections into a particular output neuron fairly constant. One of the simplest forms of weight decay was developed as early as 1968 by Grossberg[61] and was of the form

$$\frac{dw_{ij}}{dt} = \eta y_i x_j - w_{ij} \qquad (2.18)$$

It is clear that the weights will be stable (when $\frac{dw_{ij}}{dt} = 0$) at the points where $w_{ij} = \eta E(y_j x_i)$ where $E(.)$ indicates an ensemble average. Using a similar type of argument to that employed for simple Hebbian learning, we see that at convergence we must have $\eta C\mathbf{w} = \mathbf{w}$. Thus $\mathbf{w}$ would have to be

---

[1] This will certainly happen if simple Hebbian learning is used

an eigenvector of the correlation matrix of the input data with corresponding eigenvalue $\frac{1}{\eta}$. We shall be interested in a somewhat more general result.

Grossberg went on to develop more sophisticated learning equations which use weight decay e.g. for his instar coding, [63] he has used

$$\frac{dw_{ij}}{dt} = \eta\{y_i - w_{ij}\}x_j \qquad (2.19)$$

where the decay term is gated by the input term x_j and for outstar coding

$$\frac{dw_{ij}}{dt} = \eta\{x_j - w_{ij}\}y_i \qquad (2.20)$$

where the decay term is gated by the output term y_i. These rules, while still falling short of the decay in which we will be interested, show that researchers of this era were beginning to think of both differentially weighted decay terms and allowing the rate of decay to depend on the statistics of the data presented to the network.

2.4.1 Principal Components and Weight Decay

Miller and MacKay [132] have provided a definitive study of the results of a decay term on Hebbian learning. They suggest an initial distinction between Multiplicative Constraints and Subtractive Constraints. They define Multiplicative Constraints as those satisfying

$$\frac{d}{dt}\mathbf{w}(t) = C\mathbf{w}(t) - \gamma(\mathbf{w})\mathbf{w}(t)$$

where the decay in the weights is governed by the product of a function of the weights, $\gamma(\mathbf{w})$, and the weights, $\mathbf{w}(t)$, themselves. The decay term can be viewed as a feedback term which limits the rate of growth of each weight in proportion to the size of the weight itself while the first term defines the Hebbian learning itself.

Subtractive Constraints are satisfied by equations of the form

$$\frac{d}{dt}\mathbf{w}(t) = C\mathbf{w}(t) - \epsilon(\mathbf{w})\mathbf{n}$$

where the decay in the weights is governed by the product of a function of the weights , $\epsilon(\mathbf{w})$, and a constant vector, $\mathbf{n}$, (which is often $\{1, 1, ..., 1\}^T$). They prove that

- Hebbian rules whose decay is governed by Multiplicative Constraints will, in cases typical of Hebbian learning, ensure that the weights will converge to a stable point.
- This stable point is a multiple of the principal eigenvector of the covariance matrix of the input data.

- Hebbian rules governed by Subtractive Constraints will tend to lead to saturation of the weights at their extreme permissible values[2].
- Under Subtractive Constraints, there is actually a fixed point within the permitted hypercube of values, but this is unstable and is only of interest in anti-Hebbian learning (see below).
- If specific limits (w^+ and w^-) do not exist, weights under Subtractive Constraints will tend to increase without bound.

In summary then, Subtractive Constraints offer little that cannot be had from simple clipping of the weights at preset upper and lower bounds. Multiplicative Constraints, however, seem to give us not just weights which are conveniently small, but also weights which are potentially useful since

$$y_i = \sum_j w_{ij} x_j = \mathbf{w}_i . \mathbf{x}$$

where $\mathbf{w}_i$ is the vector of weights into neuron y_i and $\mathbf{x}$ is the vector of inputs. But

$$\mathbf{w}_i . \mathbf{x} = |\mathbf{w}_i| |\mathbf{x}| \cos \theta$$

where $|\mathbf{d}|$ is the length of $\mathbf{d}$ and θ is the angle between the two vectors.

This is maximised when the angle between the vectors is 0. Thus, if $\mathbf{w}_1$ is the weight into the first neuron which converges to the first Principal Component, the first neuron will maximally transmit information along the direction of greatest correlation, the second along the next largest, etc. In Section 2.3, we noted that these directions were those of greatest variance which from Section 2.2, we are equating with those of maximal information transfer through the system.

Given that there are statistical packages which find Principal Components, we should ask why it is necessary to reinvent the wheel using artificial neural networks. There are two major advantages to PCA using ANNs:

1. Traditional statistical packages require us to have available, prior to the calculation, a batch of examples from the distribution being investigated. While it is possible to run the ANN models with this method ("batch mode") ANNs are capable of performing PCA in real-time i.e. as information from the environment becomes available we use it for learning in the network. We are, however, really calculating the Principal Components of a sample, but since these estimators can be shown to be unbiased and to have variance which tends to zero as the number of samples increases, we are justified in equating the sample PCA with the PCA of the distribution. The adaptive/recursive methodology used in ANNs is particularly important if storage constraints are important.

---

[2] Such values may be partially determined by the eigenvalues of the covariance matrix but are not, in general, multiples of the eigenvectors.

2. Strictly, PCA is only defined for stationary distributions. However, in realistic situations, it is often the case that we are interested in compressing data from distributions which are a function of time; in this situation, the sample PCA outlined above is the solution in that it tracks the moving statistics of the distribution and provides as close to a local PCA as possible in the circumstances. However, most proofs of PCA ANNs convergences require the learning rate to converge to 0 in time and, in practice, it is the case that convergence is often more accurate when the learning rate tends to decrease in time. This would preclude an ANN following a distribution's statistics, which is an example of the well-known trade-off between tracking capability and accuracy of convergence.

We now look at several ANN models which use weight decay with the aim of capturing Principal Components. We will make no attempt to be exhaustive since that would in itself require a book; we do, however, attempt to give the most important examples of current network types and, in particular, four main models which illustrate the historical development of PCA models in unsupervised Artificial Neural Networks.

2.5 ANNs and PCA

The importance of the work of Oja cannot be overstated in this context. Thus we begin with his one-neuron model [137], followed by his subspace model [138] and his weighted subspace model [142, 143] before introducing Sanger's deflationary network [160]. We include no proofs in this section since these models are well known in the literature.

2.5.1 Oja's One Neuron Model

Oja [137] proposed a model which extracts the largest principal component from the input data. He suggested a single-output neuron which sums the inputs in the usual fashion

$$y = \sum_{i=1}^{m} w_i x_i$$

His variation on the Hebb rule contains a decay term

$$\Delta w_i = \eta(x_i y - y^2 w_i)$$

Note that this is a rule defined by Multiplicative Constraints ($y^2 = \gamma(w)$) and so will converge to the principal eigenvector of the input covariance matrix. The weight decay term has the simultaneous effect of making $\sum w_i^2$ tend towards 1 i.e. the weights are normalised.

However, this rule will find only the first eigenvector (that direction corresponding to the largest eigenvalue) of the data. It is not sufficient to simply throw clusters of neurons at the data since all will find the same (first) Principal Component; in order to find other PCs, there must be some interaction between the neurons.

2.5.2 Oja's Subspace Algorithm

Oja's Subspace Algorithm [138] provided a major step forward. The network has N output neurons, each of which learns using a Hebb-type rule with weight decay. Note, however, that it does not guarantee finding the actual directions of the Principal Components; the weights *do* however converge to an orthonormal basis of the Principal Component Space. We will call the space spanned by this basis the Principal Subspace. The learning rule is

$$\Delta w_{ij} = \eta \left(x_j y_i - y_i \sum_k w_{kj} y_k \right) \tag{2.21}$$

which has been shown to force the weights to converge to a basis of the Principal Subspace[3].

One advantage of this model compared with some other networks (e.g. [160]) is that it is completely homogeneous i.e. the operations carried out at each neuron are identical. This is essential if we are to take full advantage of parallel processing.

The major disadvantage of this algorithm is that it finds only the Principal Subspace of the eigenvectors, not the actual eigenvectors themselves.

2.5.3 Oja's Weighted Subspace Algorithm

The final stage is the creation of algorithms which find the actual Principal Components of the input data. In 1992, Oja et al [142, 143] recognised the importance of introducing asymmetry into the weight decay process in order to force weights to converge to the Principal Components. The algorithm is defined by the equations

$$y_i = \sum_{i=1}^n w_{ij} x_j$$

where a Hebb-type rule with weight decay modifies the weights according to

$$\Delta w_{ij} = \eta y_i \left(x_j - \theta_i \sum_{k=1}^N y_k w_{kj} \right)$$

---

[3] In this case $\gamma(w_{ij}) = y_i^2$. However, the additional weight decay constraints from the other outputs $y_i \sum_{k \neq i} w_{kj} y_k$ force decay in the directions of other eigenvectors. Therefore, the total of the decay parameters only forces weight convergence to the subspace.

Ensuring that $\theta_1 < \theta_2 < \theta_3 < \cdots$ allows the neuron whose weight decays proportional to θ_1 (i.e. whose weight decays least quickly) to learn the principal values of the correlation in the input data. That is, this neuron will respond maximally to directions parallel to the principal eigenvector, i.e. to patterns closest to the main correlations within the data. The neuron whose weight decays proportional to θ_2 cannot compete with the first, but it is in a better position than all of the others and so can learn the next largest chunk of the correlation, and so on. Empirically, it has been found that it is essential that the values of θ_i do not stray too far from 1: values of 0.9, 1.0 and 1.1 give convergence but 0.1, 0.2 and 0.3 do not.

It can be shown that the weight vectors will converge to the principal eigenvectors in the order of their eigenvalues. The algorithm clearly satisfies Miller and Mackay's definition of Multiplicative Constraints with $\gamma(w_i) = \theta_i \sum_k y_k w_{kj} x_j$.

2.5.4 Sanger's Generalized Hebbian Algorithm

Sanger [160] has developed a different algorithm (which he calls the "Generalized Hebbian Algorithm") which also finds the actual Principal Components. He also introduces asymmetry in the decay term of his learning rule:

$$\Delta w_{ij} = \eta \left(x_j y_i - y_i \sum_{k=1}^{i} w_{kj} y_k \right) \qquad (2.22)$$

Note that the crucial difference between this rule and Oja's Subspace Algorithm is that the decay term for the weights into the ith neuron is a weighted sum of the first i neurons' activations. Sanger's algorithm can be viewed as a repeated application of Oja's One Neuron Algorithm by writing it as

$$\Delta w_{ij} = \eta \left([x_j y_i - y_i \sum_{k=1}^{i-1} w_{kj} y_k] - y_i^2 w_{ij} \right) \qquad (2.23)$$

We see that the central term comprises the residuals after the first $i - 1$ Principal Components have been found, and therefore the rule is performing the equivalent of One Neuron learning on subsequent residual spaces. However, note that the asymmetry which is necessary to ensure convergence to the actual Principal Components, is bought at the expense of requiring the ith neuron to "know" that it is the ith neuron by subtracting only i terms in its decay. It is Sanger's contention that all true PCA rules are based on some measure of deflation such as shown in this rule. We will discuss deflationary networks in other contexts later in this book.

2.6 Anti-Hebbian Learning

All the ANNs we have so far met have been feedforward networks in that activation has been propagated only in one direction. However, many real biological networks are characterised by a plethora of recurrent connections. This has led to increasing interest in networks which, while still strongly directional, allow activation to be transmitted in more than one direction i.e. either laterally or in the reverse direction from the usual flow of activation. One interesting idea is to associate this change in direction of motion of activation with a minor modification to the usual Hebbian learning rule called anti-Hebbian learning (a definitive analysis of anti-Hebbian learning is given in [150]).

If inputs to a neural net are correlated, then each contains information about the other. In information theoretical terms, there is redundancy in the inputs ($I(x; y) > 0$).

Anti-Hebbian learning (Fig. 2.2) is designed to decorrelate neurons' outputs. The intuitive idea behind the process is that more information can be passed through a network when the nodes of the network are all dealing with different data. The less correlated the neurons' responses, the less redundancy is in the data transfer. Thus, the aim is to produce neurons which respond to different signals. If two neurons respond to the same signal, there is a measure of correlation between them and this is used to affect their responses to future similar data. Anti-Hebbian learning is sometimes known as lateral inhibition as this type of learning is generally used between members of the same layer and not between members of different layers. The basic model is defined by

$$\Delta w_{ij} = -\eta y_i y_j$$

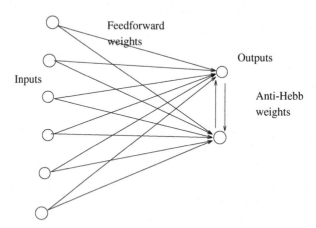

Fig. 2.2. Anti-Hebbian weights. Negative decorrelating weights between neurons in the same layer are learned using an "anti-Hebbian" learning rule.

Therefore, if initially, y_i and y_j are highly correlated then the weights between them will grow to a large negative value and each will tend to turn the other off.

It is clear that there is no need for weight decay terms or limits on anti-Hebbian weights as they are automatically self-limiting, provided decorrelation can be attained:

$$(E(y_i.y_j) \to 0) \Longrightarrow (\Delta w_{ij} \to 0) \tag{2.24}$$

i.e. weight change stops when the outputs are decorrelated. Success in decorrelating the outputs results in weights being stabilised.

It has been shown [159] that not only does anti-Hebbian learning force convergence in the particular case of a deflationary algorithm but that the lateral connections do indeed vanish.

The method is valid for all deflationary networks.

Several authors have developed Principal Component models using a mixture of one of the above PCA methods (often Oja's One Neuron Rule) and Anti-Hebbian weights between the output neurons e.g. [17, 18, 149, 158, 182].

We first note a similarity between the aims of PCA and anti-Hebbian learning: the aim of anti-Hebbian learning is to decorrelate neurons' responses. If a set of neurons performs a Principal Component Analysis, their weights form an orthogonal basis of the space of principal eigenvectors. Thus, both methods perform a decorrelation of the neurons' responses.

Further, in information-theoretic terms, decorrelation ensures that the maximal amount of information possible for a particular number of output neurons is transferred through the system. We will consider only noise-free information-transfer since if there is some noise in the system, some duplication of information may be beneficial to optimal information transfer [118].

2.7 Independent Component Analysis

There is however a more recent strand of research using artificial neural networks on the problem of separating out a single signal from a mixture of signals. This second strand deals mainly with continuous signals and has its roots in the world of signal processing. The problem is generally known as the "blind separation of sources" or sometimes "the cocktail party problem". The latter name is a reference to the human ability to extract a single voice from a mixture of voices: there is no simple algorithmic solution to this problem yet people have no difficulty following a conversation even when the conversation is embedded in multiple other conversations. The former name is in more general use: we wish to separate out a single source signal from a mixture of sources and/or noise. The problem is known as "blind" since we make (almost) no assumptions about the signals. A good reference is [85]. We have included some examples of methods for performing ICA in Appendix C.

We will consider only linear mixtures of signals.

The problem may be set up as follows: let there be N independent non-Gaussian signals $(s_1, s_2, \cdots, s_N)$ which are mixed using a (square) mixing matrix A to get N samples, x_i, each of which is an unknown mixture of the independent signals,

$$\mathbf{x} = A\mathbf{s} \tag{2.25}$$

There may in addition be noise added to the mixing process, but we shall ignore that for the time being. Then the aim is to use an artificial neural network to retrieve the original input signals when the only information presented to the network is the unknown mixture of the signals. The weights in the network will be W such that

$$\mathbf{y} = W\mathbf{x} \tag{2.26}$$

where the elements of $\mathbf{y}$ are the elements of the original signal *in some order*, i.e. we are not insisting that the first output of our neural network is equal to the first signal, the second equal to the second signal, and so on. We merely insist that neuron i's output is one of the N signals uncontaminated by any of the other signals. Neural and quasineural methods of performing this task are known as Independent Component Analysis networks (ICA) and are often thought of as extensions of PCA networks.

However, we did make one assumption when we defined the problem, which was that the signals should be non-Gaussian. The reason for this is that if we add together two Gaussian signals we simply get a third Gaussian signal. Therefore if two or more of our signals (or noise sources) are Gaussian distributed there is no way to disentangle them. This is less an assumption than an incontrovertible fact which cannot be side-stepped.

A final limit to our capabilities is with respect to scale. If we multiply one column of the mixing matrix by a and divide the amplitude of the corresponding signal by a, we get the same vector, $\mathbf{x}$. Thus, since the problem is truly blind (we are only given $\mathbf{x}$), we always have a scale ambiguity with which to contend.

2.7.1 A Restatement of the Problem

Let us take another look at the problem. In Fig. 2.3, we show two-dimensional data points, each of which were drawn independently from the uniform distribution within the parallelogram. The first Principal Component is the direction with greatest spread, the long axis of the parallelogram. The second PC is of necessity perpendicular to that; we have no choice with two dimensional data; the second PC must be perpendicular to the first and so in a plane (with 2D data) we must draw the second PC as shown. Now we are in a situation where knowledge of the value of the first Principal Component gives some information about the second: if the projection on the first principal component is near the origin, the second principal component can take a wide range of values, whereas if the first principal component is far from the origin, the second projection is tightly constrained.

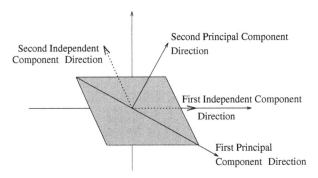

Fig. 2.3. The data were points drawn independently from the uniform parallelogram distribution shown. The first Principal Component is the direction with greatest spread, the long axis of the parallelogram. The second is of necessity perpendicular to that. The independent component directions, however, are parallel to the sides of the parallelogram.

The independent component directions, however, are parallel to the sides of the parallelogram. Knowing the first Independent Component of a data point gives us no information about the second; they truly are independent. Each, however, finds the underlying causes of the distribution in that each finds the independent directions of the uniform two-dimensional distribution.

There are two major methods used to solve this problem: one uses information theory while the other uses the higher-order moments of the data. We have already used the first two moments of a set of data:

1. The first moment is the mean. The mean can be calculated from

$$\mu = E(X) = \int p(x)x dx \qquad (2.27)$$

2. The second moment is the variance. The variance can be calculated from

$$\sigma^2 = E((X - \mu)^2) = \int p(x)(x - \mu)^2 dx \qquad (2.28)$$

For a Gaussian distribution, that is all there is to know about the distribution. For other distributions, you may well be interested in higher moments:

- The third moment measures skewness (see Figure 2.4) in a distribution:

$$E((X - \mu)^3) = \int p(x)(x - \mu)^3 dx \qquad (2.29)$$

If a distribution is perfectly symmetrical, this will evaluate to 0.
- The fourth moment measures the kurtosis of a distribution. This is a measure of the proportion of the distribution which is in the tails of the distribution compared with the proportion in the centre of the distribution:

$$E((X - \mu)^4) - 3 = \int p(x)(x - \mu)^4 dx - 3 \qquad (2.30)$$

The term "– 3" is added to ensure that a Gaussian distribution has 0 kurtosis.

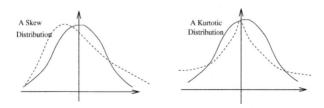

Fig. 2.4. Deviations from Gaussian distributions: the dotted line on the left represents a negatively skewed distribution; that on the right represents a positively kurtotic distribution; in each case, the solid line represents a Gaussian distribution

It can be shown that if two distributions are independent, then their higher moments satisfy the same constraint that we saw with the second-order statistics when we decorrelate the distributions:

$$E((XY)^p) = E(X^p).E(Y^p), \forall p \qquad (2.31)$$

This fact is used in some algorithms for ICA.

2.7.2 A Neural Model for ICA

Just as Oja was pivotal in the creation of neural models for PCA, so Jutten and Herault [97] have been instrumental in initiating interest in neural models for ICA. Jutten and Herault proposed a neural network architecture (Fig. 2.5) in which feedforward of activation and the lateral inhibition are defined by

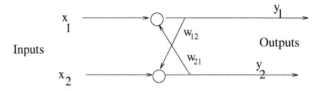

Fig. 2.5. Jutten and Herault's model.

$$y_i = x_i - \sum_{j=1}^{n} w_{ij} y_j$$

The learning rules are

$$\Delta w_{ij} = -\eta f(y_i) g(y_j), \text{ for } i \neq j \qquad (2.32)$$

which is clearly an extension of anti-Hebbian learning.

This model and a few of the most important other ICA models are discussed in Appendix C. We specifically mention Jutten and Herault's model since it has been so influential and also since it incorporates negative feedback, the subject of this book.

2.8 Conclusion

In this chapter, we have set the scene for the remainder of the book. We have discussed Hebbian learning and Principal Component Analysis and the conjunction of these two will form one of the themes of the early part of this book. We will see that the negative feedback neural network can be used with simple Hebbian learning to perform PCA. We have also discussed Information Theory and Independent Component Analysis. In Chapters 6 and 8, in particular, we will be interested in creating artificial neural networks (all of which use negative feedback) which approximate the search for independence.

In discussing these topics, we have, of necessity, been brief. While favorite textbooks are a subjective opinion, it is the author's belief that the following books provide good references to these topics:

- For Principal Component Analysis, see [93].
- For Information Theory, see [31].
- For Independent Component Analysis, see [85].
- For Artificial Neural Networks, see [71].

3

The Negative Feedback Network

3.1 Introduction

In this chapter, we will develop and investigate the negative feedback network. We will, in fact, develop an extremely simple and effective Principal Component network which needs no weight decay in its learning rule: because of the negative feedback of activation, we can use simple Hebbian learning which will not cause instability in the weight growth process and which moreover causes the weights to converge to the Principal Components of the input data. We will show that the network can be used to extract both principal and minor components.

We will note that the basic network is biologically plausible and we will investigate several modifications to the basic network while still attempting to remain within the space of models which seem possible for real neurons. Other negative feedback artificial neural networks are discussed in Appendix A; one of these is the original network of Plumbley [153] which provided the first insight into the strength of the combination of Hebbian learning and negative feedback networks.

3.1.1 Equivalence to Oja's Subspace Algorithm

Many contemporary artificial neural network models are unidirectional: an input pattern is presented to the network; the activation is propagated "forward" through the network until all neurons have had an opportunity to respond to the input pattern; finally, the weights are updated according to a learning rule. It is well known, however, that in real biological systems, activation is passed forward, backward and laterally (between neurons in the same layer). We will, in this chapter, consider a simple network composed of one layer of neurons to which the input pattern will be presented – the input layer – and one layer of neurons which we will describe as the output layer. There is therefore only a single layer of weights between input and output values but crucially, before the weights are updated we allow activation to pass forward and backward within the neural network.

We have, over the last few years, investigated a negative feedback implementation of PCA defined by (3.1) – (3.3). Let us have an N-dimensional input vector, $\mathbf{x}$, and an M-dimensional output vector, $\mathbf{y}$, with W_{ij} being the weight linking the jth input to the ith output. The learning rate, η, is a small value which will be annealed to zero over the course of training the network. The activation passing from input to output through the weights is described by (3.1). The activation is then fed back though the weights from the outputs and the residual, e, calculated for each input dimension. Finally, the weights are updated using simple Hebbian learning.

$$y_i = \sum_{j=1}^{N} w_{ij} x_j, \forall i \tag{3.1}$$

$$e_j = x_j - \sum_{i=1}^{M} w_{ij} y_i \tag{3.2}$$

$$\Delta w_{ij} = \eta e_j y_i \tag{3.3}$$

There is no explicit weight decay, normalisation or clipping of weights in the model. The subtraction of the weighted sum of the output values acts like anti-Hebbian learning. We will consider the network as a transformation from inputs x to outputs y; substituting (3.2) into (3.3), we see that the resultant network is equivalent to Oja's Subspace Algorithm (Chapter 2) and thus causes convergence to the principal component filters of the data. We have

$$\Delta w_{ij} = \eta e_j y_i = \eta \left(x_j - \sum_{k=1}^{M} w_{kj} y_k \right) y_i \tag{3.4}$$

This last formulation of the learning rule (3.4) is exactly the learning rule for the Subspace Algorithm [138], (2.21). A more formal analysis is given in Section 3.1.5

In an earlier formulation, (3.2) was discussed as

$$x_j(t+1) = x_j(t) - \sum_{i=1}^{M} w_{ij} y_i \tag{3.5}$$

while (3.3) was

$$\Delta w_{ij} = \eta x_j(t+1) y_i \tag{3.6}$$

Thus we envisage activation passing forward from inputs (x-values) to outputs (y-values) and subsequent (negative) feedback to the input neurons. This is a more usual convention in which to discuss Hebbian learning but (3.1)-(3.3) is also very convenient to emphasise the properties of the residual. We will mainly use the convention (3.1) to (3.3) in this book but will allude to the convention (3.5) and (3.6) when we wish to emphasise the Hebbian nature of the learning.

In order to compare this network with Oja's Subspace Algorithm, simulations were carried out on similar data[1] to that which Oja et al [141] used to compare the Subspace and Weighted Subspace Algorithms. The results shown in Table 3.1 are from a network with five inputs each of zero mean random Gaussians, where x_1's variance is largest, x_2's variance is next largest, etc..

Therefore, the largest eigenvalue of the input data's covariance matrix comes from the first input, x_1, the second largest comes from x_2 and so on. The advantage of using such data is that it is easy to identify the principal eigenvectors (and hence the principal subspace). There are three outputs in the network and it can be seen that the three-dimensional subspace corresponding to the first three principal components has been identified by the weights. There is very little of each vector outside the principal subspace i.e. in directions 4 and 5. The left matrix represents the results from the negative feedback network, the right shows Oja's results.

Table 3.1. Results from the simulated network and the reported results from Oja et al. The left matrix represents the results from the negative feedback network, the right from Oja's Subspace Algorithm. Note that the weights are very small outside the principal subspace and that the weights form an orthonormal basis of this space. Weights above 0.1 are shown in bold font.

W			W		
0.249	**0.789**	**0.561**	**0.207**	**−0.830**	**0.517**
0.967	**−0.234**	**−0.100**	**−0.122**	**0.503**	**0.856**
−0.052	**−0.568**	**0.821**	**0.970**	**0.241**	−0.003
0.001	0.002	0.016	-0.001	0.001	0.001
−0.001	0.009	0.005	0.000	0.000	−0.001
$W^T W$			$W^T W$		
1.001	0.000	0.000	**1.000**	0.000	0.000
0.000	**1.000**	0.000	0.000	**1.000**	0.000
0.000	0.000	**1.000**	0.000	0.000	**1.000**

The lower $(W^T W)$ section shows that the weights form an orthonormal basis of the space and the upper (W) section shows that this space is almost entirely defined by the first three eigenvectors. The negative feedback network also maintains the advantages of homogeneity and locality of computation (indeed, it is difficult to imagine a computationally simpler model). Note that while we report, in general, on simulations run on this very special type of input data, all the networks developed in this book perform excellently on all types of data.

---

[1] I did not have the value of the variances Oja used and therefore used variances of 5, 4, 3, 2 and 1.

3.1.2 Algorithm for PCA

While the above networks may be adequate for biological information processors, a more precise engineering requirement is that of finding the actual Principal Components.

Recall that Oja et al [141] amended the Subspace Algorithm by proposing the following modification to the learning rule

$$\Delta w_{ij} = \eta y_i \left(x_j - \theta_i \sum_{k=1}^{N} y_k w_{kj} \right)$$

Ensuring that $\theta_1 < \theta_2 < \theta_3 < \cdots$ allows the neuron whose weight decays proportional to θ_1 (i.e. whose weight decays least quickly) to capture the principal component of the variance. The second captures the next largest component, and so on. The crucial point is the introduction of asymmetry into the learning algorithm.

This algorithm is local and homogeneous in that each neuron knows only its own value of θ_i. Analysis of the negative feedback learning rule shows that, to simply insert a parameter, θ_i, would require computation at the level of the synapse. While this may be biologically feasible and algorithmically simple to implement, a different algorithm is developed here which uses the fact that the proposed network already incorporates subtraction of values.

We propose a very simple algorithm in this section, partly to keep our attention on the negative feedback network in all its simplicity and partly to introduce the concept of deflation in finding subsequent filters. The algorithm is: the system is created with one output; this output finds the first principal component using the above learning rule. It then loses its plasticity i.e. its weights will not subsequently change. We then create a second output neuron. Since the first neuron has found and subtracted the first principal component, the second neuron will find the largest remaining principal component. It, too, now loses its plasticity. Then the third output neuron is created, etc.. Therefore, we have introduced our asymmetry in the time dimension; note that whereas to do so with e.g. Oja's single neuron network, would have required the introduction of an extra mechanism – that of subtracting the projection of the data onto the subspace already found – we do not require this here as the network automatically finds and subtracts this subspace.

To compare the results with Oja's Weighted Subspace Algorithm, we repeated the above experiment with the algorithm. Oja's simulation was carried out for 40 000 iterations. The simulation allowed each output to learn in 13 000 iterations. The first output learned during the first 13 000 iterations, the second learned during the next 13 000 and the third learned during the last 13 000 iterations. The results are shown in Table 3.2; the left set is from the negative feedback network, the right from Oja et al [141]. Clearly both methods find the principal eigenvectors. We note that the negative feedback results have the advantage of equally weighting each eigenvector, which all therefore have length one.

Table 3.2. Results from the negative feedback network (left) and from Oja (right). Both methods find the principal eigenvectors of the input data covariance matrix. The negative feedback algorithm has the advantage that each vector is equally weighted.

W			W		
1.000	− 0.036	−0.008	**1.054**	− 0.002	− 0.002
0.036	**0.999**	−0.018	0.002	**1.000**	0.001
0.010	0.018	**1.000**	0.003	−0.002	**0.954**
− 0.002	− 0.002	0.016	− 0.001	0.001	− 0.002
0.010	0.003	0.010	0.001	−0.001	0.000
$W^T W$			$W^T W$		
1.001	0.000	0.000	**1.111**	0.000	0.000
0.000	**1.000**	0.000	0.000	**1.000**	0.000
0.000	0.000	**1.000**	0.000	0.000	**0.909**

The algorithm retains the advantages of homogeneity and locality of computation. A more analytical proof of the convergence of the algorithm is developed in Section 3.1.5.

3.1.3 Plasticity and Continuity

In this section, we investigate empirically some of the emergent properties of the negative feedback network. We view these properties as emergent properties as we do not believe that they could be expected a priori to exist, i.e. without a detailed investigation of the network.

The results reported in the last section were based on a model which suggested that only a new output could learn. The underlying assumptions are:

- an output neuron can only learn during a special period of its existence;
- only one output neuron can learn at any instant in time.

These are clearly not good properties for biological learners to have; we do not wish to have new learning remove the hard-won gains already achieved from previous learning; but equally, we do not wish to have to specify in advance how much time each neuron will have to learn. Further, in setting a specific time period during which learning will take place, we are providing the system with a form of meta information.

To test the effects of allowing neurons to continue to learn even after other new neurons were created, two more simulations were carried out. In the first, the neurons lost their plasticity gradually and there was an overlap in the times when two or more neurons were learning; in the second, neurons kept their plasticity throughout.

Thus, in this last model, the first neuron learns from its creation until the end of the simulation; the second neuron learns from its creation at iteration

Table 3.3. Results from the negative feedback network in which each output neuron stopped learning as a new one was created (left) and from the network in which each output neuron continued to learn (right).

Disjoint Learner Model			Contin. Learner Model		
W			W		
1.000	−0.036	−0.008	**1.000**	−0.015	−0.014
0.036	**0.999**	−0.018	0.014	**0.999**	0.045
0.010	0.018	**1.000**	−0.015	−0.046	**0.999**
−0.002	−0.002	0.016	−0.015	0.006	−0.021
0.010	0.003	0.010	0.003	−0.007	0.010
$W^T W$			$W^T W$		
1.001	0.000	0.000	**1.000**	0.000	0.000
0.000	**1.000**	0.000	0.000	**1.000**	0.000
0.000	0.000	**1.000**	0.000	0.000	**1.000**

13 000 until the end of the simulation; and the last neuron learns from iteration 26 000 till the end of simulation.

Only the results of the last model are reported, as the conclusions are identical: we do not have to postulate that neuron weights lose their plasticity. The left matrix of Table 3.3 repeats the results from the negative feedback model described in the previous section; the results from neurons which continue learning are shown on the right. The table shows that the neurons can retain their plasticity without there being a major loss of precision in finding the actual principal components.

We suggest that this model then represents a more plausible model of the form of learning which takes place in biological learners and further, that in most cases of unsupervised learning, the continuing learning model is preferred.

3.1.4 Speed of Learning and Information Content

One of the most interesting aspects of the proposed model is its reaction to statistical data which have inherently differing amounts of information. One might hope that a model would react to data which has more information more quickly than it does to data with less. This, in fact, happens.

We showed in Chapter 2 that the entropy [168] of a Gaussian random variable X with variance σ^2 is given by

$$h(X) = \frac{1}{2}\{1 + \log(2\pi\sigma^2)\} \tag{3.7}$$

which is a mathematical formulation of the fact that there is more information in random variables with large variance than in random variables with small variance. It would seem plausible to argue that an organism which can quickly identify data-sources with large information content would have an advantage over an organism which does not have this ability. This is, in fact, an emergent property of the model.

Therefore, in the current set of experiments, there is more information in x_1 than in x_2 and so on. That is $h(x_1) > h(x_2) > h(x_3) > h(x_4) > h(x_5)$. We find that, in the above experiment, x_1 is learned quickest, x_2 next most quickly and so on. Additional experiments to ensure that this rate was not merely a function of the order of the neuron's learning confirm that data with larger variances is learned more quickly. That this is not a necessary property of PCA networks is shown in ([41], Figure 2.7) where the network takes longest to converge to the first eigenvector.

3.1.5 Analysis of Convergence

This section provides an analytical investigation of the algorithm which causes the negative feedback network's weights to converge to the principal components of the input data's covariance matrix.

The proof of the algorithm follows closely the methods developed by Oja and Karhunen (e.g. [139]) over the last decade; the proof is in three parts each of which refers to the negative feedback learning rules. In the first section we show that the weights of a single output will converge to an eigenvector of the co-variance matrix; in the second, we show that these weights in fact converge to the principal eigenvector; in the third, we show that the algorithm ensures that the ith output's weights converge to the ith eigenvector.

Theorem 3.1 *The weights, W, of a single output with the above learning rules converges to an eigenvector of the input data covariance matrix.*

Let w_i be the weight of the connection between x_i and y.

If the weights of a single output converges to a limit, the expected weight change over a sufficiently long time will tend to zero. Given some assumptions[2], particularly regarding the learning rate η and the nature of the distribution of $\mathbf{x}$, and using $E(x)$ to indicate the expected value of x with respect to the distribution from which it is drawn,

$$E(\Delta w_i) = 0 \Longleftrightarrow E(\eta e_i y) = 0$$
$$\Longleftrightarrow E(e_i y) = 0$$
$$\Longleftrightarrow E\left[(x_i - w_i y) y\right] = 0$$
$$\Longleftrightarrow E\left[\left(x_i - w_i \sum_k w_k x_k\right) \sum_l w_l x_l\right] = 0$$
$$\Longleftrightarrow E\left(\sum_l w_l x_l x_i - w_i \sum_{kl} w_k x_k x_l w_l\right) = 0 \qquad (3.8)$$
$$\Longleftrightarrow \sum_l w_l C_{li} - w_i \sum_{kl} w_k C_{kl} w_l = 0 \qquad (3.9)$$

---

[2] This will be discussed later.

where C_{ij} is that element of the covariance matrix showing the covariance between the ith and j^{th} elements of the input data $\mathbf{x}$. If the weights of the network are to converge, then the above must be true for all values of w_i. Therefore the above may be written in matrix notation as

$$E(\Delta \mathbf{w}) = 0 \iff C\mathbf{w} - (\mathbf{w}^T C\mathbf{w})\mathbf{w} = \mathbf{0}$$
$$\iff C\mathbf{w} = (\mathbf{w}^T C\mathbf{w})\mathbf{w}$$

Now it is a standard result that the covariance matrix C is positive-semidefinite; and hence

$$\mathbf{w}^T C\mathbf{w} = \lambda \geq \mathbf{0}$$

where λ is a non-negative real number. Hence,

$$C\mathbf{w} = \lambda\mathbf{w}$$

Therefore, $\mathbf{w}$ converges to an eigenvector of C.

Theorem 3.2 *The weights, W, of a single output with the above learning rules converge to the eigenvector with the largest eigenvalue of the input data covariance matrix.*

Proof

The proof is by contradiction. Assume that $\mathbf{w}$ converges to an eigenvector $\mathbf{c}^*$ of C with corresponding eigenvalue λ^*. Then, we will show that if there exists an eigenvector $\mathbf{c}^1$ of C with corresponding eigenvalue $\lambda^1 > \lambda^*$ a small perturbation in the direction of $\mathbf{c}^1$ will cause $\mathbf{w}$ to be unstable i.e. convergence will not take place.

Let $\mathbf{w}$ have converged to a direction close to $\mathbf{c}^*$ but to have a component ϵ in the direction of $\mathbf{c}^1$. Then,

$$
\begin{aligned}
E(\Delta \mathbf{w}) &= C\mathbf{w} - (\mathbf{w}^T C\mathbf{w})\mathbf{w} \\
&= C(\mathbf{c}^* + \epsilon) - ((\mathbf{c}^{*T} + \epsilon^T)C(\mathbf{c}^* + \epsilon))(\mathbf{c}^* + \epsilon) \\
&= C\mathbf{c}^* + C\epsilon - (\mathbf{c}^{*T}C\mathbf{c}^*)\mathbf{c}^* - (\mathbf{c}^{*T}C\mathbf{c}^*)\epsilon - (\mathbf{c}^{*T}C\epsilon)\mathbf{c}^* - (\epsilon^T C\mathbf{c}^*)\mathbf{c}^* + \mathbf{O} \\
&= \lambda^*\mathbf{c}^* + \lambda^1\epsilon - \lambda^*\mathbf{c}^* - \lambda^*\epsilon - \mathbf{c}^{*T}\mathbf{C}(\epsilon\mathbf{c}^*) - C(\epsilon)^T\mathbf{c}^*\mathbf{c}^* + O(\epsilon^2) \\
&= \lambda^1\epsilon - \lambda^*\epsilon - (\lambda^1\epsilon^T\mathbf{c}^*)\mathbf{c}^* + O(\epsilon^2) \\
&= \lambda^1\epsilon - \lambda^*\epsilon + O(\epsilon^2)
\end{aligned}
$$

where we have used the facts that $C^T = C$ and that its eigenvectors are mutually orthogonal.

So, ignoring terms of $O(\epsilon^2)$, if $\lambda^1 > \lambda^*$, a perturbation in the direction of $\mathbf{c}^1$ will always be unstable. Therefore, c^* is the principal eigenvector corresponding to the largest eigenvalue of the covariance matrix.

Theorem 3.3 *If output i is installed in the network at time t_i, where $t_1 < t_2 < t_3 < \cdots$, and if the weights into the first $i - 1$ outputs have already* *converged to the first $i - 1$ eigenvectors, the weights of the ith output will converge approximately to the ith eigenvector of the input data's covariance matrix, where such eigenvectors are ordered such that the eigenvalue of vector 1 is the largest, that of vector 2 is next largest and so on.*

Proof

Let outputs $1,...,M - 1$ be already connected to the network. We assume that their weights have already converged to the subspace of the first M-1 eigenvectors, and show that the weights of output M (where $M > 1$) will converge approximately[3] to the Mth eigenvector of the covariance matrix C.

In this proof, let $\mathbf{w}_p$ be the weight vector associated with the pth output. Then,

$$
\begin{aligned}
E(\Delta \mathbf{w}_M)/\eta &= E(\mathbf{e} y_M) \\
&= E\left[(\mathbf{x} - W\mathbf{y}) y_M\right] \\
&= E\left[\left(\mathbf{x} - \sum_{k=1}^{M} y_k \mathbf{w}_k\right) y_M\right] \\
&= E\left[\left(\mathbf{x} - \sum_{k=1}^{M-1} y_k \mathbf{w}_k - y_M \mathbf{w}_M\right) y_M\right] \\
&= E\left[\left(\mathbf{x} - \sum_{k=1}^{M-1} (\mathbf{w}_k^T \mathbf{x})\mathbf{w}_k - y_M \mathbf{w}_M\right) y_M\right] \\
&= E\left[\left(\mathbf{x}_{M-1}^{\perp} - y_M \mathbf{w}_M\right) y_M\right]
\end{aligned}
$$

where $\mathbf{x}_{M-1}^{\perp}$ is the projection of $\mathbf{x}$ onto the subspace of possible values orthogonal to the first $M - 1$ eigenvectors.

Consider the application of this equivalence to the ith component of $\mathbf{w}_M$, i.e. w_{Mi}, the weight on the connection between e_i and y_M. Then, denoting the ith component of $\mathbf{x}_{M-1}^{\perp}$ by p_i,

$$
\begin{aligned}
E(\Delta w_{Mi}) = 0 &\iff E[(p_i - y_M w_{Mi})y_M] = 0 \\
&\iff E\left[\left(p_i - w_{Mi}\sum_k w_{Mk}x_k\right)\sum_l w_{Ml}x_l\right] = 0 \\
&\iff E\left[\sum_l w_{Ml}x_l p_i - w_{Mi}\sum_{kl} w_{Mk}x_k x_l w_{Ml}\right] = 0 \quad (3.10)
\end{aligned}
$$

We note the similarity between this equation and equation (3.8) in Theorem 3.1. For values of x_l within the subspace $\mathbf{x}_{M-1}^{\perp}$, the first term of equation (3.10) acts exactly like $p_l p_i$ and so the remainder of Theorems 1 and 2 hold for values

---

[3] Approximately, since the proof really requires an infinite convergence time for each weight vector. For a stationary source, $\mathbf{x}$, the finite time intervals used are close to perfect, but we can only claim "approximately" here.

of $\mathbf{x}$ restricted to this subspace. For values of $\mathbf{x}$ outside this subspace, the first term is 0 (x_l is in the subspace whose basis is the first $M - 1$ eigenvectors, p_i is in the orthogonal projection of this space) and the second term causes the weights to decrease to zero (recall that $\mathbf{w}^T C \mathbf{w} = \lambda$ is a scalar).

Therefore we can apply Theorems 1 and 2 to this subspace to show that the Mth output's weights will converge to the eigenvector corresponding to the largest eigenvalue of this subspace. This eigenvector has eigenvalue smaller than those of the $M - 1$ eigenvectors already allocated to weight vectors $w_1, \cdots, w_{M-1}$, but is larger than any other. Hence this eigenvalue is the Mth largest eigenvalue of the covariance matrix of the original input vector, $\mathbf{x}$.

Therefore, if the result is true for $M - 1$ outputs, it is true for M outputs. We know (Theorem 1) that it is true for one output. Therefore, the algorithm will force the weights to converge as required.

Assumptions in the Proofs of Convergence

The proof given above is based on a proof developed by Oja and Karhunen [139] and by Oja et al. [141, 142] for their feedforward networks. The major difficulty with the proof is the step from the stochastic equations (3.8) which are used in an empirical algorithm to the ordinary differential equations (3.9) which are solvable as seen above. Denoting by C_k the covariance matrix of the input data after k presentations of input vectors from the distribution, the proof given in [139] makes four critical assumptions:

1. Each C_k is almost surely bounded and symmetric and the C_k are mutually statistically independent with $E(C_k) = C$ for all k.
2. The eigenvalues of C have unit multiplicity.
3. $\eta_k \geq 0, \sum \eta_k^2 < \infty, \sum \eta_k = \infty$.
4. Each C_k has a probability density which is bounded away from zero uniformly in k in some neighbourhood of C in R^{n*m}.

The first constraint is easiest to satisfy since by taking k large enough we can sample the distribution sufficiently often so that the condition is almost surely satisfied.

The second assumption cannot be guaranteed for every distribution, however not satisfying it will only result in (a pair of) neurons converging to the subspace spanned by the eigenvectors with equal eigenvalues.

The third assumption is the difficult one to satisfy in any particular stochastic realisation of the algorithm: we are constraining the learning rate in a way that will not be practicable to sustain in any actual simulation; not only must the learning rate converge to zero (which is easy to manage), but it must do so sufficiently slowly that $\sum_j \eta_j$ is infinite. This leads to a long simulation! In practice, it has been found that slow annealing of the learning rate will, under a wide range of annealing schedules, cause the weights to converge to the principal components.

Another way to regard the problem is to say that we have not proved convergence; we have only proved that if the weights converge, they do so in a specific direction. We know also that if the weights reach this direction, they will be stable there but we have not proved that, in any single simulation, they must reach this direction. The proof that they would so converge with probability one uses the fact that each point in the neighbourhood of the attractor is sampled infinitely often.

3.2 The VW Model

The models discussed until now have one major drawback when considered as a model of biological systems: the weights of the connections from the output neuron, y, to the input neuron, x, are assumed to be identical to those from the input neuron, x, to the output neuron, y. This is biologically implausible: information flow in the neuron is unidirectional. This leads us to propose a model where these weights are initially different:

$$\mathbf{y} = W\mathbf{x} \tag{3.11}$$

$$\mathbf{e} = \mathbf{x} - V^T\mathbf{y} \tag{3.12}$$

$$\Delta W = \alpha_w \mathbf{e}\mathbf{y}^T \tag{3.13}$$

$$\Delta V = \alpha_v \mathbf{e}\mathbf{y}^T \tag{3.14}$$

where the initial values of both V and W are small random numbers not correlated in any way with each other. Note that both learning rules for W and V are identical up to the learning rate and use only simple Hebbian learning.

The convention we will use here is that w_{ij} is the weight of the connection from x_j to y_i; similarly, v_{ij} is the weight of the connection from y_i to x_j. Unless specifically stated otherwise, we shall be interested in the vectors to and from the output neurons. Therefore we take the vectors $\mathbf{v_i}$ to be the weight vector from the ith output neuron, i.e. to be the vector of form $\{v_{ik}\}$ for all k; similarly we take the vector $\mathbf{w_i}$ to be the vector of weights into the ith output neuron i.e. to be the vector $\{w_{ik}\}$ for all k; we note here that $\mathbf{v_i}$ corresponds to a column of the matrix V of weights while $\mathbf{w_i}$ is a row of W. Both are vectors of length n where n is the number of input neurons.

It is shown in Section 3.2.2, that, for a single output network, $\mathbf{v}$ and $\mathbf{w}$ converge to an eigenvector and that, at equilibrium, the weights $\mathbf{v}$ and $\mathbf{w}$ converge to the eigenvector, $\mathbf{c_1}$ with the largest eigenvalue such that

$$\mathbf{v} = \frac{1}{|\mathbf{w}|^2}\mathbf{w} \tag{3.15}$$

We then show that it is possible to apply the further analysis developed for the WW network and hence show that the ith output neuron converges to the ith eigenvector of the covariance matrix.

Table 3.4. Results from the negative feedback network (left) with symmetric weights,W and for the V and W vectors from the VW model (see text).

WW model			VW Model					
W			W			V		
1.000	−0.036	−0.008	**0.985**	−0.041	−0.003	**1.013**	−0.017	−0.024
0.036	**0.999**	−0.018	−0.019	**1.033**	0.031	−0.027	**0.965**	0.032
0.010	0.018	**1.000**	0.022	−0.032	**1.028**	0.020	−0.017	**0.969**
−0.002	−0.002	0.016	−0.024	−0.041	0.038	−0.007	−0.034	0.037
0.010	0.003	0.010	0.098	−0.007	−0.011	0.010	0.000	0.002

Experimental results, shown in Table 3.4 confirm this. It can be seen that both **v** and **w** converge to the same eigenvector, although the results are slightly less accurate than in the previous algorithm. However, given the simplicity of this biologically inspired model, the results are extremely clear: any entity which used such a method would be able to extract the greatest amount of information from its environment with a minimal number of output neurons using a very simple learning rule.

3.2.1 Properties of the VW Network

The motivation for the introduction of the VW model is that it removes a constraint from the network builder: in the WW model, the weights into and out of each neuron must be the same and so must be known in a meta-sense i.e. outwith the learning space. One feature of symmetry still remaining in the network is the equivalence of the learning rates in the V and W weights.

Experimental results show that, when **v** and **w** learn with different rates, the angle between **v** and **w** converges as quickly as before but the weight, **v** or **w**, with the larger learning rate acquires a larger length than the other. Indeed the result of the last theorem still applies.

While most of the emergent properties of the symmetric (WW) network still are found with the VW network, there is one property which this network does not have: the neurons cannot retain their plasticity when new neurons are created.

There always remains a slight angle between **v** and **w**; although this can be made arbitrarily small, it is sufficient to destabilise the output neurons' weights. It is not possible for the weights **v** and **w** to be both exactly orthogonal to any new neuron's weights; therefore the new neuron will destabilise the weights of existing neurons. The interaction between **v** and **w** will further move the weights away from the eigenvector and so the weights will be rotated in the principal subspace. This is also an empirical finding.

Therefore, for VW neurons, each neuron's weights must be allowed to converge to the eigenvector but must then lose their plasticity. This is algorithmically easy to implement, but the need to take this action has led to a search for other algorithms.

3.2.2 Theoretical Discussion

Lemma 1

*If the weights, **w**, of a single output neuron with the above learning rules converge to an eigenvector of the input data covariance matrix, then the weights **w** and the weights **v** converge to the same eigenvector.*

Proof

Let w_i be the weight of the connections from x_i to y and and v_i be that from y to x_i.

If the weights of a single output neuron converge to a limit, the expected weight change over a sufficiently long time will tend to zero. Given the usual approximations, particularly regarding the learning rate η, and using $E(x)$ to indicate the average value of x over the time period,

$$E(\Delta w_i) = 0 \iff E(\eta x_i y) = 0$$
$$\iff E(x_i y) = 0$$
$$\iff E((x_i - v_i y)y) = 0$$
$$\iff E\left[\left(x_i - v_i \sum_k w_k x_k\right) \sum_l w_l x_l\right] = 0$$
$$\iff E\left[\sum_l w_l x_l x_i - v_i \sum_{k,l} w_k x_k x_l w_l\right] = 0$$
$$\iff \sum_l w_l C_{li} - v_i \sum_{k,l} w_k C_{kl} w_l = 0$$

where C_{ij} is that element of the covariance matrix of the input data, $\mathbf{x}$, showing the covariance between the ith and j^{th} elements. We note that the same criterion may be deduced from $E(\Delta v_i) = 0$. If the weights of the output neurons are to converge, then the above must be true for all values of w_i. Therefore it may be written in matrix notation as

$$E(\Delta \mathbf{w}) = 0 \iff C\mathbf{w} - (\mathbf{w}^T C \mathbf{w})\mathbf{v} = 0$$
$$\iff C\mathbf{w} = (\mathbf{w}^T C \mathbf{w})\mathbf{v}$$

Now it is a standard result that the covariance matrix C is positive-semidefinite; and hence

$$\mathbf{w}^T C \mathbf{w} = \gamma \geq 0 \tag{3.16}$$

where γ is a non-negative real number. Hence,

$$C\mathbf{w} = \gamma \mathbf{v} \tag{3.17}$$

Therefore, if $\mathbf{w}$ converges to an eigenvector of C (see below), then $C\mathbf{w} = \lambda \mathbf{w}$ for some real number, λ, and so $\mathbf{v} = \alpha \mathbf{w}$, where α is a scalar; that is, $\mathbf{v}$ and $\mathbf{w}$ converge to the same eigenvector. Therefore, it is possible to apply the further analysis developed for the WW network and hence show that the ith output neuron converges to the ith eigenvector of the covariance matrix.

However there remains the possibility that the weight $\mathbf{w}$ will not converge to an eigenvector.

Theorem 3.4 *If the weights, $\mathbf{w}$, of a single output neuron with the above learning rules converge, then the weights $\mathbf{w}$ and the weights $\mathbf{v}$ converge to the same eigenvector of the input data's covariance matrix.*

From the lemma, we know that the weights converge as stated if they converge to an eigenvector. Therefore, we must prove that if $\mathbf{w}$ converges, it does so to an eigenvector of C. We use a contradiction argument.

Assume that there is a solution of

$$C\mathbf{w} = \gamma\mathbf{v} \qquad (3.18)$$

where $\mathbf{w}$ is not an eigenvector nor the degenerate solution, $\mathbf{w} = \mathbf{0}$.

Let the eigenvectors of C be $\mathbf{c}_1, \mathbf{c}_2, ..., \mathbf{c}_n$. Then

$$\mathbf{w} = \sum_{i=1}^{n} w_i \mathbf{c}_i$$

Since $\mathbf{w} \neq \mathbf{0}$, there exists a direction $\mathbf{c}_b$, such that $w_b \neq 0$. Since $\mathbf{w}$ is not an eigenvector, there exists one other direction $\mathbf{c}_a$ with a nonzero component, which we denote as w_a.

Then

$$\mathbf{w} = w_a\mathbf{c}_a + w_b\mathbf{c}_b + \sum_{i\neq a,b} w_i\mathbf{c}_i$$

where $1 \leq a, b \leq n, a \neq b$, and

$$\mathbf{v} = v_a\mathbf{c}_a + v_b\mathbf{c}_b + \sum_{i\neq a,b} v_i\mathbf{c}_i$$

and from Equation (3.18),

$$\lambda_b w_b = \gamma v_b$$

$$\lambda_a w_a = \gamma v_a$$

Consider a disturbance of magnitude $\epsilon > 0$ in the direction of $\mathbf{c}_a$ i.e. a disturbance of $\epsilon_{\mathbf{a}}$. Then if $\mathbf{w}$ is a stable point of convergence of the weights, the expected change in the weights over time is zero. Therefore,

$$E(\Delta\mathbf{w}) = 0$$

$$\Longleftrightarrow C\mathbf{w} - (\mathbf{w}^T C\mathbf{w})\mathbf{v} = 0$$

$$\Longleftrightarrow C\left(w_a\mathbf{c}_a + w_b\mathbf{c}_b + \epsilon_a + \sum_{i\neq a,b} w_i\mathbf{c}_i\right) - \gamma'\left(v_a\mathbf{c}_a + v_b\mathbf{c}_b + \sum_{i\neq a,b} v_i\mathbf{c}_i\right) = 0$$

$$\Longleftrightarrow \lambda_a w_a \mathbf{c}_a + \lambda_b w_b \mathbf{c}_b + \lambda_a \epsilon_{\mathbf{a}} + \sum_{i \neq a,b} \lambda_i w_i \mathbf{c}_i - \gamma' v_a \mathbf{c}_a$$

$$-\gamma' v_b \mathbf{c}_b - \gamma' \sum_{i \neq a,b} v_i \mathbf{c}_i = 0$$

where $\gamma' = (\mathbf{w} + \epsilon_a)^T C (\mathbf{w} + \epsilon_a) \geq 0$ since C is a positive semidefinite matrix.

Now, considering the components of the transformation in the direction of $\mathbf{c}_b$,

$$\lambda_b w_b - \gamma' v_b = 0$$

$$\text{Then,} \quad \gamma v_b - \gamma' v_b = 0$$

Therefore, $\gamma = \gamma'$ since $v_b \neq 0$. Now, considering the components of the transformation in the direction of $\mathbf{c}_a$,

$$\lambda_a w_a + \lambda_a \epsilon - \gamma' v_a = 0$$

$$\gamma v_a - \gamma v_a + \lambda_a \epsilon = 0$$

$$\lambda_a \epsilon = 0$$

which is a contradiction. Hence there does not exist a nonzero, noneigenvector solution to equation (3.18).

Theorem 3.5 *At equilibrium, the weights* $\mathbf{v}$ *and* $\mathbf{w}$ *converge to the same eigenvector,* $\mathbf{c}_a$ *with*

$$\mathbf{v} = \frac{1}{|\mathbf{w}|^2}\mathbf{w} \tag{3.19}$$

Proof

At equilibrium,

$$C\mathbf{w} = (\mathbf{w}^T C \mathbf{w})\mathbf{v} = \gamma \mathbf{v}$$

and, by Theorem 4, $\mathbf{w}$ is an eigenvector of C, $\mathbf{c}_a$. Therefore,

$$C\mathbf{w} = \lambda_a \mathbf{w}$$

where λ_a is the eigenvalue corresponding to eigenvector $\mathbf{c}_a$.

$$\text{Therefore,} \quad \lambda_a \mathbf{w} = \gamma \mathbf{v}$$

$$\text{Therefore,} \quad \mathbf{v} = \frac{\lambda_a}{\gamma}\mathbf{w} = \frac{\lambda_a}{\mathbf{w}^T C \mathbf{w}}\mathbf{w}$$

Now, $\mathbf{w}^T C \mathbf{w}$ is a scalar; hence,

$$\mathbf{w}^T C \mathbf{w} = |\mathbf{w}^T C \mathbf{w}| = |\mathbf{w}||C\mathbf{w}| = |\mathbf{w}||\lambda_a \mathbf{w}| = \lambda_a |\mathbf{w}|^2$$

Therefore,

$$\mathbf{v} = \frac{1}{|\mathbf{w}|^2}\mathbf{w}$$

Note: The theorems in this section imply that the further analysis of this negative feedback network is identical to that performed previously for the WW network. In other words, a negative feedback network with asymmetric weights, $\mathbf{w}_i$ and $\mathbf{v}_i$, will calculate the principal components if the output neurons are created in the network as in the previous section.

3.3 Using Distance Differences

Another possible model is suggested by the innate asymmetry in real biological neural networks in terms of the distances between neurons. This will manifest itself as different times to respond to a signal depending on the distance which the signal must travel (assuming that there is some uniformity in the speed of information transfer).

This differential is used in a new model where different output neurons take different lengths of times to respond to the input signal $\mathbf{x}$. Therefore while the activation from the input neurons is transmitted *to* all output neurons at the same time, each output neuron's response takes a different length of time to feedback to the input neurons. Thus, the negative feedback is felt and used in a phased manner, and learning takes place immediately when the returned signal is received. Therefore, we embed the learning process in the feedback loop, so that we now postulate a learning and activation-transmission process which takes place in the order in which the following equations are given:

$$\text{Initial value of } \mathbf{e}(0) = \mathbf{x} \tag{3.20}$$

$$\mathbf{y} = W\mathbf{x} \tag{3.21}$$

$$\text{For each output in turn } \mathbf{e}(i) = \mathbf{e}(i-1) - \mathbf{v}_i y_i \tag{3.22}$$

$$\Delta \mathbf{w}_i = \eta y_i \mathbf{e}^T(i) \tag{3.23}$$

$$\Delta \mathbf{v}_i = \eta y_i \mathbf{e}^T(i) \tag{3.24}$$

Other than the first two steps,(the acceptance of the initial activation $\mathbf{x}$ and its forward transmission to the output neurons), the process [defined by (3.22), (3.23) and (3.24)] is repeated for each output neuron in turn. This corresponds to the feedback from the output neurons being received at different times (perhaps depending on the physical distance which the activation must traverse, perhaps depending on the efficiency of transmission of the output neuron). This process results in the weights of the first (fastest) output neuron learning the first Principal Component; the second fastest output neuron learns the second Principal Component, and so on. Experimental results from a network with five inputs and three output neurons are given in Table 3.5. In order to demonstrate the effect of the network, we have carried out our simulations on the same type of data as previously. Clearly the first three principal components have been found by the three output neurons.

Table 3.5. Results of the differential distance model; each column shows the converged weights between one output neuron and the input neurons after learning on data from independent zero mean Gaussians with descending variances.

V			W		
1.000	0.006	−0.010	**1.000**	0.006	−0.010
−0.000	**−1.000**	0.013	−0.000	**−1.000**	0.013
0.012	0.023	**1.000**	0.012	0.023	**1.000**
0.000	−0.003	0.004	0.000	−0.002	0.004
−0.002	−0.004	−0.001	−0.002	−0.004	−0.001

Note that the crucial difference between this model and previous models is the embedding of the learning process in the activation reception process. When this is done, the resulting network is more similar to a Sanger-type [160] network rather than an Oja-type network. The kth output neuron is learning to extract the maximum amount of information which is left after the previous $(k - 1)$ output neurons have extracted their information.

3.3.1 Equivalence to Sanger's Algorithm

Sanger's algorithm has, as a learning rule,

$$\Delta w_{ij} = \eta y_i \left(x_j - \sum_{k=1}^{i} y_k w_{kj} \right)$$

in a totally feedforward architecture, where the outputs at y are given by

$$y_i = \sum_j w_{ij} x_j$$

We can show that the output neuron network using the rules determined by (3.20) to (3.24) is equivalent to Sanger's algorithm when we use the WW model in which the feedforward and feedback weights are identical.

Let the **e** values be indexed with the time of feedback from the output neurons. Then,

- $e_j(0)$ is the initial value of e_j at time 0 i.e. $e_j(0) = x_j$.
- $e_j(1)$ is the value of e_j after receiving the feedback activation from the first (and hence closest) output neuron i.e. $e_j(1) = e_j(0) - v_{1j}y_1$. Note that the time values are only ordinal indices; they do not imply equal intervals between feedback activations.

Similarly, if $e_j(2)$ is the value of e_j after receiving feedback from the first two output neurons, then

$$e_j(2) = e_j(1) - v_{2j}y_2 = e_j(0) - \sum_{k=1}^{2} v_{kj}y_k \qquad (3.25)$$

In general, if $e_j(i)$ is the value of e_j after receiving feedback from the first i output neurons,

$$e_j(i) = e_j(0) - \sum_{k=1}^{i} v_{kj} y_k = x_j - \sum_{k=1}^{i} v_{kj} y_k \qquad (3.26)$$

Therefore,

$$\Delta v_{ij} = \Delta w_{ij} = \eta e_j(i) y_i$$

$$= \eta \left(e_j(0) - \sum_{k=1}^{i} v_{kj} y_k \right) y_i$$

$$= \eta y_i \left(x_j - \sum_{k=1}^{i} v_{kj} y_k \right)$$

which is exactly Sanger's formulation (see Chapter 2).

3.4 Minor Components Analysis

Whereas a Principal Component Analysis finds the eigenvectors of the covariance matrix with greatest eigenvalues, a Minor Component Analysis (MCA) finds those eigenvectors with smallest eigenvalues. We will perform MCA with the same network as before but reverse the learing rule so that $\Delta w_i = -\alpha y_i \mathbf{e}$.

We showed above that all eigenvectors are stationary points of the learning rule, $\Delta w_i = \alpha y_i \mathbf{e}$ i.e. where $E(\Delta w_i) = 0$, and similarly we may show that eigenvectors are also solutions to $\Delta w_i = -\alpha y_i \mathbf{e} = 0$. We must still prove that only the eigenvector with the minimum eigenvalue can be stable.

Assume that $\mathbf{w}$ converges to an eigenvector $\mathbf{c}^*$ of C with corresponding eigenvalue λ^*. Then, we will show that if there exists an eigenvector $\mathbf{c}^1$ of C with corresponding eigenvalue $\lambda^1 < \lambda^*$ a small perturbation in the direction of $\mathbf{c}^1$ will cause $\mathbf{w}$ to be unstable i.e. convergence will not take place.

Let $\mathbf{w}$ have converged to a direction close to $\mathbf{c}^*$ but to have a component ϵ in the direction of $\mathbf{c}^1$. Then,

$$E(\Delta \mathbf{w}) = -C\mathbf{w} - (\mathbf{w}^T C \mathbf{w})\mathbf{w}$$

$$= -\{C(\mathbf{c}^* + \epsilon) - ((\mathbf{c}^{*T} + \epsilon^T)C(\mathbf{c}^* + \epsilon))(\mathbf{c}^* + \epsilon)\}$$

$$= -\{C\mathbf{c}^* + C\epsilon - (\mathbf{c}^{*T}C\mathbf{c}^*)\mathbf{c}^* - (\mathbf{c}^{*T}C\mathbf{c}^*)\epsilon - (\mathbf{c}^{*T}C\epsilon)\mathbf{c}^* - (\epsilon^T C\mathbf{c}^*)\mathbf{c}^* +$$

$$= -\{\lambda^*\mathbf{c}^* + \lambda^1\epsilon - \lambda^*\mathbf{c}^* - \lambda^*\epsilon - \mathbf{c}^{*T}C(\epsilon\mathbf{c}^*) - C(\epsilon)^T\mathbf{c}^*\mathbf{c}^* + O(\epsilon^2)\}$$

$$= -\{\lambda^1\epsilon - \lambda^*\epsilon - (\lambda^1\epsilon^T\mathbf{c}^*)\mathbf{c}^* + O(\epsilon^2)\}$$

$$= -\lambda^1\epsilon + \lambda^*\epsilon + O(\epsilon^2)$$

where we have used the facts that $C^T = C$ and that its eigenvectors are mutually orthogonal.

So, ignoring terms of $O(\epsilon^2)$, if $\lambda^1 < \lambda^*$, a perturbation in the direction of $\mathbf{c}^1$ will always be unstable. Therefore, directions other than the eigenvector corresponding to the smallest eigenvalue of the covariance matrix are unstable.

Note that if the covariance matrix is not of full rank, i.e. $\exists \mathbf{c}_i : \lambda_i = 0$, then the weights converge to the projection of the initial values of the weights onto the subspace with zero eigenvalue.

Xu et al. [186] have shown that the Total Least Squares (TLS) fitting problem can be solved by performing a Minor Component Analysis of the data: i.e. finding those projections which instead of containing maximum variance of the data contain minimum variance. In the next section, we first review regression and the TLS solution and then show how Minor Component analysis can be used to solve it.

3.4.1 Regression

Regression comprises finding the best estimate of a dependent variable, y, given a vector of predictor variables, $\mathbf{x}$. Typically, we must make some assumptions about the form of the predictor surface e.g. that the surface is linear or quadratic, smooth or disjoint, etc.. The accuracy of the results achieved will test the validity of our assumptions.

This can be more formally stated as: let (X, Y) be a pair of random variables such that $X \in R^n, Y \in R$. Standard regression aims to estimate the response surface

$$f(x) = E(Y|X = x) \tag{3.27}$$

from a set of p observations, $\mathbf{x}_i, y_i, i = 1, ..., p$.

The usual method of forming the optimal surface is the Least (Sum of) Squares Method which minimises the Euclidean distance between the actual value of y and the estimate of y based on the current input vector, $\mathbf{x}$). Formally, if we have a function, $f()$, which is an estimator of the predictor surface, and an input vector, $\mathbf{x}$, then our best estimator of y is given by minimising

$$E = \min_f \sum_i^N (y_i - f(\mathbf{x}_i))^2 \tag{3.28}$$

i.e. the aim of the regression process is to find that function $f()$ which most closely matches y with the estimate of y based on using $f()$ on the predictor, $\mathbf{x}$, for all values $(y, \mathbf{x})$.

For a linear function of a scalar x, we have $y = mx + c$, and so the search for the best estimator, $f()$, is the search for those values of m and c which minimise

$$E_1 = \min_{m,c} \sum_i (y_i - mx_i - c)^2$$

For each sample point in Fig. 3.1, this corresponds to finding that line which minimises the sum of the vertical lengths such as PR from all actual y-values to the best-fitting line, $y = mx + c$.

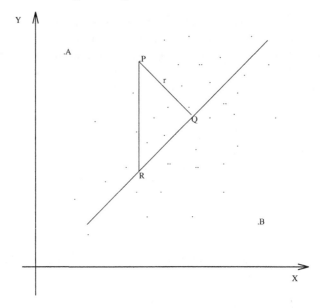

Fig. 3.1. The vertical lines will be minimised by the Least Squares method. The shortest distances, r_i, will be minimised by the Total Least Squares method.

However, in minimising this distance, we are making an assumption that only the y-values contain errors while the x-values are known accurately. This is often not true in practical situations in which, for example, which variable constitutes the response variable and which the predictor variables is often a matter of choice rather than being a necessary feature of the problem. Therefore, the optimal line will be that which minimises the distance, PQ=r, i.e. which minimises the shortest distance from each point, (x_i, y_i) to the best fitting line. Obviously, if we know the relative magnitude of the errors in x and y, we will incorporate that into the model; however here we assume no fore-knowledge of the magnitudes of errors. Thus, we are seeking those values of m and c which minimise

$$E_2 = \min_{m,c} \sum_i r_i^2 = \min_{m,c} \sum_i \frac{(y_i - mx_i - c)^2}{1 + m^2}$$

This is the so-called Total Least Squares method. Because of the additional computational burden introduced by the nonlinearity in calculating E_2, TLS is less widely used than LS although the basic idea has been known for approximately a centuary.

3.4.2 Use of Minor Components Analysis

We will solve the TLS fitting problem by performing a Minor Component Analysis of the data. The basic idea is that the noise in the data will typically contain less variance than the spread along the regression line. Therefore, the regression line must be orthogonal to the minor component of the data so we select the line in this direction which goes through the centre of the data. Since there may be errors in both y and x we do not differentiate between them and indeed incorporate y into the input vector $\mathbf{x}$. Therefore we reformulate the problem as: find the direction $\mathbf{w}$ such that we minimise E_2 i.e.

$$E_2 = \min_{\mathbf{w}} \frac{(\mathbf{w}.\mathbf{x} + c)^2}{\mathbf{w}^2} \text{ over all inputs } \mathbf{x}$$

$$= \min_{\mathbf{w}} \sum_{i=1}^{N} \frac{(\mathbf{w}.\mathbf{x}_i + c)^2}{\mathbf{w}^2}$$

$$= N \min_{\mathbf{w}} \frac{\mathbf{w}^T R \mathbf{w} + 2c\mathbf{w}^T \mathbf{E}(\mathbf{x}) + c^2}{\mathbf{w}^T \mathbf{w}}$$

where $R = \frac{1}{N} \sum_{i=1}^{N} \mathbf{x}_i \mathbf{x}_i^T$, the autocorrelation matrix of the data set and $E(\mathbf{x}) = \frac{1}{N} \sum_{i=1}^{N} \mathbf{x}_i$, the mean vector of the data set. Since, at convergence, $\frac{dE_2}{d\mathbf{w}} = 0$, we must have

$$R\mathbf{w} + cE(\mathbf{x}) - \lambda\mathbf{w} = 0 \tag{3.29}$$

where $\lambda = \frac{\mathbf{w}^T R \mathbf{w} + 2c\mathbf{w}^T E(\mathbf{x}) + c^2}{\mathbf{w}^T \mathbf{w}}$. Now we wish to find a hyperplane of the form

$$\mathbf{w}.\mathbf{x} + c = 0$$

So, taking expectations of this equation we have $c = -\mathbf{w}.\mathbf{E}(\mathbf{x})$ which we can substitute

$$C\mathbf{w} - \lambda\mathbf{w} = 0 \tag{3.30}$$

where now $\lambda = \frac{\mathbf{w}^T C \mathbf{w}}{\mathbf{w}^T \mathbf{w}}$ where C is the covariance matrix. From this we can see that every eigenvector is a solution of the minimisation of E_2.

We apply the method of Xu et al. [186][4] to the negative feedback network, to get the learning rule

$$\Delta\mathbf{w}_i = -\alpha y_i \mathbf{x} \tag{3.31}$$

which causes convergence to the eigenvector with the smallest eigenvalue, provided such eigenvalue is strictly smaller than all other eigenvalues.

As an example of the network in operation, we show in the first line of Table 3.6 the converged values of the weights of an MCA network when sample points are drawn from the line and both x and y coordinates are subject to noise. Clearly the algorithm has been successful. However, the robustness of regression solutions has generated a distinct area of research in the statistics literature. In the next section, we review some of this and consider the effect of noise on TLS solutions.

---

[4] Xu et al. derived the algorithm with a feedforward network using Hebbian learning with weight decay.

3.4.3 Robustness of Regession Solutions

We consider firstly the effect of white noise on all measurements and then consider the effect of substantial outliers on the convergence of the algorithm.

Table 3.6. Directions converged to when the points from the distribution were disturbed by noise drawn from N(0,0.05).

Actual Distribution	Direction Found	Outliers
$3x + 2y = 10$	$0.300x + 0.200y = 1$	None
$3x + 2y = 1$	$2.970x + 2.018y = 1$	None
$3x + 2y = 0.1$	$24.3x + 23.7y = 1$	None
$3x + 2y = 1$	$3.424x + 1.714y = 1$	1% in y direction
$3x + 2y = 1$	$2.459x + 2.360y = 1$	1% in x direction

For the lines shown in Table 3.6, points were drawn uniformly from only the first (both x and y positive) quadrant of the distribution determined by the line in each case. The first three lines show the direction to which the network converged when the distribution was affected by only white noise in both x and y direction drawn from N(0,0.05). Clearly the degree of accuracy of the convergence depends very greatly on the relative proportion of the amount of variance due to the length of the distribution from which points were drawn and the white noise. In the third case, the noise was of the same order as the variance due to the spread of points on the line and the convergence was severely disrupted.

As an example of the effect of outliers, we repeated the experiment with MCA regression to the line $2x + 3y = 1$ with 1% outliers (100 points out of 10000). The error in each outlier was $O(1)$. The results are shown in the last two lines of Table 3.6. Clearly the effect, even at this level of intensity, is much more severe when outliers occur in the x than the y direction.

Xu et al. suggest amending (3.31) to (our notation)

$$\Delta \mathbf{w}_i = -\alpha \beta y_i \mathbf{x} \qquad (3.32)$$

where $\beta = \frac{1}{|\mathbf{x}^2 - y_i^2|}$, to give a robust regression procedure. Clearly, for an outlier, β will be very small *provided* the weights $\mathbf{w}$ have already converged towards the direction indicated by the cloud of data. However, a network which meets a so-called leverage point early in learning will be just as liable to converge to a false direction and will treat the "good" points as before as outliers. Thus we see that in line 4 of Table 3.6, there is some disruption caused by 1% of the sample being disturbed by noise in the y-direction; but there is virtually total loss of regression parameters when the noise is in the x-direction (line 5).

Our overall conclusion, therefore, is that the method of MCA provides a valid method of calculating the Total Least Squares Minimum Error of a data set, but it is not correct to say that the method is robust to outliers.

3.4.4 Application to ICA

An example of blind source separation is the cocktail party problem, where a number of people at a party all talk simultaneously. A set of microphones are then placed throughout the room to receive the sound of everyone talking simultaneously. Because the microphones are placed in different positions, each microphone picks up slightly different mixtures. Using the different mixtures from the microphones we then try to isolate the individual speaker's speech signals.

Table 3.7. The relative power of each signal in each of the three mixes. Signal 3 is very much the weakest yet is readily recovered.

	Signal 1	Signal 2	Signal 3
Mixture 1	$2.8 * 10^{11}$	$0.3 * 10^{11}$	1
Mixture 2	$1.1 * 10^{11}$	$0.2 * 10^{11}$	1
Mixture 3	$2.5 * 10^{11}$	$1.1 * 10^{11}$	1

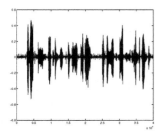 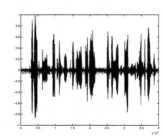

Fig. 3.2. Left: The original signal. Right: The recovered signal using Minor Component Analysis.

We can use MCA to extract a signal from a mixture of signals: we use three speech signals (each speaker said "perhaps the most frequent use of ICA is in the extraction of independent causes") and mix them so that the power of the weakest signal constitutes a small fraction of the total power in the mixes. Table 3.7 gives an example of the relative power of each of the signals in each of the three mixtures. Fig. 3.2 (left) shows the weakest power signal while Fig. 3.2 (right) shows the recovered output: the low power signal has been recovered.

We emphasise that we are using a linear network (and hence the second-order statistics of the data set) to extract the low-power voice. The mixture is such that humans cannot hear the third voice at all in any of the three mixtures and yet the MCA method reliably finds that signal with least power.

We note that a limitation of this method as a biological model might seem to be that the signal to be recovered must be swamped by the noise: if there is a component of the noise which is lower power than the signal, it will be recovered. However there is one situation (and perhaps a frequent one given the symmetry of our ears and surroundings) in which MCA may be useful: if the mixing matrix is ill-conditioned. Consider the mixing matrix

$$A = \begin{pmatrix} 0.8 & 0.4 \\ 0.9 & 0.4 \end{pmatrix} \tag{3.33}$$

whose determinant is -0.04. Its eigenvalues are 1.2325 and -0.0325. The major principal component will be the term which is constant across the two signals and so the smaller component will be the residual when this has been removed. Two voice signals were mixed using this matrix and one recovered by the MCA method is shown in Fig. 3.3 which should be compared with the original signal shown in Fig. 3.2. Most interesting is that the more ill-conditioned the mixing matrix, the better is the recovery of the lesser-amplitude voice. Given the closeness of our ears compared with the area from which signals are liable to be emitted, this suggests that this model might be extremely possible in a biological implementation: the fact that the data is ill-conditioned might actually be of help in disentangling different signals.

3.5 Conclusion

We have, in this Chapter, discussed a negative feedback implementation of a network which has been shown to be capable of performing Principal Component Analysis. We have extended the basic algorithm in several ways which may be of biological interest. Specifically:

- We have shown that we may have different weights feeding back from those feeding forward.
- We may feed back the outputs and have the subsequent weight change in a specific order in order to perform an actual Principal Component Analysis.
- If we reverse the direction of the weight change, we create a network which performs Minor Component Analysis.

This last method is mentioned for completeness but it should be mentioned that there are serious questions to be raised with respect to the convergence and stability of Minor Component Analysis methods. There is also the question of the magnitude of the weights from online algorithms which perform

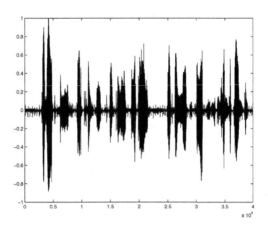

Fig. 3.3. The signal recovered from ill-conditioned mixing. Compare with Fig. 3.2 .

Minor Components Analysis which is not discussed herein. A good recent review of MCA algorithms and a discussion of these problems is given in [133].

In this chapter, we have seen that the algorithm may be approximately derived from maximisation of variance. In Chapter 5, we show how the equivalent nonlinear algorithm may be derived from minimisation of the mean square error of the low-dimensional representation; the network of this chapter is a special case of that algorithm and so we may equivalently show that this network performs the best linear compression of a data set where "best" is defined in terms of minimum mean squared error.

However, since we have shown that the network is algorithmically equivalent to Oja's or Sanger's algorithms, we must ask if we are gaining anything by phrasing the operation as a negative feedback operation rather than as a simple feedforward network with weight decay in the Hebbian learning rule. We believe that phrasing the operation as a three phase operation – feedforward, feedback and then weight change – gives us a mental model which permits changes which would not have been possible with a two-phase network with weight decay. We have already met one of these – the VW network; others which will be met in this book are the introduction of competition at the outputs before feedback (Chapter 7) or optimising the network learning with respect to the probability density functions of the residuals after feedback (Chapter 8). Such innovations would simply not be possible if we use a totally feedforward artificial neural network with negative feedback only within the learning rule.

Before we investigate these issues, we show in the next chapter how lateral weights may be used (before the feedback operation) to force convergence to the Principal Component directions. That chapter and the subsequent ones in Part 1 can be read almost in any order though it is necessary that the reader at least skims Chapter 6 before reading Chapter 8.

4

Peer-Inhibitory Neurons

At the start of the previous chapter, we discussed how, in biological systems, activation was passed forward, backward and also laterally through networks of neurons. The properties of the system which used the first two of these directions for information passing were discussed fully in that chapter; however, we did not consider any lateral interactions between the neurons. We take this approach in the current chapter.

Four factors make the negative feedback network especially exciting as a PCA network:

Simplicity: There are no logistic or hyperbolic functions to be calculated; there is no additional computation within the learning rule; there is no sequential passing back of errors or decay terms.

Homogeneity: Every output neuron is performing exactly the same calculation as its neighbours.

Locality of Information: Each output neuron uses only the information which it receives from its own connections; similarly with the input neurons which calculate the residuals.

Parallelism: Each operation at each output neuron is independent of what is happening at any other output neuron; similarly with the input neurons.

However, the phased creation of neurons described in the previous chapter does not utilise the inherent potential of this network for parallel information processing. We now develop learning algorithms which do this while retaining as much as possible of the other features.

Thus, in this chapter, we create the entire network at one instant in time and train all weights simultaneously. We investigate lateral trainable connections which learn using simple Hebbian learning to self-organise. We analyse these connections to find under what conditions they may be used to force a totally parallel network to learn the actual Principal Components themselves. We amend the basic network by allowing the inhibitory effect of each output neuron to act on the other output neurons as well as the input neurons. Two methods will be used with this amended network in order to create the necessary asymmetry: in the first, we will allow the network weights to be upgraded at different rates; in the second, we will use different activation functions to force convergence to the Principal Components.

The first type of network will be characterised by

$$\text{Feedforward:} \quad \mathbf{y}' = W\mathbf{x} \tag{4.1}$$

$$\text{Lateral action:} \quad \mathbf{y} = \mathbf{y}' - U\mathbf{y}' \tag{4.2}$$

$$\text{Feedback:} \quad \mathbf{e} = \mathbf{x} - V^T\mathbf{y} \tag{4.3}$$

$$\text{Weight change:} \quad \Delta W = \eta_w \mathbf{y}\mathbf{e}^T \tag{4.4}$$

$$\text{Weight change:} \quad \Delta V = \eta_v \mathbf{y}\mathbf{e}^T \tag{4.5}$$

$$\text{Weight change:} \quad \Delta U = \gamma \mathbf{y}\mathbf{y}^T \tag{4.6}$$

where $\mathbf{y}'$ is the initial activation of the output neuron before receiving the lateral inhibition from other output neurons, and U is the matrix of weights between the output neurons. We do not, however, allow self-connections from output neurons to themselves.

We note that we have now a 3-phase operation:

1. The activation is fed forward from the input neurons to the output neurons
2. The output neurons feed their activation to their peers and recalculate their activations
3. The activation is fed back to the input neurons from the output neurons

While this is more computationally complex than before, we only require $O(m^2)$ additional calculations, where m is the number of output neurons. Further, all learning processes continue to use simple Hebbian learning.

We will introduce a matrix $G(x) = (I - U)W(x)$,[1] which represents the forward function from $\mathbf{x}$ to $\mathbf{y}$. G is an integral part of the mathematical model which we will use for understanding the network, but it makes no overt contribution to the development of the network in the real, stochastic world. The actual learning in the network, i.e. the weight updates, is accomplished by updating the actual weights U, V and W although we will discuss $\frac{dG}{dt}$ as though it were being performed in the same sense that, e.g. $\frac{dW}{dt}$ is performed.

We can prove (an obvious special case of Theorem 4.2) that the learning rules detailed above are equivalent to

$$\frac{dV}{dt} = \frac{dW}{dt} = (I - U)WC - (I - U)WCW^T(I - U)^TV \tag{4.7}$$

$$\frac{dU}{dt} = (I - U)WCW^T(I - U)^T \tag{4.8}$$

$$\frac{dG}{dt} = (I - U)\frac{dW}{dt} - \frac{dU}{dt}W \tag{4.9}$$

---

[1] I being the identity matrix.

where G is the forward function relating $\mathbf{x}$ and $\mathbf{y}$ and C is the covariance matrix of the input data.

We will show, as with other models with lateral inhibition, that $U = 0$ is a stable stationary point of the system.

$$\text{Now, } G = (I - U)W$$

$$\text{and so } \frac{dG}{dt} = (I - U)\frac{dW}{dt} - \frac{dU}{dt}W$$

$$= (I - U)\{(I - U)WC - (I - U)WCW^T(I - U)^TV\}$$

$$-(I - U)WCW^T(I - U)^TW$$

$$= (I - U)\{GC - GCG^TV\} - GCG^TW$$

$$\rightarrow GC - GCG^TV - GCG^TW$$

$$\text{as U} \rightarrow 0$$

Now $G \rightarrow W$ as $\mathrm{U} \rightarrow 0$ and so $\frac{dG}{dt} \rightarrow WC - 2WCW^TW$ using the fact that $V = W$.

It can be seen that the necessary asymmetry between the Hebbian learning term and the weight decay term has not been achieved; however, the important point to note is that part of the weight decay term comes from the $\frac{dU}{dt}$ term which we can manipulate independently of the $\frac{dW}{dt}$ term in order to create the necessary asymmetry.

In summary, this chapter will show how it is possible to manage the lateral connections to force the weights to converge to Principal Component filters (rather than just to identify the Principal Subspace). We can do this by having the lateral weights learn at different rates or by having different weights of activation functions in certain models. We discuss the type of models in which this method does not work.

4.1 Analysis of Differential Learning Rates

Let us consider the system of equations:

$$\mathbf{y}' = W\mathbf{x} \tag{4.10}$$

$$\mathbf{y} = \mathbf{y}' - U\mathbf{y}' \tag{4.11}$$

$$\mathbf{e} = \mathbf{x} - V^T\mathbf{y} \tag{4.12}$$

$$\Delta W = \eta_W \mathbf{y}\mathbf{e}^T \tag{4.13}$$

$$\Delta V = \eta_V \mathbf{y}\mathbf{e}^T \tag{4.14}$$

$$\Delta U = \Gamma \mathbf{y}\mathbf{y}^T \tag{4.15}$$

Let us review our naming conventions: the convention we will use is that w_{ij} is the weight of the connection from x_j to y_i; similarly, v_{ji} is the weight of the connection from y_i to e_j; u_{ij} is the weight of the connection from y_j to

y_i. Unless specifically stated otherwise, we shall be interested in the vectors to and from the output neurons. Therefore, we take the vectors $\mathbf{v}_i$ to be the weight vector into the ith output neuron, i.e. to be the vector of form $\{v_{ki}\}$ for all k; similarly we take the vector $\mathbf{w}_i$ to be the vector of weights from the ith output neuron i.e. to be the vector $\{w_{ik}\}$ for all k. Both are vectors of length n where n is the number of input neurons. Note that the learning rates of U values are different for different output neurons as we wish to force the first output neuron to learn the first principal component, the second the next and so on. Thus we have a diagonal matrix, Γ .

Since we ensure that there are no self-connections, the main diagonal of U is composed of zeros. Also note that Γ is the matrix $\mathrm{diag}\{\gamma_1, \gamma_2, \cdots, \gamma_m\}$ where m is the number of output neurons and γ_i is the learning rate for the U weights of the ith output neuron such that $\gamma_1 < \gamma_2 < ... < \gamma_m$. We allow all learning rates to decrement to zero as time tends to infinity.

As introduced in the previous section, $G(x) = (I - U)W(x)$ is the forward function from $\mathbf{x}$ to $\mathbf{y}$. We will assume that, if $\gamma_i(t)$ is the value of γ_i during time interval t, $\lim_{t \to 0} \frac{\gamma_i(t)}{\eta_w(t)}$ exists and is positive. This assumption will be discussed in Section 4.1.1.

Theorem 4.1 $\mathbf{v}_i$ *converges if and only if* $\mathbf{w}_i$ *converges, where* $\mathbf{v}_i$ *is the weight vector from the* i^{th} *output neuron and* $\mathbf{w}_i$ *is the weight vector into the* i^{th} *output neuron. Further,*

$$\mathbf{v}_i = a\mathbf{w}_i + \mathbf{p}$$

where $a = \lim_{t \to 0} \frac{\eta_V(t)}{\eta_W(t)}$,
$\eta_V(t)$ *is the value of* η_V *during time interval* t,
and $\mathbf{p}$ *is a vector depending on the initial conditions of* $\mathbf{v}_i$ *and* $\mathbf{w}_i$.

Proof
At time B, we have

$$w_{ij}(B) = w_{ij}(B - 1) + \eta_{ij}(B)e_j(B)y_i(B)$$

If we start from time 0, we can equate the continuous time point T with the sum of the discrete intervals η_{ij}:

$$T = \sum_{p=0}^{B} \eta_{ij}(p)$$

Thus, we are breaking up continuous time into discrete time steps η_{ij} . Now the convergence, if it exists, must be taking place simultaneously over all weights. Therefore, we must ensure that $\eta_{ij} = \eta$ for all values of i, j. In order to have no limit on continuous time, we must have

$$\sum_{p=0}^{\infty} \eta(p) = \infty$$

If we further assume that $\eta(p) > 0$ for all p, then we have

$$\Delta w_{ij} = \eta e_j y_i$$

$$= \eta \left(x_j - \sum_s v_{sj} y_s \right) y_i$$

$$= \eta \left(x_j - \sum_s v_{sj} \left(\sum_l w_{sl} x_l - \sum_p u_{sp} \sum_l w_{pl} x_l \right) \right) \times$$

$$\left(\sum_t w_{it} x_t - \sum_q u_{iq} \sum_r w_{qr} x_r \right)$$

Or, in matrix terms,

$$\frac{W(B) - W(B-1)}{\eta(B)} = (I - U(B))W(B)\mathbf{x}(B)\mathbf{x}(B)^T$$

$$-(I - U(B))W(B)\mathbf{x}(B)\mathbf{x}(B)^T W(B)^T (I - U(B))^T V(B)$$

$$(4.16)$$

If we also assume that

$$\lim_{p \to \infty} \eta(p) = 0$$

then the sequence of $w_{ji}(T)$ asymptotically approaches a continuous-time function and the left-hand side of (4.16) approaches its derivative. Then we can replace (4.16) with the corresponding averaged differential equation

$$\frac{dW}{dt} = (I - U)WC - (I - U)WCW^T (I - U)^T V \qquad (4.17)$$

where C is the covariance matrix of the stationary distribution producing the x_k values. Now, under the same assumptions as in the previous chapter about the rate η it can be shown that the solution of the stochastic algorithm approaches the solution of the differential equation (4.17) with probability 1.

Now consider $\mathbf{v}$'s learning.

$$v_{ij}(B) = v_{ij}(B-1) + \eta_V(B)e_j(B)y_i(B)$$

Therefore,

$$\frac{v_{ij}(B) - v_{ij}(B-1)}{\eta(B)}$$

$$= \frac{\eta_V(B)}{\eta(B)} [\sum_l w_{li}(B)x_l(B)x_j(B) \left(1 - \sum_q u_{iq} \right)$$

$$- \sum_k v_{kj}(B) \left(1 - \sum_q u_{iq} \right) \sum_{p,l} w_{pk}(B)x_p(B)x_l(B)w_{li}(B) \left(1 - \sum_p u_{sp} \right)]$$

Given the same assumptions as before and making the additional assumption that

$$\lim_{p \to 0} \frac{\eta_V(p)}{\eta(p)} = a > 0 \text{ i.e. the limit exists and is positive.}$$

we have the corresponding differential equation,

$$\frac{dV}{dt} = a((I - U)WC - (I - U)WCW^T(I - U)^T V) \tag{4.18}$$

Therefore,

$$\frac{dV}{dt} = a\frac{dW}{dt}$$

Therefore, W converges to a solution (where $\frac{dW}{dt} = 0$) if and only if V converges to a solution.

$$\text{Now, } \frac{dW}{dt} = (I - U)WC - (I - U)WCW^T(I - U)^T V = f(W, V)$$

$$\text{Let } F(W, V) = \int_0^\infty f(W, V)dt$$

$$\text{Then, } V = aF(W, V) + aK \text{ and } W = F(W, V) + K$$

where K is a function of the initial values of V and W.

Thus, $\mathbf{v_i} = \mathbf{w_i} + \mathbf{p}$, where $\mathbf{p}$ is a vector depending only on the initial values of the system.

Thus if $\mathbf{v_i}$ and $\mathbf{w_i}$ converge, they do so simultaneously and close to the same vector.

Note 1 For the remainder of this section we will assume that $a=1$. i.e. the learning rates for V and W are equal.

Note 2 We note that the vectors $\mathbf{v}_i$ and $\mathbf{w}_i$ may be made arbitrarily close by limiting the original size of vectors $\mathbf{v}_i(0)$ and $\mathbf{w}_i(0)$. i.e. $\mathbf{p}$ may be made arbitrarily small by appropriate initial choice of $\mathbf{v}$ and $\mathbf{w}$. Hence we are able to assume that $\mathbf{v}_i \approx \mathbf{w}_i$.

Theorem 4.2 *The learning rules detailed above are equivalent to*

$$\frac{dV}{dt} = \frac{dW}{dt} = (I - U)WC - (I - U)WCW^T(I - U)^T V \tag{4.19}$$

$$\frac{dU}{dt} = A(I - U)WCW^T(I - U)^T \tag{4.20}$$

$$\frac{dG}{dt} = (I - U)\frac{dW}{dt} - \frac{dU}{dt}W \tag{4.21}$$

where G is the forward function relating $\mathbf{x}$ and $\mathbf{y}$ and A is the matrix $diag\{a_1, a_2, ..., a_m\}$ with $a_i = \lim_{t \to 0} \frac{\gamma_i(t)}{\eta_W(t)}$ with $\gamma_i(t)$ being the value of γ_i during the time interval t.

Proof

With the same assumptions as before, we can write

$$\frac{W(B) - W(B-1)}{\eta(B)} = (I - U(B))W(B)\mathbf{x}(B)\mathbf{x}(B)^T$$

$$-(I - U(B))W(B)\mathbf{x}(B)\mathbf{x}(B)^T W(B)^T (I - U(B))^T V(B) \tag{4.22}$$

If we also assume that

$$\lim_{p \to \infty} \eta(p) = 0$$

then the sequence of $w_{ij}(T)$ asymptotically approaches a continuous-time function and the left-hand side of (4.22) approaches its derivative. Then we can replace (4.22) with the corresponding averaged differential equation

$$\frac{dW}{dt} = (I - U)WC - (I - U)WCW^T(I - U)^T V \tag{4.23}$$

where C is the covariance matrix of the stationary distribution producing the x_k values. Now, under certain assumptions about the rate η it can be shown that the solution of the stochastic algorithm approaches the solution of the differential equation (4.23) with probability 1.

Similarly, for the weight updates of the U weights,

$$u_{ij}(B) = u_{ij}(B-1) + \gamma_i(B)y_i(B)y_j(B)$$

Therefore,

$$\frac{U(B) - U(B-1)}{\eta(B)} = \frac{\gamma_i(B)}{\eta(B)}y_iy_j$$

If we make the further assumption that $\lim_{B \to 0} \frac{\gamma_i(B)}{\eta(B)}$ exists, we can take the limit of the above stochastic equation giving

$$\frac{dU}{dt} = AQ$$

where $A = \text{diag} \{a_1, a_2, ..., a_m\}$ with $a_i = \lim_{t \to 0} \frac{\gamma_i(t)}{\eta(t)} > 0$, and Q the $m * m$ matrix with elements $q_{ij} = E(y_iy_j), i \neq j$, and $q_{ii} = 0$ for all i, j. $E()$ indicate an ensemble average. We will, for the time being, assume that the a_i values are constant during the learning process. We will return to this assumption in Section 4.1.1 Now,

$$Q = E(\mathbf{y}\mathbf{y}^T) \tag{4.24}$$

$$= E((I - U)W\mathbf{x}\mathbf{x}^T W^T(I - U)^T) \tag{4.25}$$

$$= (I - U)WCW^T(I - U)^T \tag{4.26}$$

where C_{ij} is $E(x_ix_j)$ for all i, j. Hence,

$$\frac{dU}{dt} = A(I - U)WCW^T(I - U)^T$$

The transform from $\mathbf{x}$ to $\mathbf{y}$ is G where $G(t) = (I - U(t))W(t)$ where $U(t)$ is the value of U at time t, etc. Then,

$$\frac{dG}{dt} = (I - U(t))\frac{dW(t)}{dt} - \frac{dU(t)}{dt}W(t) \tag{4.27}$$

$$= (I - U)\frac{dW}{dt} - \frac{dU}{dt}W \tag{4.28}$$

Theorem 4.3 $U = 0$, the $m * m$ zero matrix, is a solution of $\frac{dG}{dt} = 0$, and $\mathbf{g_i} = \mathbf{w_i} = \frac{1}{\sqrt{1+a_i}}\mathbf{c_i}$ is the corresponding solution for G and W where $\mathbf{c_i}$ are the eigenvectors of the covariance matrix of the input data in order. i.e. if $\gamma_{u_1} < \gamma_{u_2} < \cdots < \gamma_{u_m}$, then $\mathbf{w}_i = \frac{1}{\sqrt{1+a_i}}\mathbf{c}_i$ where $\mathbf{c}_i$ is that eigenvector with corresponding eigenvalue λ_i where $\lambda_1 > \lambda_2 > \cdots > \lambda_m$ and, as before, $a_i = \lim_{t\to0}\frac{\gamma_i(t)}{\eta_W(t)}$

Proof

$$\frac{dG}{dt} = (I - U)\frac{dW}{dt} - \frac{dU}{dt}W$$

First we note that as $U \to 0$,

$$\frac{dG}{dt} = \frac{dW}{dt} - \frac{dU}{dt}W$$
$$= (I - U)WC - (I - U)WCW^T(I - U)^TV - A(I - U)WCW^T(I - U)^TW$$
$$= GC - GCG^TV - AGCG^TW$$
$$\to WC - (I + A)WCW^TW$$

at the point of convergence of V and W.

Note the similarity between these equations and those required for Oja's Weighted Subspace Theorem [142, 143]; therefore, we conjecture that a solution of $\frac{dG}{dt} = 0$ at $U = 0$ is $\mathbf{g}_i = \mathbf{w}_i = \frac{1}{\sqrt{1+a_i}}\mathbf{c}_i$ the ith eigenvector of C in normal order. Here we show that the stated values are solutions; stability will be proved later.

$$\frac{dG}{dt} = (I - U)\frac{dW}{dt} - \frac{dU}{dt}W$$
$$= (I - U)((I - U)WC - (I - U)WCW^T(I - U)^TV)$$
$$\quad - A(I - U)WCW^T(I - U)^TW$$
$$\to WC - WCW^TV - AWCW^TW$$
$$= \Lambda W - KV - AKW$$

where K is the diagonal matrix whose $(i, i)^{th}$ element is $\lambda_i|\mathbf{w_i}|^2$ with λ_i the ith eigenvalue and Λ is the diagonal matrix whose $(i, i)^{th}$ element is λ_i.

Then taking $\mathbf{g}_i$ as the ith vector of G i.e. going into the ith output neuron and using the fact that $\mathbf{w}_i = \mathbf{v}_i$, we have

$$\frac{d\mathbf{g}_i}{dt} = \lambda_i \mathbf{w}_i - k_i \mathbf{w}_i - a_i k_i \mathbf{w}_i$$

$$= \left(\frac{\lambda_i}{\sqrt{1+a_i}} - \frac{\lambda_i}{1+a_i} \frac{1}{\sqrt{1+a_i}} - \frac{a_i \lambda_i}{1+a_i} \frac{1}{\sqrt{1+a_i}} \right) \mathbf{c}_i$$

$$= \left(\frac{\lambda_i(1+a_i) - \lambda_i - a_i \lambda_i}{\sqrt{1+a_i}(1+a_i)} \right) \mathbf{c}_i = 0$$

So the stated values are stationary points of the system.

Note: We can in fact go further, in that if $U = 0$ then

$$\frac{d\mathbf{g}_i}{dt} = \lambda_i \mathbf{w}_i - k_i \mathbf{w}_i - a_i k_i \mathbf{w}_i$$

$$= (\lambda_i - \lambda_i |\mathbf{w}_i|^2 - a_i \lambda_i |\mathbf{w}_i|^2) \mathbf{w}_i$$

$$= \lambda_i (1 - |\mathbf{w}_i|^2 - a_i |\mathbf{w}_i|^2) \mathbf{w}_i$$

so if $\frac{d\mathbf{g}_i}{dt} = 0$ then $|\mathbf{w}_i|^2(1+a_i) = 1$ i.e. $|\mathbf{w}_i| = \pm \frac{1}{\sqrt{1+a_i}}$.

Theorem 4.4 *At the solutions* $U = 0$, $\mathbf{w}_i = \frac{1}{\sqrt{1+a_i}} \mathbf{c}_i$ *of* $\frac{dG}{dt} = 0$, *then*

$$\frac{du_{ij}}{dt} = 0 \text{ for all } i,j$$

if $\lambda_i \neq 0$ *i.e. the* i^{th} *eigenvalue is not zero.*

Proof

$$\frac{dU}{dt} = A(I - U)WCW^T(I - U)^T$$

$$= AWCW^T$$

Now $\mathbf{w}_i C \mathbf{w}_j^T = 0$ for all $i \neq j$ and $\mathbf{w}_i C \mathbf{w}_i^T = \lambda_i |\mathbf{w}_i|^2$. Therefore, WCW^T is a diagonal matrix of the form $\text{diag}\{k_1, k_2, \cdots, k_m\}$ where $k_i = \lambda_i |\mathbf{w}_i|^2$, λ_i being the ith eigenvalue. Then

$$\frac{dU}{dt} = AK = \text{diag}\{a_1 k_1, a_2 k_2, \cdots, a_m k_m\}$$

Therefore, $\frac{du_{ij}}{dt} = 0$ for all $i \neq j$.

Theorem 4.5 *The solutions* $u_{ij} = 0$, $\mathbf{w}_i = \frac{1}{\sqrt{1+a_i}} \mathbf{c}_i$ *for all* i,j *of* $\frac{dG}{dt} = 0$ *ensure all variables are stationary at this point.*

Proof

At the stated points of solution of $\frac{dG}{dt} = 0$ then $\frac{dU}{dt} = 0$. But

$$\frac{dG}{dt} = (I - U)\frac{dW}{dt} - \frac{dU}{dt}W$$
$$= \frac{dW}{dt}$$

So $\frac{dG}{dt} = \frac{dU}{dt} = \frac{dW}{dt} = \frac{dV}{dt} = 0$.

Theorem 4.6 *If $\mathbf{c}_i$ are the unit length eigenvectors of C, the solutions*

$$\mathbf{v}_i = \mathbf{w}_i = \frac{1}{\sqrt{1 + a_i}}\mathbf{c}_i$$

$$\mathbf{u}_i = \mathbf{0}, \text{ where } \mathbf{0} \text{ is the } m*1 \text{ zero vector}$$

of the equations governing the dynamics of this network, are asymptotically stable for all i.

Proof

First consider the $\mathbf{u}_i$. We have already shown that $U = 0 \implies \frac{dU}{dt} = 0$. Consider a disturbance of ϵ in $U = 0$. We have

$$\frac{dU}{dt} = A(I - (0 + \epsilon))WCW^T(I - (0 + \epsilon))^T$$
$$= AK - A\epsilon WCW^T - AWCW^T\epsilon + O(\epsilon^2)$$
$$= -A\epsilon WCW^T - AWCW^T\epsilon + O(\epsilon^2) \text{ (off diagonal)}$$
$$= -A\epsilon K - AK\epsilon + O(\epsilon^2)$$

Since A and K are both diagonal matrices with entries ≥ 0, if $\epsilon > 0$, the rate of change of U is negative i.e. U must decrease. If $\epsilon < 0$, the rate of change of U is positive i.e. U will increase.

Now consider the W weights. We have proved that the stated values are solutions; we must still prove asymptotic stability. Note that at the stated points of convergence,

$$G = (I - U)W = W$$
$$\frac{dG}{dt} = (I - U)\frac{dW}{dt} - \frac{dU}{dt}W$$

Now since $G = W$ at the points stated, any instantaneous disturbance in W will have an equal instantaneous effect on G. Therefore, we will investigate the effect of a disturbance in W on G in order to derive the asymptotic stability of W. We do this through investigating the effects of the disturbance on U and W. Let there be a disturbance of E in the converged weights W. Then,

$$\frac{dU}{dt} = A(W + E)C(W + E)^T$$
$$= AWCW^T + AECW^T + AWCE^T + AECE^T$$
$$\approx (AWCW^T) + AECW^T + AWCE^T$$

ignoring terms of $O(E^2)$. Thus,

$$\frac{dU}{dt}(W + E) = (AWCW^T + AECW^T + AWCE^T)(W + E)$$
$$\approx AWCW^TW + AECW^TW + AWCE^TW + AWCW^TE$$
$$= (AWCW^TW) + AECW^TW + AWCE^TW + AKE$$

Similarly,

$$\frac{dW}{dt} = (W + E)C - (W + E)C(W + E)^TV$$
$$\approx WC + EC - WCW^TV - ECW^TV - WCE^TV$$
$$= (WC - WCW^TV) + EC - ECW^TV - WCE^TV$$

So still ignoring terms of $O(E^2)$,

$$\frac{dG}{dt} = \frac{dW}{dt} - \frac{dU}{dt}W$$
$$= (WC - WCW^TV + AWCW^TW) + EC - ECW^TV - WCE^TV$$
$$\quad -AECW^TW - AWCE^TW - AWCW^TE$$
$$= EC - ECW^TV - WCE^TV - AECW^TW - AWCE^TW - AKE$$
$$= EC - AKE - (I + A)(ECW^T + WCE^T)W$$

Now, considering a disturbance of ϵ in the direction of $\mathbf{c}_j$ of the weight $\mathbf{w}_i$,(i.e. a disturbance of ϵ_j) we note first that the matrix$(I + A)(ECW^T + WCE^T)$is a diagonal matrix with its jth element $(1 + a_j)\frac{2\lambda_j\epsilon}{\sqrt{1+a_j}}$. So considering the rate of change of $\mathbf{g}_i$ in the direction of $\mathbf{c}_j$,

$$\frac{d\mathbf{g}_i}{dt} = \lambda_j\epsilon_j - a_jk_j\epsilon_j - (1 + a_j)\frac{2\lambda_j\epsilon}{\sqrt{1+a_j}}w_j$$
$$= \left(\lambda_j\epsilon - a_j\frac{\lambda_j}{1+a_j}\epsilon - (1 + a_j)\frac{2\lambda_j\epsilon}{\sqrt{1+a_j}}\frac{1}{\sqrt{1+a_j}}\right)\mathbf{c}_j$$
$$= -\lambda_j\left(1 + \frac{a_j}{1+a_j}\right)\epsilon\mathbf{c}_j \qquad (4.29)$$

Since C is symmetric, $\lambda_j > 0$. Further, the learning rates γ_j were such that $a_j > 0$. Then, equation (4.29) shows that if $\epsilon > 0$, which would cause G to grow, the system will self-organise to cause G to shrink; if $\epsilon < 0$ the system will self-organise to cause G to grow. Since we have shown that $U = 0$ is a stable solution, then the solutions of the W vectors must also be stable.

Now consider the $\mathbf{v}_i$. The proof that the stated values are solutions is implicit in the section above. To show asymptotic stability, let there be a disturbance of $\epsilon > 0$ in V. Then

$$\frac{dV}{dt} = (I - U)WC - (I - U)WCW^T(I - U)^T(V + \epsilon)$$
$$= WC - WCW^T(V + \epsilon)$$
$$= WC - WCW^T V - WCW^T \epsilon$$
$$= -WCW^T \epsilon$$
$$= -K\epsilon < 0$$

since every element of K is greater than 0. Similarly, if $\epsilon < 0$, we have $\frac{dV}{dt} > 0$. Thus all the stated values are stable points of the system.

4.1.1 The GW Anomaly

There is an apparent anomaly in the above equations. The solution of $\frac{dG}{dt} = 0$ is $\mathbf{g}_i = \mathbf{w}_i = \frac{1}{\sqrt{1+a_i}}\mathbf{c}_i$ whereas the solution of $\frac{dW}{dt} = 0$ occurs at $\mathbf{w}_i = \mathbf{c}_i$. Further, as $U \to 0$, $G \to W$. This suggests a less stable system than before, and this is indeed the case. Thus, in order to minimise instability, it is necessary to ensure that the a_i values are low. Experimental results suggest a value of 0.1 is sufficient to ensure stable convergence to Principal Components. However, we note that, at $\mathbf{w}_i = \frac{1}{\sqrt{1+a_i}}\mathbf{c}_i$, $U = 0$ and so

$$\frac{dW}{dt} = (I - U)WC - (I - U)WCW^T(I - U)^T W^T V$$
$$\to WC - WCW^T V$$
$$\text{and so} \quad \frac{d\mathbf{w}_i}{dt} = \frac{\lambda_i}{\sqrt{1+a_i}}\mathbf{c}_i - \frac{\lambda_i}{1+a_i}\frac{1}{\sqrt{1+a_i}}\mathbf{c}_i$$
$$= \frac{\lambda_i}{\sqrt{1+a_i}}\frac{a_i}{1+a_i}\mathbf{c}_i$$

Therefore, for any $a_i \neq 0$, there will be a tendancy for the weights to grow away from the global optimum. However, as seen in the equations governing $\frac{dG}{dt}$, this cause instantaneous change in U, which will drive the W weights in the opposite direction. In order to produce a damped system, the values of a_i should be small. One possible response to this anomaly is to insist that as we are taking a limit to infinity, the a_i values can only be ≥ 0 i.e. to allow equality. However, this is not the experimental situation where a strict ratio is maintained as the terms decrease to 0 nor does it help the analysis as we then have a diagonal matrix which is not of full rank and would not then provide the differential decay necessary for convergence to the Principal Components.

The approach chosen here is to choose the values of a_i appropriately small so that the term $\frac{1}{\sqrt{1+a_i}} \approx 1$. Under this constraint the system has been found experimentally to be stable.

The final point to note is that in this system the decay of the learning rate to 0 may be essential to the fixed stability of the system; if the learning rates are not allowed to decay to zero, the very dynamical nature of the convergence will continue.

4.1.2 Simulations

The results of a typical experiment on the same type of data as in the previous chapters are indicated in Table 4.1. Here, the first output neuron has the smallest learning rate i.e. $\alpha_{u_1} < \alpha_{u_2} < \alpha_{u_3}$. Further, while the initial values of the **w** and **v** weights were $O(0.0001)$ those of the **u** weights were $O(0.00001)$.

In order to show that it is the different learning rate which causes the convergence to Principal Components, the same experiment was rerun with all the U weights having the same learning rate; the results of this are shown in Table 4.2. While there may appear to be a soft PCA taking place, this effect vanishes in larger networks. This effect – that increased size removes the tendency to perform a "soft" PCA – has been found in other models in this section, and therefore slightly larger networks have been used in obtaining other corroborative empirical results.

Table 4.1. Results show the weights of a typical set of V and W weights from the parallel learning algorithm with the angle in radians between the **v** and **w** vectors. No. of iterations=40 000. Initially, $\alpha_w = \alpha_v = 0.0001; \alpha_{u_1} = 0.000005; \alpha_{u_2} = 0.00001; \alpha_{u_3} = 0.000015$.

V			W		
1.000	0.012	0.000	**1.000**	0.012	−0.000
0.030	**1.000**	0.008	0.030	**1.000**	0.009
0.005	−0.003	**0.994**	0.004	−0.003	**0.994**
0.004	0.007	−0.023	0.004	0.007	−0.023
−0.002	0.000	−0.000	−0.002	0.000	0.000

Output Neuron No.:	1	2	3
Angle (radians)	0.0011	0.0006	0.0004

Table 4.2. Results of the same network as before with homogeneous U learning rates.

V			W		
0.651	0.188	1.0312	0.650	0.188	1.031
−0.151	0.984	−0.092	−0.150	0.984	−0.092
0.824	0.049	−0.305	0.842	0.049	−0.305
−0.019	0.006	0.006	−0.018	0.006	0.006
−0.001	−0.001	−0.001	−0.001	−0.001	−0.002

4.2 Differential Activation Functions

In this section we investigate three models of peer inhibitory output neurons which use activation functions instead of learning rate to break the symmetry of the system. We will not repeat the explicit derivations of the last section for each of the three models as the mathematics is usually very similar; however specific points of interest will be identified and analysed. First each of the three models is introduced and experimental results are given; points of interest in each are identified. Then a comparison of the models is made and additional models which have similar properties are outlined. Unless stated otherwise, the empirical data is obtained from a network with 12 inputs, 7 output neurons and input data with variance of $x_1 >$ variance of $x_2 > \cdots$. A is a diagonal matrix with $A_{11} > A_{22} > \cdots$. The simulated network in this section is slightly larger than in previous sections in order to highlight various interesting empirical results which are not so obvious in smaller networks.

Restricting ourselves to models where only the output neurons use activation functions and restricting such activation functions to multiplicative factors (so that we still have a linear system), there are several possible models; we will identify three separate classes of models by determining the characteristics of three of these models. We will use the same conventions in naming vectors as before. Note, in particular, that there are still no self-connections for the output neurons i.e. the main diagonal of U is composed of zeros. In this section, all $\mathbf{u}$ weights will learn at the same rate, η_U, but there will be differential activation functions (multiplicative factors) on the output neurons. For simplicity, we assume that $\eta_W = \eta_V = \eta_U = \eta$. (This does not affect our results and provides a simpler mathematical model).

4.2.1 Model 1: Lateral Activation Functions

The first model is governed by the system of equations:

$$\mathbf{y}' = W\mathbf{x} \tag{4.30}$$

$$\mathbf{y} = \mathbf{y}' - AU\mathbf{y}' \tag{4.31}$$

$$\mathbf{e} = \mathbf{x} - V^T\mathbf{y} \tag{4.32}$$

$$\Delta W = \eta \mathbf{y} \mathbf{e}^T \tag{4.33}$$

$$\Delta V = \eta \mathbf{y} \mathbf{e}^T \tag{4.34}$$

$$\Delta U = \eta \mathbf{y} \mathbf{y}^T \tag{4.35}$$

Then, omitting details, we have $G = (I - AU)W$ and

$$\frac{dW}{dt} = (I - AU)WC - (I - AU)WCW^T(I - AU)^TV \tag{4.36}$$

$$\frac{dU}{dt} = (I - AU)WCW^T(I - AU)^T \tag{4.37}$$

$$\frac{dG}{dt} = (I - AU)\frac{dW}{dt} - A\frac{dU}{dt}W \tag{4.38}$$

Then if U converges to 0 at which point $G = W = V$,

$$\frac{dG}{dt} = (I - AU)\{(I - AU)WC - (I - AU)WCW^T(I - AU)^TV\}$$
$$-A\{(I - AU)WCW^T(I - AU)^T\}W$$
$$= (I - AU)\{GC - GCG^TV - AGCG^TW\}$$
$$\rightarrow GC - (I + A)GCG^TW \text{ as } U \rightarrow 0$$

at convergence, which should be compared with the equations in the previous section.

However, the dynamics of the two models should not be assumed to be the same; note, for example, the different format of the equations governing the behaviour of the U values. This will be shown to be important in the investigation below. The $\mathbf{w}_i$ weights (almost) converge to the eigenvectors of the input data's covariance matrix. The underlying rationale for this network is that each output neuron has different susceptibility to the inhibition from its peers. The results shown in Table 4.3 are from a 12-input, 7 output-neuron

Table 4.3. Model 1 results. The results are from a network with 12 input neurons and 7 output neurons; each column is the vector of weights into each output neuron. In most cases the actual Principal Components have been identified.

	1	2	3	4	5	6	7
1	−0.005	−0.001	0.004	0.006	−0.023	−0.025	**0.999**
2	0.001	0.005	0.012	0.008	−0.014	**0.999**	0.029
3	−0.011	0.011	0.022	−0.045	**−0.998**	−0.023	−0.031
4	−0.003	0.021	0.058	**−0.999**	0.024	0.010	−0.003
5	0.001	**0.146**	**−0.996**	−0.031	0.019	−0.002	0.009
6	0.040	**0.998**	**0.119**	−0.004	−0.012	−0.019	−0.002
7	**0.998**	0.029	0.023	0.004	−0.016	0.013	0.014
8	−0.016	−0.013	0.004	0.004	0.003	0.011	−0.007
9	0.004	0.006	−0.003	0.004	−0.003	−0.001	−0.20
10	0.013	0.005	−0.001	0.002	−0.001	0.003	−0.005
11	−0.001	−0.004	−0.001	−0.001	−0.003	0.001	0.004
12	0.001	−0.001	−0.003	−0.003	−0.004	−0.002	0.001

network, with $a_i = 1.58 - 0.2 * (i - 1)$ for $i = 1,...,7$. We note that while almost all the Principal Components have been certainly identified, the second and third output neurons have not identified precisely their respective Principal Components. The vectors seem to be almost correct and to satisfy $\mathbf{w}_2.\mathbf{w}_3 = 0$ (the vectors are mutually normal) yet are not in the direction of the eigenvectors themselves. In fact by appropriate choice of the parameters a_i, this effect can be eliminated; however:

1. We wish to develop a network which will not require any fine tuning as it is used in different situations.

2. The analysis of this fault provides insight into the network behaviour.

The reason for this fault lies in the convergence of the U values. In the model of the last section the learning rule for the U values was shown to be

$$\frac{dU}{dt} = A(I - U)WCW^T(I - U)^T$$

Notice that as $U \to 0$, this learning rule continues to be dominated by the A matrix whereas, in Model 1, the effect of the A matrix vanishes as $U \to 0$ (see (4.37)).

The importance of $\frac{dU}{dt}$ is due to the fact that the value of $\frac{dU}{dt}$ is a major component in the decay term in $\frac{dG}{dt}$. Thus, as U tends to zero and hence $\frac{dU}{dt} \to 0$, the decay term tends to zero. In the previous model this decay term maintained its differential effect as it decreased, but in Model 1, as $U \to 0$, the decay loses its directional impact – it becomes homogeneous. More formally, consider the convergence to a solution of the system which is not an eigenvector. Let $\mathbf{w}_i$ have converged to $a\mathbf{c}_i + b\mathbf{c}_j, a, b \neq 0$ and let $\mathbf{w}_j$ have converged to $c\mathbf{c}_i + d\mathbf{c}_j, c, d \neq 0$. Both c and d are necessarily not zero as $\mathbf{w}_i.\mathbf{w}_j = 0$. Then we can show that the $(i, j)^{th}$ element of WCW^T can be shown to be

$$\mathbf{w}_i C\mathbf{w}_j^T = (\lambda_i a\mathbf{c}_i + \lambda_j b\mathbf{c}_j).(c\mathbf{c}_i + d\mathbf{c}_j)$$
$$= \lambda_i ac + \lambda_j bd$$
$$\text{Similarly, } \mathbf{w}_i C\mathbf{w}_i^T = \lambda_i a^2 + \lambda_j b^2$$
$$\mathbf{w}_j C\mathbf{w}_i^T = \lambda_i ac + \lambda_j bd$$
$$\mathbf{w}_j C\mathbf{w}_j^T = \lambda_i c^2 + \lambda_j d^2$$

Now the ith row of $(I - AU)$ is $[\; -a_i u_{i1} \cdots 1 \cdots -a_i u_{ij} \cdots -a_i u_{im} \;]$ where the 1 is in the ith position. Similarly with the jth row. Then,

$$\frac{du_{ij}}{dt} = (I - AU)_i WCW^T(I - AU)_j^T$$
$$= u_{ij}^2(a_i a_j(\lambda_i ac + \lambda_j bd)) - u_{ij}(a_j(\lambda_i a^2 + \lambda_j b^2)$$
$$+ a_i(\lambda_i c^2 + \lambda_j d^2)) + (\lambda_i ac + \lambda_j bd)$$

Thus $\frac{du_{ij}}{dt}$ at $u_{ij} = 0$ is equal to $(\lambda_i ac + \lambda_j bd)$ which is exactly zero for $b = c = 0$, i.e. the eigenvectors.

But, as $u_{ij} \to 0$, a situation arises where there is no particular impulse for the change of u_{ij} in any particular direction provided the constraint $ac + bd = 0$ is satisfied. The symmetry of the formula shows that $\frac{du_{ij}}{dt} = \frac{du_{ji}}{dt}$; thus the differential term in $\frac{dg_i}{dt}$ also vanishes at this point and so the weights, having approached the eigenvectors, need not converge precisely to any eigenvector – the driving force of differential weight decay has vanished.

For the system analysed in the previous section, we have

$$\frac{du_{ij}}{dt} = a_i u_{ij}^2(\lambda_i ac + \lambda_j bd) - u_{ij}(a_i(\lambda_i a^2 + \lambda_j b^2) + a_j(\lambda_i c^2 + \lambda_j d^2)) + a_i(\lambda_i ac + \lambda_j bd)$$

Note the asymmetry in this rule in that even as $u_{ij} \to 0$, $\frac{du_{ij}}{dt} \neq \frac{du_{ji}}{dt}$. Therefore this system will maintain a preferred degree of slope no matter how small U becomes. As noted earlier, the value of $\frac{du_{ij}}{dt}$ is precisely the value of the decay term in $\frac{d\mathbf{g}_i}{dt}$ and so W will continue to converge, taking account of the network's asymmetry, no matter how small U becomes.

It is possible to somewhat circumvent this problem by choosing values of a_i which are sufficiently different to make the term containing u_{ij} significant until U is closer to zero; however, this is an heuristic and an a priori decision on the size of the a_i cannot be made.

A further difficulty with this model is that we now have the activations fed to the output neurons being processed differently depending on their origins: all output neurons are responding equally to the fed-forward activations from the input neurons but are responding differentially to the activations from their peers. This seems unrealistic for a biological model and requires an engineered model to have meta-information (as to whether to use an activation function or not). Model 2 is designed to rectify this.

4.2.2 Model 2: Lateral and Feedforward Activation Functions

Our equations are almost the same as in the last section, but note that every input to Z carries an activation function times the weighted inputs. The rationale behind this model is a belief that all inputs to an output neuron should be treated equally.

$$\mathbf{y}' = AW\mathbf{x} \tag{4.39}$$

$$\mathbf{y} = \mathbf{y}' - AU\mathbf{y}' \tag{4.40}$$

$$\mathbf{e} = \mathbf{x} - V^T\mathbf{y} \tag{4.41}$$

$$\Delta W = \eta \mathbf{y} \mathbf{e}^T \tag{4.42}$$

$$\Delta V = \eta \mathbf{y} \mathbf{e}^T \tag{4.43}$$

$$\Delta U = \eta \mathbf{y} \mathbf{y}^T \tag{4.44}$$

Then we have

$$\mathbf{y} = (I - AU)\mathbf{y}' = (I - AU)AW\mathbf{x}$$

Therefore

$$G = (I - AU)AW$$

$$\frac{dG}{dt} = (I - AU)A\frac{dW}{dt} - A\frac{dU}{dt}AW$$

As before, we can show that

$$\frac{dW}{dt} = (I - AU)AWC - (I - AU)AWC((I - AU)AW)^T V \quad (4.45)$$

$$\frac{dU}{dt} = (I - AU)AWC((I - AU)AW)^T \quad (4.46)$$

and so that

$$\frac{dG}{dt} = (I - AU)A(I - AU)AWC$$
$$-(I - AU)A(I - AU)AWC((I - AU)AW)^T V$$
$$-A(I - AU)AWC((I - AU)AW)^T AW$$
$$\rightarrow A^2WC - A^2WC(AW)^T V - A^2WC(AW)^T AW \text{ as } U \rightarrow 0$$
$$= A\{(AW)C - (AW)C(AW)^T V - (AW)C(AW)^T AW\} \quad (4.47)$$

The factor A, which multiplies the whole of the right side of (4.47) acts on the whole of that side equally i.e. does not have the differential decay effect necessary to force convergence to Principal Components. It does however have the effect that the first vector $\mathbf{w}_1$ has the highest learning rate and so will tend to adapt to those directions which contain the greatest variance before the others do. This results in a "fuzzy" PCA. The first and last terms are precisely those of the Subspace Algorithm [138] i.e. will cause (AW), and hence W, to converge to the Principal Subspace though not to the Principal Components themselves. The results of a simulation based on the usual set of data are shown in Table 4.4.

Table 4.4. Model 2 results. Results from a network with 12 input neurons and 7 output neurons. Each vector into each output neuron (the columns above) has almost all of its weight into the first seven directions. The actual Principal Components have not, in general, been identified.

	1	2	3	4	5	6	7
1	**0.438**	−0.359	0.763	0.001	−0.039	−0.013	0.004
2	**0.661**	0.199	−0.418	−0.176	−0.070	0.017	0.008
3	0.022	**0.709**	0.275	0.478	0.112	−0.012	−0.001
4	0.076	−0.266	−0.182	**0.852**	0.273	0.006	0.047
5	0.007	0.010	0.023	−0.206	**1.063**	0.394	0.010
6	−0.027	0.009	0.016	0.060	−0.248	**1.248**	0.101
7	0.009	0.006	0.010	−0.007	−0.007	−0.062	**1.611**
8	0.003	0.006	−0.010	−0.006	0.002	−0.009	0.006
9	−0.011	0.010	−0.013	−0.002	0.001	−0.006	0.010
10	−0.003	0.004	−0.004	0.005	−0.002	0.002	0.013
11	0.003	0.001	0.003	0.002	0.001	−0.005	−0.002
12	0.000	0.001	0.002	0.002	0.001	0.002	0.002

A second drawback of this model is that the activation function in (4.40) is applied only to the effect of the other inhibitory output neurons. i.e. the output neurons are calculating their final output values after the activation function has been applied. This may not be appropriate in a biological mode.

4.2.3 Model 3: Feedforward Activation Functions

Now we only have an activation function on the first calculation of the z values:

$$\mathbf{y'} = AW\mathbf{x} \tag{4.48}$$

$$\mathbf{y} = \mathbf{y'} - U\mathbf{y'} \tag{4.49}$$

$$\mathbf{e} = \mathbf{x} - V^T\mathbf{y} \tag{4.50}$$

$$\Delta W = \eta\mathbf{y}\mathbf{e}^T \tag{4.51}$$

$$\Delta V = \eta\mathbf{y}\mathbf{e}^T \tag{4.52}$$

$$\Delta U = \eta\mathbf{y}\mathbf{y}^T \tag{4.53}$$

Then we have

$$\mathbf{y} = (I - U)\mathbf{y'} = (I - U)AW\mathbf{x}$$

Therefore

$$G = (I - U)AW$$

$$\frac{dG}{dt} = (I - U)A\frac{dW}{dt} - \frac{dU}{dt}AW$$

As before, we can show that

$$\frac{dW}{dt} = (I - U)AWC - (I - U)AWC((I - U)AW)^TV \tag{4.54}$$

$$\frac{dU}{dt} = (I - U)AWC((I - U)AW)^T \tag{4.55}$$

and thus

$$\begin{aligned}
\frac{dG}{dt} &= (I - U)A(I - U)AWC - (I - U)A(I - U)AWC((I - U)AW)^TV \\
&\quad -(I - U)AWC((I - U)AW)^TAW \\
&\to A^2WC - A^2WC(AW)^TV - AWC(AW)^TAW \text{ as } U \to 0 \\
&= A\{(AW)C - (AW)C(AW)^TV\} - (AW)C(AW)^T(AW) \\
&= A\{(AW)C - AWC(AW)^T\frac{AW}{A} - A^{-1}(AW)C(AW)^T(AW)\} \tag{4.56}
\end{aligned}$$

The rationale behind this model is that each output neuron has an equal inhibitory effect on the others but has a differential response to inputs.

The central term causes convergence to the Principal Subspace but within that subspace causes no convergence to the Principal Components themselves.

The last term is the one which causes convergence to the actual Principal Components.

In more detail, consider the system as governed by

$$\frac{dG}{dt} = (I - U)A(GC - GCG^T V - ((I - U)A)^{-1}GCG^T AW)$$

$$\rightarrow A(GC - GCG^T V - A^{-1}GCG^T G) \text{ as } U \rightarrow 0$$

We note that as A is diagonal and of full rank, it has an inverse, which is also diagonal, and each element of the inverse, $(A^{-1})_{ii} = a_i^{-1}$.

Table 4.5. Model 3: the results are from a network with 12 input neurons and 7 output neurons; each column is the vector of weights into each output neuron. In all cases the actual Principal Components have been identified. Note the different direction of "slope" of the bold figures (see text).

	1	2	3	4	5	6	7
1	**0.794**	−0.024	0.014	−0.009	−0.004	−0.002	0.001
2	0.020	**0.849**	0.003	−0.007	0.014	0.013	0.005
3	−0.027	−0.012	**0.918**	0.047	0.028	−0.008	−0.010
4	−0.005	0.010	−0.021	**1.007**	−0.015	0.018	−0.001
5	0.002	−0.002	−0.017	−0.004	**1.129**	0.016	0.013
6	−0.005	−0.018	0.011	0.000	−0.007	**1.308**	−0.040
7	0.009	0.013	0.014	−0.001	−0.025	0.021	**1.613**
8	−0.007	0.010	−0.002	−0.004	−0.001	−0.010	0.006
9	−0.017	−0.001	0.002	−0.004	−0.004	0.006	−0.013
10	−0.003	0.001	0.002	−0.003	0.001	0.006	0.003
11	0.003	−0.002	0.003	0.001	−0.003	−0.003	0.002
12	0.001	0.001	0.004	0.003	0.001	0.002	0.001

Then, as before, the central term will have no effect on convergence once the weights have converged to the Principal Subspace. Within that subspace convergence of the weights is governed by the equation

$$\frac{dG}{dt} = A(GC - A^{-1}GCG^T G)$$

This causes $\mathbf{g}_i$ to converge to $\frac{1}{\sqrt{A_{ii}^{-1}}}\mathbf{c}_k = \sqrt{a_i}\mathbf{c}_k$ where $k = m - i$. Note that if $a_i > a_j$ then $a_i^{-1} < a_j^{-1}$ and so this model causes convergence in the "opposite direction" to that normally associated with the A values. See Table 4.5. Now, $\mathbf{g}_i = (I - U)A\mathbf{w}_i \rightarrow a_i\mathbf{w}_i$; therefore,

$$a_i\mathbf{w}_i = \sqrt{a_i}\mathbf{c}_i$$

$$\text{i.e. } \mathbf{w}_i = \frac{1}{\sqrt{a_i}}\mathbf{c}_i$$

Now G is simply a mathematical construct to help us understand the model; the actual learning processes take place in the modification of the W and U weights; in particular, the values of the W weights are determined by the convergence of $\frac{dW}{dt}$. At $\mathbf{w}_i = \frac{1}{\sqrt{a_i}}\mathbf{c}_i$

$$
\begin{aligned}
\frac{d\mathbf{w}_i}{dt} &= a_i\mathbf{w}_iC - a_i\mathbf{w}_iC(a_i\mathbf{w}_i)^T\mathbf{v}_i \\
&= a_i\lambda_i a_i^{-\frac{1}{2}}\mathbf{c}_i - a_i(a_i^{-\frac{1}{2}}\mathbf{c}_i)C(a_ia_i^{-\frac{1}{2}}\mathbf{c}_i)^T(a_i^{-\frac{1}{2}}\mathbf{c}_i) \\
&= \lambda_i a_i^{\frac{1}{2}}\mathbf{c}_i - \lambda_i a_i^{\frac{1}{2}}\mathbf{c}_i.a_i^{\frac{1}{2}}\mathbf{c}_i(\lambda_i a_i^{-\frac{1}{2}}\mathbf{c}_i) \\
&= \lambda_i a_i^{\frac{1}{2}}\mathbf{c}_i - \lambda_i a_i a_i^{-\frac{1}{2}}\mathbf{c}_i \\
&= 0.
\end{aligned}
$$

In other words, that solution of the overall system dynamics, $\frac{dG}{dt} = 0$, is also a solution of $\frac{dW}{dt} = 0$. The system will converge in harmony.

4.2.4 Summary

We present a summary comparison of the models using the rate of change of the various weights to guide the comparison.

Change in W, $\frac{dW}{dt}$

Note first that in all three models, $\frac{dW}{dt}$ will cause convergence to the Principal Subspace but not to the actual Principal Components themselves. We repeat the equations here for convenience:

Model 1: $\frac{dW}{dt} = (I - AU)WC - (I - AU)WCW^T(I - AU)^TV$

Model 2: $\frac{dW}{dt} = (I - AU)AWC - (I - AU)AWC((I - AU)AW)^TV$

Model 3: $\frac{dW}{dt} = (I - U)AWC - (I - U)AWC((I - U)AW)^TV$

All three equations are of the form

$$
\frac{dW}{dt} \to GC - GCG^TV \text{ as } U \to 0
$$

and as before it can be shown that $V = KG$ for some diagonal matrix K. Therefore, all of these equations will cause the G vectors to converge to the Principal Subspace but not to the Principal Components themselves. We note that if $a_i\mathbf{w}_i$ has converged to an eigenvector, then $\mathbf{w}_i$ has converged to the same eigenvector. Further, these equations will determine the size of the W vectors; since each vector, $\mathbf{g}_i$, is of length 1, we have, noting that $\lim_{t\to\infty} u_{ij} = 0$:

Model 1: $|\mathbf{w}_i| = 1$.

Model 2: $|\mathbf{w}_i| = a_i^{-\frac{1}{2}}$.

Table 4.6. Experimental values of A_i and corresponding values of $f(A_i)$ for the functions shown.

	A_1	A_2	A_3	A_4	A_5	A_6	A_7
x	1.58	1.38	1.18	0.98	0.78	0.58	0.38
$\sqrt{x}$	1.26	1.17	1.09	0.99	0.88	0.76	0.62
$\frac{1}{\sqrt{x}}$	0.79	0.85	0.92	1.01	1.14	1.31	1.62

Table 4.7. Lengths of the relevant vectors from the three models.

	w_1	w_2	w_3	w_4	w_5	w_6	w_7
Model 1	0.998	1.000	0.998	0.999	0.999	0.999	1.000
Model 2	0.792	0.861	0.930	1.013	1.130	1.309	1.614
Model 3	0.795	0.849	0.919	1.008	1.130	1.308	1.614

Model 3: $|\mathbf{w}_i| = a_i^{-\frac{1}{2}}$.

This analysis is corroborated by Tables 4.6 and 4.7. We will demonstrate that the stated solution is correct for Model 2; the other models can be similarly[2] analysed. We have

$$\frac{dW}{dt} = (I - AU)AWC - (I - AU)AWC((I - AU)AW)^T V$$
$$\rightarrow AWC - AWC(AW)^T V \text{ as } U \rightarrow 0$$

Now let $\mathbf{w}_i = a_i^{-\frac{1}{2}}\mathbf{b}_i$ where $\mathbf{b}_i = \sum_j b_j\mathbf{c}_i$, i.e. $\mathbf{b}_i$ is a unit length combination of the vectors $\mathbf{c}_i$, with $\mathbf{c}_i$ the eigenvectors of C as usual. Then,

$$\frac{dW}{dt} = AWC - AWC(AW)^T V$$
$$\frac{d\mathbf{w}_i}{dt} = a_i(a_i^{-\frac{1}{2}}\sum_j b_j\mathbf{c}_j)C - a_i(a_i^{-\frac{1}{2}}\sum_j b_j\mathbf{c}_j)Ca_i(a_i^{-\frac{1}{2}}\sum_j b_j\mathbf{c}_j)^T(a_i^{-\frac{1}{2}}\sum_j b_j\mathbf{c}_j)$$
$$= a_i^{\frac{1}{2}}\sum_j \lambda_j b_j\mathbf{c}_j - a_i^{\frac{1}{2}}\sum_j \lambda_j b_j\mathbf{c}_j\left(a_i^{\frac{1}{2}}\sum_j b_j\mathbf{c}_j\right)\left(a_i^{-\frac{1}{2}}\sum_j b_j\mathbf{c}_j\right)$$
$$= a_i^{\frac{1}{2}}\sum_j \lambda_j b_j\mathbf{c}_j - a_i^{\frac{1}{2}}\sum_j \lambda_j b_j\mathbf{c}_j\left(\sum_j b_j\mathbf{c}_j\right)^2$$
$$= 0$$

since $\mathbf{b}_j$ is a unit length vector and so $(\sum_j b_j\mathbf{c}_j)^2 = 1$;

---

[2] Indeed, more simply since we can use the fact that these models cause convergence to the eigenvectors.

Change in U, $\frac{dU}{dt}$

Neither will $\frac{dU}{dt}$ cause convergence to the actual eigenvectors.

Model 1: $\frac{dU}{dt} = (I - AU)WCW^T(I - AU)^T$

Model 2: $\frac{dU}{dt} = (I - AU)AWC((I - AU)AW)^T$

Model 3: $\frac{dU}{dt} = (I - U)AWC((I - U)AW)^T$

Note that all equations have the general form $\frac{dU}{dt} = GCG^T$. We consider only the case $U = 0$, since it can be shown that, at $U = 0$ in all models, $\frac{dU}{dt} = 0^3$. Then since A is a diagonal matrix, convergence to nonzero diagonal elements is achieved whenever the rows and columns of W are orthogonal. (If $a_i\mathbf{w}_i \perp a_j\mathbf{w}_j$ then $\mathbf{w}_i \perp \mathbf{w}_j$). Consider Model 3, at $U = 0$, $\mathbf{w}_i = a_i^{-\frac{1}{2}}\mathbf{c}_i$; then

$$
\begin{aligned}
\frac{du_{ij}}{dt} &= (I - U)A\mathbf{w}_i C((I - U)A\mathbf{w}_j)^T \\
&\rightarrow A\mathbf{w}_i C(A\mathbf{w}_j)^T \text{ as } U \rightarrow 0 \\
&= a_i\lambda_i\mathbf{w}_i.a_j\mathbf{w}_j \\
&= \lambda_i a_i a_j\mathbf{w}_i.\mathbf{w}_j \\
&= \lambda_i a_i a_j\delta_{ij}a_i^{-\frac{1}{2}}a_j^{-\frac{1}{2}} \\
&= \lambda_i\delta_{ij}a_i^{\frac{1}{2}}a_j^{\frac{1}{2}}
\end{aligned}
$$

where δ_{ij} is the Kronecker delta. Therefore, off the main diagonal, $\frac{du_{ij}}{dt} = 0$. So the equations governing the growth of U and W merely ensure that the columns of W form an orthogonal basis of the Principal Subspace of the covariance matrix of the input data.

Change in G, $\frac{dG}{dt}$

$\frac{dG}{dt}$ is the equation which causes convergence to the Principal Components. $\frac{dG}{dt}$ is the manifestation of the interaction between the dynamical development of U and that of W. Recall that G is defined as

Model 1: $G = (I - AU)W$

Model 2: $G = (I - AU)AW$

Model 3: $G = (I - U)AW$

If we assume convergence at $U=0$, then we have

Model 1:

$$
\begin{aligned}
\frac{dG}{dt} &= (I - AU)\{GC - GCG^TV\} - AGCG^TW \\
&\rightarrow GC - (I + A)GCG^TW \\
&\approx GC - (I + A)GCG^TG
\end{aligned}
$$

---

[3] This is not to be taken that we assume that $\frac{dU}{dt} = 0 \Longrightarrow U = 0$.

This causes convergence of $\mathbf{g}_i$ to $\frac{1}{\sqrt{1+a_i}}\mathbf{c}_i$; however, note the caveats made in Section 4.2.1.

Model 2:

$$\frac{dG}{dt} = (I - AU)A\{GC - GCG^T V\} - AGCG^T(AW)$$
$$\rightarrow A\{GC - GCG^T V - GCG^T(AW)\}$$

There is no specific parameter which will force the weights to the actual Principal Components themselves; both decay terms cause convergence to the Principal Subspace but within that subspace are nondirectional.

Model 3

$$\frac{dG}{dt} = (I - U)A\{GC - GCG^T V\} - GCG^T AW$$
$$\rightarrow A\{GC - GCG^T V - A^{-1}GCG^T AW\}$$
$$\approx A\{GC - GCG^T(A^{-1}G) - A^{-1}GCG^T G\}$$

This causes convergence to $\frac{1}{\sqrt{a_i^{-1}}}\mathbf{c}_{m-i}$. The essential point to note is that the vector associated with the smallest value of A corresponds to the vector with the largest eigenvalue.

4.2.5 Other Models

Clearly the models identified above are not the only possible models; however, all models investigated have been found to be of one of the three classes defined by the above three models e.g.

Model 4: Our equations are almost the same as in Model 2, but note that the second outputs of the output neuron are calculated after the subtraction of the inputs from their peers :

$$\mathbf{y}' = AW\mathbf{x} \tag{4.57}$$
$$\mathbf{y} = A(\mathbf{y}' - U\mathbf{y}') \tag{4.58}$$
$$\mathbf{e} = \mathbf{x} - V^T\mathbf{y} \tag{4.59}$$
$$\Delta W = \eta \mathbf{y} \mathbf{e}^T \tag{4.60}$$
$$\Delta V = \eta \mathbf{y} \mathbf{e}^T \tag{4.61}$$
$$\Delta U = \eta \mathbf{y} \mathbf{y}^T \tag{4.62}$$

Then we have

$$\mathbf{y} = A(I - U)\mathbf{y}' = A(I - U)AW\mathbf{x}$$

Therefore

$$G = A(I - U)AW$$
$$\frac{dG}{dt} = A(I - U)A\frac{dW}{dt} - A\frac{dU}{dt}AW$$

As before, we can show that

$$\frac{dW}{dt} = A(I - U)AWC - A(I - U)AWC(A(I - U)AW)^T V \quad (4.63)$$

$$\frac{dU}{dt} = A(I - U)AWC(A(I - U)AW)^T \quad (4.64)$$

and so that

$$
\begin{aligned}
\frac{dG}{dt} &= A(I - U)AA(I - U)AWC \\
&\quad -A(I - U)AA(I - U)AWC(A(I - U)AW)^T V \\
&\quad -AA(I - U)AWC(A(I - U)AW)^T AW \\
&= A(I - U)A(GC - GCG^T V - (A(I - U))^{-1}GCG^T(AW)) \\
&\rightarrow A^2\{(A^2W)C - (A^2W)C(A^2W)^T V - A^{-1}(A^2W)C(A^2W)^T AW\}
\end{aligned}
$$

This model acts similarly to Model 3 in that the A^{-1} causes convergence. All effects, however, are even more pronounced: the differential learning rates of the feedforward functions g_i are even more exaggerated, and the differences in the size of vectors are larger. The rationale behind this model is that each output neuron will calculate its activation at all times based on the sum of the inputs at that time. This is possibly the most realistic biological model; it requires no meta knowledge and is local, simple and parallel.

Model 5:

$$\mathbf{y}' = W\mathbf{x} \quad (4.65)$$

$$\mathbf{y} = A(\mathbf{y}' - U\mathbf{y}') \quad (4.66)$$

$$\mathbf{e} = \mathbf{x} - V^T\mathbf{y} \quad (4.67)$$

$$\Delta W = \eta\mathbf{y}\mathbf{e}^T \quad (4.68)$$

$$\Delta V = \eta\mathbf{y}\mathbf{e}^T \quad (4.69)$$

$$\Delta U = \eta\mathbf{y}\mathbf{y}^T \quad (4.70)$$

Then we have

$$\mathbf{y} = A(I - U)\mathbf{y}' = A(I - U)W\mathbf{x}$$

Therefore

$$G = A(I - U)W$$

$$\frac{dG}{dt} = A(I - U)\frac{dW}{dt} - A\frac{dU}{dt}W$$

As before, we can show that

$$\frac{dW}{dt} = A(I - U)WC - A(I - U)WC(A(I - U)W)^T V \quad (4.71)$$

$$\frac{dU}{dt} = A(I - U)WC(A(I - U)W)^T \quad (4.72)$$

$$\frac{dG}{dt} \rightarrow A\{GC - GCG^T V - GCG^T W\} \quad (4.73)$$

This model acts like Model 2. It finds the Principal Subspace but not the Principal Components themselves as there is no differential decay in the model.

Model 6:

$$\mathbf{y}' = AW\mathbf{x} \tag{4.74}$$
$$\mathbf{y} = A(\mathbf{y}' - AU\mathbf{y}') \tag{4.75}$$
$$\mathbf{e} = \mathbf{x} - V^T\mathbf{y} \tag{4.76}$$
$$\Delta W = \eta \mathbf{y}\mathbf{e}^T \tag{4.77}$$
$$\Delta V = \eta \mathbf{y}\mathbf{e}^T \tag{4.78}$$
$$\Delta U = \eta \mathbf{y}\mathbf{y}^T \tag{4.79}$$

Then we have

$$\mathbf{y} = A(I - AU)\mathbf{y}' = A(I - AU)AW\mathbf{x}$$

Therefore

$$G = A(I - AU)AW$$
$$\frac{dG}{dt} = A(I - AU)A\frac{dW}{dt} - AA\frac{dU}{dt}AW$$

As before, we can show that

$$\frac{dW}{dt} = A(I - AU)AWC - A(I - AU)AWC(A(I - AU)AW)^T V$$
$$\frac{dU}{dt} = A(I - AU)AWC(A(I - AU)AW)^T \tag{4.80}$$
$$\frac{dG}{dt} \rightarrow A^2\{GC - GCG^T V - GCG^T AW\} \tag{4.81}$$

Again there is no asymmetry in the learning process and so the model will act like Model 2 – it finds the Principal Subspace but not the Principal Components.

4.3 Emergent Properties of the Peer-Inhibition Network

A possible criticism of envisaging biological neural nets as performing a Principal Component Analysis is that it leads to a situation whereby one neuron is in charge of all information passing in a particular direction; therefore, if it is in any way damaged, the information in that direction which should be passed on will be lost.

An interesting property of large Peer-inhibitory networks is that such so-called "grandmother" cells take a very long while to form: the network quickly self-organises until each output neuron's weights are maximally sensitive to four or five directions, but it then takes a very long while to converge to a single Principal Component. A typical set of weights is shown in Tables 4.9 to 4.12. It should be seen that each weight is gradually converging to a particular Principal Component; what is more difficult to show is that the direction of each Principal Component is maximally associated with the weights of approximately four or five output neurons after 50 000 iterations and the weights only gradually thereafter converge to a single Principal Component. The full matrix would appear as a "fuzzy" diagonal of bold-faced type.

4.4 Conclusion

In this chapter, we have used the negative feedback effect of each output neuron on the other output neurons in an attempt to ensure that the weights into each output neuron converge to the actual Principal Components themselves. We have shown that it is not enough to simply feed back the activations of the output neurons as they are calculated – some measure of asymmetry must be introduced into the network.

Two main methods of introducing such asymmetry have been shown to be successful: output neurons using different learning rates and output neurons using different activation functions. While both of these methods have been shown to be successful, the results of the analysis and experiments with different activation functions show that simply to introduce an asymmetry into the network without a theoretical understanding of the consequences could lead to unpredictable consequences: in the case of activation functions, it has been shown that the same activation function can have the desired effect, no effect or the opposite effect to that which might be predicted depending on where it is introduced.

Nevertheless, several models have been shown to be extremely successful at finding Principal Components of input data and hence of transmitting the maximum amount of information with the least possible amount of hardware. The inherent parallelism of the network should make possible a very fast implementation of the network on parallel hardware.

Table 4.8. Each row represents the first 7 components of the first 5 output neuron weights in a network of 100 inputs and 50 output neurons after 50 000 iterations; all weights not shown are less than 0.1, most considerably less...

	Input 1	Input 2	Input 3	Input 4	Input 5	Input 6	Input 7
Output neuron 1	**0.737**	**0.405**	**0.464**	0.063	**0.154**	**0.114**	−0.079
Output neuron 2	0.072	**0.323**	**−0.637**	**0.360**	**0.557**	−0.030	−0.021
Output neuron 3	0.046	**0.638**	**−0.375**	**−0.461**	**−0.429**	−0.045	**0.297**
Output neuron 4	**0.340**	**−0.259**	**−0.255**	**0.533**	**−0.475**	−0.015	**−0.440**
Output neuron 5	**0.506**	**−0.466**	**−0.366**	**−0.443**	**0.123**	**0.322**	**0.170**

Table 4.9. The same network as above after 100000 iterations...

	Input 1	Input 2	Input 3	Input 4	Input 5	Input 6	Input 7
Output neuron 1	**0.898**	**0.325**	**0.256**	−0.021	0.081	−0.025	−0.048
Output neuron 2	0.021	**0.361**	**−0.690**	**0.379**	**0.466**	−0.032	0.059
Output neuron 3	−0.095	**0.626**	**−0.319**	**−0.496**	**−0.471**	0.043	−0.091
Output neuron 4	**0.266**	**−0.289**	**−0.378**	**0.490**	**−0.653**	−0.086	**−0.226**
Output neuron 5	**0.329**	**−0.525**	**−0.488**	**−0.525**	**0.101**	**0.190**	**0.209**

Table 4.10. The same network as above after 200000 iterations...

	Input 1	Input 2	Input 3	Input 4	Input 5	Input 6	Input 7
Output neuron 1	**0.922**	0.022	**0.162**	**−0.130**	−0.032	−0.015	0.023
Output neuron 2	**0.207**	**0.191**	**−0.850**	**0.339**	**0.248**	0.083	−0.053
Output neuron 3	−0.082	**0.845**	−0.082	**−0.457**	**−0.273**	0.004	0.060
Output neuron 4	0.034	**−0.184**	**−0.257**	**0.180**	**−0.914**	**−0.121**	−0.016
Output neuron 5	−0.010	**−0.479**	**−0.398**	**−0.737**	0.005	**0.237**	**0.131**

Table 4.11. The same network as above after 300000 iterations...

	Input 1	Input 2	Input 3	Input 4	Input 5	Input 6	Input 7
Output neuron 1	**0.961**	**−0.133**	**0.139**	**−0.167**	−0.037	−0.002	0.041
Output neuron 2	**0.218**	**0.164**	**−0.902**	**0.308**	0.091	0.035	−0.015
Output neuron 3	0.063	**0.949**	0.078	**−0.265**	**−0.101**	0.056	0.016
Output neuron 4	−0.032	−0.063	**−0.102**	0.034	**−0.971**	**−0.157**	0.067
Output neuron 5	**−0.137**	**−0.219**	**−0.367**	**−0.860**	0.011	**0.190**	**0.103**

Table 4.12. At the other end of the matrix/table, the output neurons' weights are converging only slightly more slowly.

	Input 50	Input 49	Input 48	Input 47	Input 46	Input 45	Input 44
Output neuron 45	0.013	**0.273**	−0.061	**0.123**	**−0.717**	−0.077	**0.297**
Output neuron 46	**0.129**	**−0.172**	**0.122**	**0.271**	−0.017	**−0.868**	0.039
Output neuron 47	0.065	**0.317**	−0.072	**−0.859**	−0.090	**−0.313**	0.018
Output neuron 48	−0.088	0.058	**0.963**	**−0.114**	**−0.105**	**0.121**	−0.093
Output neuron 49	**0.731**	**0.558**	0.080	**0.223**	**0.270**	**0.108**	0.002
Output neuron 50	**0.617**	**−0.654**	0.007	**−0.237**	**−0.297**	**0.170**	−0.015

5

Multiple Cause Data

Barlow [5] has developed a theory of learning based on the neuron as a "suspicious coincidence detector": if input A is regularly met in conjunction with input B this represents a suspicious coincidence; there is something in the neuron's environment which is worth investigating. A crude example might be the coincidence of mother's face and food and warmth for a young animal. Field [45] has made an important distinction between compact codes and sparse distributed codes (Fig. 5.1). These types of codes are sometimes known as "factorial codes": we have lots of different symbols representing the different parts of the environment and the occurrence of a particular input is simply the product of probabilities of the individual code symbols. So if neuron 1 says that it has identified a sheep and neuron 2 states that it has identified blackness, then presenting a black sheep to the network will cause neurons 1 and 2 to both fire. Also such a coding should be invertible: if we know the code we should be able to go to the environment and identify precisely the input which caused the code reaction from the network. So when we see neurons 1 and 2 firing we know that it is due to a black sheep.

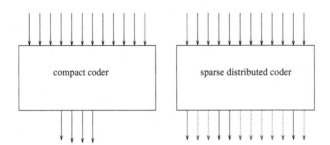

Fig. 5.1. The coder on the left is transforming the data by reducing its dimensionality. That on the right retains the dimensionality of the data but sparsifies its representation

A compact code is a code such as that formed by projecting the data onto the first few Principal Components of the data. In other words, such a code will reduce the dimensionality of the representation while, if it is to be a good code, retaining as much of the information in the data as possible. A sparse distributed code, on the other hand, retains (or perhaps even increases) the dimensionality of the representation, but in such a way that any individual code uses only a few dimensions of the channel.

However, in a sparse distributed code, while overall each individual cell may have the same probability of firing, the chances of two cells firing together are very much reduced. Thus, the chances of false "suspicious coincidences" are very much reduced. The statistics of such a code are strongly kurtotic – each code has a great number of low firing cells corresponding to the random occasional firing of neurons, while at the same time there are a few cells, those which correspond to the signal, firing very strongly.

Hertz et al. [73] point out that simple competitive learning leads to the creation of **grandmother** cells, the proverbial neuron which would fire if and only if your grandmother hove in sight. The major difficulty with such neurons is their lack of robustness: if you lose your grandmother cell, will you never again recognise your grannie. In addition, we should note that with N grandmother cells we can only recognise N categories, whereas if we are using a binary code, we could distinguish between 2^N categories.

So simple competitive learning leads to a single neuron firing in response to an input pattern. At the other extreme, when a large number of neurons are firing for each input pattern, subsequent layers have to work much harder to identify the information being presented by that layer.

Földiák [46] has suggested that an appropriate compromise between these competing constraints would be to have the neurons in a layer forming a sparse coding of the input data. i.e. each pattern is represented by a set of m firing neurons where $m > 1$ but $m << n$, the number of neurons in the layer. He believes that such a representation potentially trades off the benefits of increased representational capacity to be had with a distributed representation with the simplicity to be had with a completely local representation. It is this balance between cooperation (so that a set of output neurons can represent the input pattern which is currently being presented) and competition (so that not all outputs are used to represent all patterns) that seems necessary for the extraction of salient features of the problem. Other networks which are designed to tackle the same problem are discussed in Appendix B.

We will maintain a close connection with psychological principles which should suggest that we are using a biologically plausible rule such as the Hebbian rule. We have seen that the Hebbian rule will extract information from the environment. What we need to do is modify the Hebbian rule so that each neuron responds to a particular set of inputs which is unique to itself. One way to do this is with competitive learning; we will be more interested in a second way which uses an implicit competition. This will involve a rectification of the negative feedback PCA network (we will ensure that all outputs or weights remain positive) which we will discuss initially in terms of constrained PCA and then subsequently in terms of Factor Analysis (FA).

5.0.1 A Typical Experiment

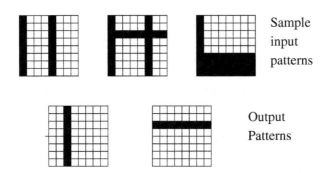

Sample input patterns

Output Patterns

Fig. 5.2. The top line shows sample data (input values) presented to the network. The second layer shows the independent sources which we hope will also be the network's response. If this is so, it has clearly identified the suspicious coincidences of lines (several squares regularly being nonzero simultaneously).

A standard data set, which we will use in this chapter, consists of a square grid of input values where $x_i = 1$ if the ith square is black and 0 otherwise (see Fig. 5.2). However the patterns are not random patterns: each input consists of a number of randomly chosen horizontal or vertical lines. The network must identify the existence of these lines. The important thing to note is that each line can be considered as an independent source of blackening a pixel on the grid: it may be that a particular pixel will be twice blackened by both a horizontal and a vertical line at the same time, but we need to identify both of these sources.

Typically, on an 8*8 grid, each of the 16 possible lines are drawn with a fixed probability of $\frac{1}{8}$ independently from each of the others. The data set then is highly redundant in that there exists 2^{64} possible patterns and we are only using at most 2^{16} of these. We will typically have 16 output neurons whose aim is to identify (or respond optimally to) one of the input lines. Thus from any pattern composed of some of the set of 16 lines, we can identify exactly which of the 16 lines were used to create the pattern. Note the factorial nature of the coding we are looking for: neurons 1, 3 and 10 will fire if and only if the input is composed of a pattern from sources 1, 3 and 10. Note also the code's reversibility: given that neurons 1, 3 and 10 are firing we can recreate the input data exactly.

If we use a principal component net on this data, our first principal component will be a small magnitude uniform vector over all 64 positions. i.e. we get a global smearing of the patterns which does not reveal how each pattern came to be formed. Subsequent principal components will not identify a unique bar.

5.1 Non-negative Weights

There is one obvious asymmetry used in nature which we have not used as yet: it is believed that signals from neurons may be excitatory or inhibitory but not both i.e. a neuron's output can excite (positively) other neurons or it can inhibit (negatively) other neurons; what cannot happen is that it switches between excitation and inhibition. The results reported in previous chapters were based on a model where the weights were allowed to take any value, positive or negative, and so every neuron could switch between excitatory and inhibitory values. If we allow only non-negative weights i.e. ensure that if a weight, while learning, never takes a negative value, we have the following interesting situation.

Assume that two weights of our converged network have values $a\mathbf{c_i}+b\mathbf{c_j}$ and $c\mathbf{c_i} + d\mathbf{c_j}$, with the same notation as before. Then, since the weights converge to an orthogonal basis of the space, $ac + bd = 0$. Now if none of the terms a, b, c or d can be negative, then at least two must be zero (one from each term ac and bd). In other words, this constraint swings the weight vectors through the weight space to the actual Principal Components themselves. Since we are not directing the process, situations where several sets of weights converge to the same Principal Component tend to appear. An extreme example is shown in Table 5.1 in which we report the results of a simulation on the same type of data as previously but where the basic VW negative feedback network was set up and the weights allowed to learn concurrently but with the constraint of non-negativity.

Table 5.1. The weights of a five input, four output neuron network with the same type of input data as previously.

	Input 1	Input 2	Input 3	Input 4	Input 5
Neuron 1	**0.552**	0	0.004	0.000	0.001
Neuron 2	**0.700**	0	0.005	0.000	0.001
Neuron 3	0.035	**0.991**	0.004	0.000	0.000
Neuron 4	**0.450**	0	0.003	0.000	0.001

Clearly, the weights of output neurons 1, 2 and 4 have all converged to the same Principal Component. Note that, at the end of the simulation, the weights marked only "0" have been stopped from becoming negative.

5.1.1 Other Data Sets

However, there is one clear difficulty with this program – if we are calculating Principal Components from a general data set, there must be a negative term in at least one of the Principal Components' coordinates. (In order to have orthogonal directions, the inner product of the components must be zero and hence there must be at least one negative component).

Table 5.2. Sample Principal Components of the data calculated using a standard statistical package.

Direction	1	2	3	4	5	Value
First PC	**0.584**	**0.811**	0.000	−0.002	−0.002	59.3
Second PC	−0.006	0.001	**−0.469**	**−0.617**	**−0.632**	33.7
Third PC	**−0.811**	**0.584**	−0.010	0.005	0.011	7.1
Fourth PC	0.012	−0.008	**−0.876**	**0.235**	**0.421**	2.4
Fifth PC	0.002	−0.001	**0.111**	**−0.751**	**0.650**	0.5

To further investigate the network's potential, data from a distribution whose Principal Components are shown in Table 5.2 was used as input to the network: it should be clear that there is a sharp division in the data between the first two directions and the last three.

It might seem to be possible for the network to converge to a mixture of the above weights e.g. the filters {0.584,0,0.469,0,0} and {0,0.811,0,0.617,0.632} span the subspace of the first two Principal Components. This does not happen; the network converges to the first two Principal Components themselves (see analysis in the next section).

It is impossible for the network using the positive weight constraint to converge to any filter containing a negative component i.e. from the third onwards. To find out how the network would respond to a situation where there were more degrees of freedom than possible directions to be found, we used the network with these five inputs and four output neurons (with the constraint that no weights are allowed to become negative). The results are shown in Table 5.3.

Table 5.3. The weights of a five input, four output negative feedback network operating on the data of the previous table.

Output neuron 1	0.005	0.000	**0.465**	**0.616**	**0.635**
Output neuron 2	**0.391**	**0.518**	0.001	0.000	0.000
Output neuron 3	**0.324**	**0.467**	0.001	0.000	0.000
Output neuron 4	**0.296**	**0.409**	0.001	0.000	0.000

It is clear that the first output neuron has found the second Principal Component while the second, third and fourth output neurons have found the first Principal Component. This is a general finding with this type of network with the non-negative weight constraints.

This form of information extraction may be important if the data has been preprocessed in order to isolate the "texture" data from the "colour" data from the "smell" data, etc.. This type of distributed data-processing is known to happen in biological neural networks. However, this type of data-processing cannot be an initial data-processing function. The information must first be differentiated into disjoint dimensions: if there is any overlap between the dimensions in which the data exists, no more than one Principal Component per data set is possible.

We note that the length of the total vector of weights into output neurons 2, 3 and 4 is one unit. This is convenient in that it dispels the end of "grandmother cells"– that elusive neuron which would recognise only your grandmother. If such recognition is spread over a group of neurons such as is shown here, this provides a robustness in the network which has been missing up till now.

Restricting ourselves to our specialised data, we can show that the principal component directions are found: in Table 5.4, we show the weights from a network with 100 inputs of the same specialised form as before and 50 output neurons. All weights not shown were under 0.015 after 100 000 iterations. We

Table 5.4. Results from a negative feedback neuron network with 100 inputs labelled 0–99, and 50 output neurons labelled 0–49, e.g. the weights into output neuron 0 have converged to (input) direction 9 and the weight in that direction was 1.000

Neuron	Input	Weight	Neuron	Input	Weight	Neuron	Input	Weight
0	9	1.000	17	23	0.999	34	10	0.374
1	19	1.000	18	3	0.308	35	20	0.999
2	33	0.998	19	4	0.455	36	16	0.999
3	8	0.554	20	28	0.998	37	25	0.999
4	12	1.000	21	5	1.000	38	6	0.999
5	1	0.370	22	27	0.998	39	0	0.641
6	1	0.490	23	35	0.993	40	18	0.999
7	0	0.611	24	26	0.998	41	31	0.998
8	2	0.999	25	30	0.999	42	22	1.000
9	32	0.997	26	14	0.999	43	7	0.674
10	15	0.999	27	3	0.951	44	0	0.142
11	8	0.481	28	10	0.870	45	10	0.320
12	29	0.998	29	11	0.797	46	21	0.999
13	34	0.995	30	17	0.999	47	7	0.738
14	0	0.441	31	8	0.679	48	11	0.602
15	13	1.000	32	1	0.451	49	1	0.648
16	4	0.890	33	24	0.999			

note that

- The weights into each output neuron converged to a single Principal Component.
- Some of the directions with largest eigenvalues, (those of the first 12 Principal Components) were covered by more than one output neuron. Maximally, directions 0 and 1 were covered by the weights of four output neurons.
- The weights in each direction still (approximately) had length 1.
- There is no half-way house with this network's converged weights – the weights into different output neurons are either totally orthogonal or in completely the same direction.

The last two points are potentially important in considering a negative feedback network as a possible explanation of biological networks' information management processes. If such recognition is spread over a group of neurons such as is shown here, this provides a robustness in the network which has been missing up until now. Further, since the total weight in any direction still has length 1, then directions which are represented by more than one output neuron are not over emphasised in any data processing.

Experiments with larger sizes of networks have shown that the above effects increases with size.

5.1.2 Theoretical Analysis

Consider a network with four inputs and two output neurons. Let the eigenvector of the input data with the largest eigenvalue be $\mathbf{a} = \{a_1, a_2, 0, 0\}$ and the second eigenvector be $\mathbf{b} = \{0, 0, b_3, b_4\}$. Then, in the situation described in the last section, if $\mathbf{w}_i$ is the vector of weights into output neuron i, then $\mathbf{w}_1$ converges to $\mathbf{a}$ and $\mathbf{w}_2$ to $\mathbf{b}$, or vice versa. We show that this is a stable solution.

The expected input is

$$E(x) = k_a\mathbf{a} + k_b\mathbf{b} = \{k_aa_1, k_aa_2, k_bb_3, k_bb_4\} \tag{5.1}$$
$$E(y_1) = k_a(a_1^2 + a_2^2) \tag{5.2}$$
$$E(e_1) = k_aa_1 - k_aa_1(a_1^2 + a_2^2) \tag{5.3}$$

So, the expected change in the weight between x_1 and y_1 is

$$E(\Delta w_{11}) = \eta E(e_1y_1)$$
$$= \eta\{(a_1(1 - (a_1^2 + a_2^2))k_a^2(a_1^2 + a_2^2)\} \tag{5.4}$$

So, since $a_i \neq 0, \forall i$ and $k_a > 0$, $E(\Delta w_{11}) = 0 \iff a_1^2 + a_2^2 = 1$. Since the eigenvector $\mathbf{a}$ has length 1, the converged weights are stable.

Now consider a network whose weights have converged to values incorporating both eigenvectors e.g. let $\mathbf{w}_1$ have converged to $\{a_1, 0, b_3, 0\}$ and $\mathbf{w}_2$

have converged to $\{0, a_2, 0, b_4\}$. Then a similar argument to the above leads to

$$E(\Delta w_{11}) = (k_a a_1^2 + k_b b_3^2)a_1(k_a - (k_a a_1^2 + k_b b_3^2))$$

So $E(\Delta w_{11}) = 0 \iff k_a = k_b \frac{b_3^2}{1-a_1^2} = k_b \frac{b_3^2}{a_2^2}$. This equation imposes constraints on the input data relating the internal proportion of each eigenvector in each direction to the relative size of each eigenvalue.

Thus, while it is possible to construct data to satisfy these criteria, it is not generally the case that data sets will comply with the constraints.

Further, note that this equation is only one of four derivable from the system. We can show that the system requires $\frac{k_a}{k_b} = \frac{b_3^2}{a_2^2} = \frac{b_4^2}{a_1^2}$ for stability. This will not generally be true.

5.1.3 Conclusion

We have shown that the constraint of not allowing the weights to become negative has an interesting effect on our artificial data sets: the actual Principal Components (rather than the Principal Subspace) is found in a totally parallel network with no lateral connections. The experiment with the data whose Principal Components, of necessity, contained a negative weight value showed one limitation of this network and so it cannot be used with general data sets to find all principal components. However, the experiments do suggest that the network can extract interesting factors which underlie a data set. It is from this perspective that we now examine this network.

5.2 Factor Analysis

A standard method of finding independent sources in a data set is the statistical technique of Factor Analysis (FA). It is in the nature of human existence that we find raw data far less interesting than data which has been defined in some structured model: we may have raw crime statistics showing criminals ages, history, background, family circumstances, employment and so on, but it is only when we structure the data that we gain information. We need not necessarily be stating that a specific factor causes the criminality, merely that it helps to explain the data. In doing so we are performing an elementary factor analysis. Various early neural network models which approach Factor Analysis are discussed in Appendix B.

PCA and FA are closely related statistical techniques both of which achieve an efficient compression of the data but in a different manner. They can both be described as methods to explain the data set in a smaller number of dimensions, but FA is based on assumptions about the nature of the underlying data whereas PCA is model free.

We can also view PCA as an attempt to find a transformation from the data set to a compressed code, whereas in FA we try to find the linear transformation which takes us from a set of hidden factors to the data set. Since PCA is model free, we make no assumptions about the form of the data's covariance matrix. However, FA begins with a specific model which is usually constrained by our prior knowledge or assumptions about the data set. The general FA model can be described by the following relationship:

$$\mathbf{x} = \Lambda \mathbf{f} + \mathbf{u} \tag{5.5}$$

where $\mathbf{x}$ is a vector representative of the data set, $\mathbf{f}$ is the vector of factors, Λ is the matrix of factor loadings, and $\mathbf{u}$ is the vector of specific (unique) factors.

The usual assumptions built into the model are that:

• $E(\mathbf{f}) = 0$, $\text{Var}(\mathbf{f}) = I$ i.e. the factors are zero mean, of the same power and uncorrelated with each other.
• $E(\mathbf{u}) = 0$, $\text{Cov}(\mathbf{u}_i, \mathbf{u}_j) = 0$, $\forall i, j$, i.e the specific factors are also zero mean and uncorrelated with each other.
• $\text{Cov}(\mathbf{f}, \mathbf{u}) = 0$ i.e. the factors and specific factors are uncorrelated.

Let $\Sigma = E(\mathbf{x}\mathbf{x}^T)$ be the covariance matrix of $\mathbf{x}$ (again assuming zero mean data). Then $\Sigma = \Lambda \Lambda^T + \Phi$ where Φ is the covariance matrix of the specific factors, $\mathbf{u}$, and so Φ is a diagonal matrix, $\text{diag}\{\Phi_{11}, \Phi_{22}, , \Phi_{MM}\}$. Now whereas PCA attempts to explain Σ without a specific model, FA attempts to find parameters Λ and Φ which explain Σ and only if such models can be found will a Factor Analysis be successful.

Estimations of the Factor loading is usually done by means of one of two methods – Maximum Likelihood Estimation or Principal Factor Analysis [125]. Since Principal Factor Analysis is a method which uses the covariance matrix of the input data, this is the method in which we shall be interested in this chapter.

5.2.1 Principal Factor Analysis

We can expand the main diagonal of $\Sigma = E(\mathbf{x}\mathbf{x}^T)$ as

$$\sigma_{ii}^2 = \sum_j \lambda_{ij}^2 + \Phi_{ii}^2 = h_i^2 + \Phi_{ii}^2 \tag{5.6}$$

i.e. the variance of the data set can be broken into two parts, the first of which is known as the communality and is the variance of $\mathbf{x}_i$ which is shared via the factor loadings with the other variables. The second is the specific or unique variance associated with the ith input.

In Principal Factor Analysis (PFA), an initial estimate of the communalities is made. This is inserted into the main diagonal of the data covariance matrix and then a PCA is performed on the "reduced correlation matrix". A commonly used estimate of the communalities is the maximum of the square of the multiple correlation coefficient of the ith variable with every other variable. Then PCA is used to give the eigenvector–eigenvalue decomposition of the matrix $\Sigma - \hat{\Phi} = \sum_{i=1}^{M} \alpha_i \mathbf{c}_i \mathbf{c}_i^T$ where $\hat{\Phi}$ is the estimated specific variances and α_i is the ith eigenvalue corresponding to the ith eigenvector $\mathbf{c}_i$. The estimated factor loadings are given by $\hat{\lambda}_i = \alpha_i^{\frac{1}{2}} \mathbf{c}_i$, where $\hat{\lambda}_i$ is the column vector associated with the ith factor. Now estimates of the specific variances can be calculated using

$$\hat{\Phi}_{ii}^2 = \sigma_{ii}^2 - \sum_{j=1}^{M} \hat{\lambda}_{ij}^2 \tag{5.7}$$

Note, however, that this method will only provide a feasible factor analysis if all of the $\hat{\Phi}_{ii}^2 = \sigma_{ii}^2 - \sum_{j=1}^{M} \hat{\lambda}_{ij}^2$ are non-negative since these values represent the specific variances. This fact can be used to determine the number of factors which the model permits.

5.2.2 The Varimax Rotation

In PCA the orthogonal components are arranged in descending order of importance and a unique solution is always possible. The factor loadings in FA are not unique and there are likely to be substantial loadings on more than one factor that may be negative or positive. This often means that the results in standard FA are difficult to interpret. To overcome these problems it is possible to perform a rigid rotation of the axes of the factor space and so identify a more simplified structure in the data that is more easily interpretable. One well-known method of achieving this is the Varimax rotation [125]. This has as its rationale that factors should be formed with a few large loadings and as many near-zero loadings as possible, normally achieved by an iterative maximization of a quadratic function of the factor loadings. It is worth noting that the Varimax rotation aims for a sparse response to the data and this was acknowledged as an efficient form coding.

5.2.3 Relation to Non-negativity

We consider the output vector $\mathbf{y}$ to be a vector of factors and the weight matrix, W, to be the factor loadings. In constraining the weight vectors, we are making an assumption about the form of the model. The learning rules for the Principal Subspace network have been shown [101] to be an approximation to the learning rules necessary to minimise $F(W, \mathbf{x}) = ||(\mathbf{x} - WW^T\mathbf{x})||^2$. By adding the constraint that all weights must be non-negative, we are creating a constrained optimisation problem which can be solved by the method of Lagrange multipliers:
Minimise $L(W, \mathbf{x}) = ||(\mathbf{x} - WW^T\mathbf{x})||^2 + \Lambda g(W)$ where Λ is the matrix of Lagrange multipliers and $g(W) = W$ in this case. Now at the optimal point,

the weights, W, and the Lagrange multiplier matrix, Λ , are known to satisfy the Kuhn–Tucker equations:

$$\frac{\partial F}{\partial w_{ij}^*} + \lambda_{ij}^* \frac{\partial g(W^*)}{\partial w_{ij}^*} = 0$$

$$g(W^*) \geq 0$$

$$\lambda_{ij}^* \leq 0$$

$$\lambda_{ij}^* g(W^*) = 0$$

which, in the special case $g(W) = W$ gives

$$\frac{\partial F}{\partial w_{ij}^*} + \lambda_{ij}^* = 0$$

$$w_{ij}^* \geq 0$$

$$\lambda_{ij}^* \leq 0$$

$$\lambda_{ij}^* w_{ij}^* = 0$$

where we have used the asterisk to identify the optimal point.

Now this last equation $\lambda_{ij}^* w_{ij}^* = 0$ means that this additional constraint is removed for any nonzero weight i.e. if $w_{ij}^* \neq 0$ then λ_{ij}^* must be 0. So for positive weights, a standard PCA is being performed i.e. the first condition becomes $\frac{\partial F}{\partial w_{ij}^*} = 0$. The additional constraint only applies to those weights which have gone to zero through rectification. Thus a (depleted) PCA is being formed – each neuron is attempting to extract the maximum variance from the depleted input vector. So this method is similar to PFA in that the model is constrained before performing a PCA. This method has been shown to be effective in finding the factors which underlie a data set.

5.2.4 The Bars Data

In this section, we illustrate the network's performance with the benchmark "bars data". The networks are trained over 50 000 presentations of the data with a learning rate of 0.05 which is annealed linearly to zero over the training period. The squares in the following figures are individual weight vectors each connected to one output, arranged 2-dimensionally for convenience when viewing the results. The diameter of circles within the squares represent the individual weight values where black is a positive weight and white is a negative weight.

We use the more difficult form of this data in that horizontal and vertical bars may appear together. Each bar (horizontal or vertical) may appear at the input to the network with a probability of $1/8$ as described previously.

The converged weights of our network when using the straightforward rectification of the outputs are shown in Figure 5.3.

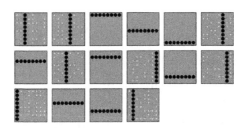

Fig. 5.3. Converged weights of the nonlinear PCA network with the straightforword rectification $[y]^+$ with 16 outputs. Each square represents the weights into a single output neuron and these weights have been arranged so that horizontal and vertical bars are identifiable by eye.

5.2.5 Continuous Data

As an extension to the factor analysis on discrete data, we use five mixtures of sine waves as input data to our network so that

$$x_0 = \sin(t) + \sin(2t)$$
$$x_1 = \sin\left(t + \frac{\pi}{8}\right) + \sin\left(2t + \frac{\pi}{4}\right)$$
$$x_2 = \sin\left(3t + \frac{3\pi}{7}\right)$$
$$x_3 = \sin\left(4t + \frac{4\pi}{3}\right) + sin(5t)$$
$$x_4 = 2\sin(5t)$$

The first two mixtures, x_0 and x_1, are identical but slightly out of phase, the third is a totally independent sine wave and the last two contain the same sine wave, however, one has another sine wave mixed with it. Therefore the relationship between the outputs of the sources is straightforward in the case of x_3 and x_4 but time-varying in the case of x_0 and x_1 where the underlying source is emitting different phase signals (Figure 5.4 might give a clearer idea of the relationships). Results are shown in Table 5.5: the first neuron is identifying x_3 and x_4 while the second identifies x_0 and x_1; both ignore x_3.

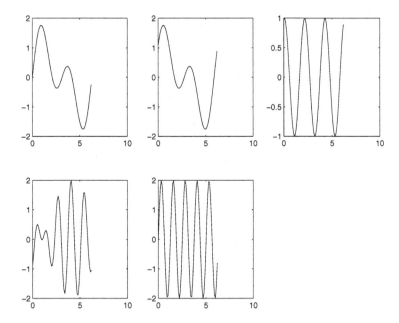

Fig. 5.4. The five dimensional data set: the first two are the same signal but with slightly diffferent phases; the last two both contain a signal of frequency sin(5t).

Table 5.5. The converged weights of the network showing that the underlying sources have been found.

	Output 1	Output 2
Input 1	0.000755	**0.708634**
Input 2	0.001283	**0.705238**
Input 3	0.021043	0.021338
Input 4	**0.708208**	0.002265
Input 5	**0.705604**	0.001692

5.2.6 Generalised PCA

Not unnaturally, the topic of "nonlinear PCA" receives a great deal of attention from the neural net community e.g. [169, 32, 144, 100, 103, 140, 101, 99]. The impetus for such a development is the recognition that neural networks are ideally suited to nonlinear adaptation because of their incremental methods of learning: while closed-form solutions may exist for linear processes such as PCA, such methods are simply not possible for non-linear algorithms.

Karhunen and Joutsensalo [101] derived from Oja's Subspace Algorithm [138]

$$\Delta W(t) = \eta(t)[I - W(t)W^T(t)]\mathbf{x}(t)\mathbf{x}^T(t)W(t)$$

three nonlinear equations by introducing a nonlinear function $f()$ in one of three ways:

$$\Delta W(t) = \eta(t)[\mathbf{x}(t)\mathbf{x}^T(t)W(t) - W(t)f(W^T(t)\mathbf{x}(t))\mathbf{x}^T(t)W(t)] \qquad (5.8)$$
$$\Delta W(t) = \eta(t)[I - W(t)W^T(t)]\mathbf{x}(t)f(\mathbf{x}^T(t)W(t)) \qquad (5.9)$$
$$\Delta W(t) = \eta(t)[\mathbf{x}(t)f(\mathbf{x}^T(t)W(t)) - W(t)f(W^T(t)\mathbf{x}(t))f(\mathbf{x}^T(t)W)] (5.10)$$

(5.8) has a first term identical to the PCA rule; it can be shown to converge to the Principal Components and, it is claimed in [101], does so more robustly than the totally linear algorithm. (5.9) was derived in from a constrained optimisation criterion and will be used in Exploratory Projection Pursuit in Chapter 6. We now follow [101] in deriving (5.10).

First consider a negative feedback network which calculates a (nonlinear) function, $f()$, of the weighted activations passed to it and then returns (via the same weights) and subtracts this function from the input neurons. i.e.

$$y_i = f(a_i) = f\left(\sum_{j=1}^{N} w_{ij}x_j\right) \qquad (5.11)$$

$$e_j = x_j - \sum_{k=1}^{M} w_{kj}y_k \qquad (5.12)$$

$$\Delta w_{ij} = \eta_t y_i e_j \qquad (5.13)$$

$$= \eta_t f\left(\sum_{k=1}^{N} w_{ik}x_k\right)\left\{x_j - \sum_{l=1}^{M} w_{lj}f\left(\sum_{k=1}^{N} w_{lk}x_k\right)\right\} \qquad (5.14)$$

Consider the optimisation criterion that the reconstruction error, $\mathbf{e}$, of the N-dimensional input vector $\mathbf{x}$ is made as small as possible after activation is returned from M output neurons. i.e. we wish to minimise

$$J(W) = \frac{1}{2}\mathbf{1}^T E(\mathbf{e}^2|W) = \frac{1}{2}\mathbf{1}^T E[(\mathbf{x} - Wf(W^T\mathbf{x}))^2|W] \qquad (5.15)$$

where $\mathbf{1}$ is the vector of 1s. We note that Karhunen and Joutsensalo have suggested that, instead of the mean-square error which we have used, we may use any even, monotonic, nonconstant, nonnegative, continuously differentiable cost function which has a minimum at $e = 0$. They also note that the function $f()$ must be an odd function in order that the feedback process stabilises.

Now consider the reconstruction error e_j at the jth input neuron.

$$e_j = x_j - \sum_{i=1}^{M} w_{ij}f(\mathbf{w}_i.\mathbf{x}) \qquad (5.16)$$

where, as before, $\mathbf{w_i}$ is the vector of weights into the ith output neuron. Then we wish to find stationary point(s) of the derivative of $J(W)$ i.e. where

$$\frac{\partial J(W)}{\partial \mathbf{w}_m} = \sum_{j=1}^{M} e_j \frac{\partial e_j}{\partial \mathbf{w}_m} = 0 \qquad (5.17)$$

Now,

$$\frac{\delta e_j}{\delta \mathbf{w}_m} = -w_{mj} f'(\mathbf{w}_m.\mathbf{x}) - f(\mathbf{w}_m.\mathbf{x})[0,0,...,1,0,..0]^T \qquad (5.18)$$

where $f'()$ denotes the derivative of $f()$ w.r.t $\mathbf{w}_m$ and the last vector has a 1 in only the jth position. Then,

$$\frac{\partial J(W)}{\partial \mathbf{w}_m} = -\sum_{j=1}^{M} \left(x_j - \sum_{i=1}^{M} w_{ij} f(\mathbf{w}_i.\mathbf{x}) \right).$$

$$(w_{mj} f'(\mathbf{w}_m.\mathbf{x})\mathbf{x} + f(\mathbf{w}_m.\mathbf{x})[0,0,..,1,0..,0])$$
$$= -(\mathbf{x} - W^T f(W\mathbf{x}))W f'(W\mathbf{x})\mathbf{x} - (\mathbf{x} - W^T f(W\mathbf{x}))f(W\mathbf{x}) \qquad (5.19)$$

This can be used in the usual way in the gradient ascent algorithm

$$\Delta W \propto \frac{\partial J(W)}{\partial W}$$

to give a learning rule

$$\Delta \mathbf{w}_m = (\mathbf{x} - W^T f(W\mathbf{x}))W f'(W\mathbf{x})\mathbf{x} + (\mathbf{x} - W^T f(W\mathbf{x}))f(W\mathbf{x}) \qquad (5.20)$$

Now the first term affects the weight update on an element–by–element basis while the second term affects each element uniformly; so the driving force of weight convergence comes from the second term.

Therefore the algorithm given by (5.10) may be thought of as an approximation of an algorithm to minimise the reconstruction error at the summing neurons. We note that the linear case as developed in previous chapters is a special case of this algorithm with $f(\mathbf{y}) = \mathbf{y}$ and Xu [185] has shown in this case that the direction of the rate of change of weights using the subspace learning algorithm is on average the same as the derived direction found by the derived algorithm (5.20).

5.2.7 Non-negative Outputs

A constraint which has the same effect of enforcing non-negativity on the weights is to only allow the outputs to be non-negative: this can be shown to have the same effect of forcing the weights to learn the individual factors underlying a data set. Thus we are using a simple nonlinear function in (5.11) of

$$y_i = f(a_i) = \begin{cases} a_i, & \text{if } a_i > 0 \\ 0, & \text{if } a_i \leq 0 \end{cases} \qquad (5.21)$$

Though this works perfectly well in identifying underlying causes, it has the disadvantage from a theoretical perspective that it is a discontinuous (and hence nondifferentiable) function. However, we can smooth the effect rather than have a strict cut-off by creating a function of the activation which gives a very small value when the weights are negative. This allows us to use all the theory associated with nonlinear Principal Component networks [101, 102]. The algorithm then becomes

$$\Delta w_{ij} = \eta\{y_i x_j - y_i \sum_k w_{kj} y_k\}$$

$$\text{where e.g. } y_i = \exp\left(\sum_j w_{ij} x_j\right)$$

$$\text{or } y_i = \frac{1}{1 + \exp(\{-\sum_j w_{ij} x_j\} + a)}$$

The last of these we will call the soft threshold function since the value of a can be chosen to vary the response of the function to different data sets.

All of these methods used with the negative feedback rule in parallel, cause convergence of the weights to find the individual bars. Figure 5.5 shows the converged weights when using the threshold implementation of the network. We use the *soft threshold* function with the artificial data here as it is very flexible to work with and forgiving of nonoptimal network parameters. The other threshold functions from this family once optimised, however, yield results that are virtually indentical.

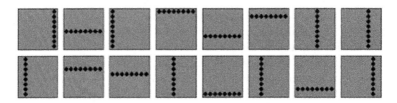

Fig. 5.5. Trained threshold network – 16 outputs.

5.2.8 Additive Noise

However, the outputs of the network as described up until now will learn partial bars when there are more outputs than causes, i.e. the bars are shared between the outputs. If there are sixteen causes in the input space then only sixteen outputs should be used to code these individual causes regardless of the dimensionality of the output space. We do not wish to require to have this prior knowledge of the number of sources hidden in the signals; we wish the network to tell us in some way how many signals there are in the mixture and stop learning when only these signals have been learned. We will show that, by adding noise to the network after the application of the nonlinearity, we can ensure that only as many outputs respond to the data as there are causes. An added bonus is that this may be interesting from a biological perspective as real neurons tend to operate in a noisy environment. There are already a number of a biologically plausible aspects to the network such as Hebbian learning, local learning, sparse coding, as well as thresholding and, as we shall see, the possibility of topographical mapping.

First we discuss a basis in which solely non-negative coordinates can determine the position of every point.

5.2.9 Dimensionality of the Output Space

By enforcing positive–only output values it may be said that we are searching for positive–only codes on the outputs of the network that will represent the data. Another way that this may be expressed is that we are looking for a set of positive only coordinates that will describe each data point. It is now shown that the following is true: for every n–dimensional Euclidean space, there exists an $(n+1)$ basis in which every point in the space may be expressed with nonnegative coordinates. We first illustrate this for a two–dimensional space and then discuss the general case.

Let $\mathbf{e}_1, \mathbf{e}_2$, be a basis of a two–dimensional space. Then every point, P, may be expressed as $P = (p_1, p_2)$ which is equivalent to $\vec{OP} = p_1\mathbf{e}_1 + p_2\mathbf{e}_2$ where O is the origin.

Let $\epsilon = -(\mathbf{e}_1 + \mathbf{e}_2)$. Then we will show that every point $P(p_1, p_2)$ can be expressed as $P' = (p'_1, p'_2, p')$ in the basis $(\mathbf{e}_1, \mathbf{e}_2, \epsilon)$ where $p'_1, p'_2, p' \geq 0$.

This is trivially true if $p_1, p_2 > 0$ since $P' = (p_1, p_2, 0)$ in the basis $(\mathbf{e}_1, \mathbf{e}_2, \epsilon)$.

Let $p_1 < 0, p_2 > 0$, then $P(p_1, p_2) = P'(0, p_2 - p_1, -p_1)$, since

$$\vec{OP'} = 0\mathbf{e}_1 + (p_2 - p_1)\mathbf{e}_2 + p_1(\mathbf{e}_1 + \mathbf{e}_2)$$
$$= p_1\mathbf{e}_1 + p_2\mathbf{e}_2 = \vec{OP}$$

Finally, each of $0, p_2 - p_1, -p_1 \geq 0$, since $p_1 < 0$ and $p_2 > 0$. Similarly if $p_2 < 0$.

If both $p_1, p_2 < 0$, $P'(-p_2, -p_1, -(p_1 + p_2))$ has positive coordinates and

$$\vec{OP'} = -p_2\mathbf{e}_1 - p_1\mathbf{e}_2 - (p_1 + p_2)(-(\mathbf{e}_1 + \mathbf{e}_2))$$
$$= p_1\mathbf{e}_1 + p_2\mathbf{e}_2 = \vec{OP}$$

Thus the assertion is true for every two–dimensional space.

Now let $\mathbf{e}_1, ..., \mathbf{e}_n$, be a basis for an n-dimensional space. Let $\epsilon_n = -\sum_{i=1}^{n} \mathbf{e}_i$; then we must prove that every point $P(p_1, .., p_n)$ in the $\mathbf{e}_1, ..., \mathbf{e}_n$

basis can be represented as $P'(p_1', ..., p_n')$ in the $\mathbf{e}_1, ..., \mathbf{e}_n, \epsilon_n$ basis with non-negative coordinates.

Let A be the subset of indices in $U = \{1, ..., n\}$ which are such that $p_i < 0$. Then

$$\vec{OP} = \sum_{i=1}^{n} p_i \mathbf{e}_i = \sum_{i \in (U-A)} p_i \mathbf{e}_i + \sum_{i \in A} p_i \mathbf{e}_i$$

$$= \sum_{i \in (U-A)} p_i \mathbf{e}_i - \sum_{i \in (U-A)} \left(\sum_{j \in A} p_j \right) \mathbf{e}_i - \sum_{i \in A} \left(\sum_{j \in A, j \neq i} p_j \right) \mathbf{e}_i - \sum_{j \in A} p_j \epsilon_n$$

$$= \sum_{i \in (U-A)} \left(p_i - \sum_{j \in A} p_j \right) \mathbf{e}_i - \sum_{i \in A} \left(\sum_{j \in A, j \neq i} p_j \right) \mathbf{e}_i - \sum_{j \in A} p_j \epsilon_n$$

where we have used $U - A$ to denote the complement of A in U. Now each coordinate is non-negative in the new basis.

This result merely shows that is possible to code an n–dimensional space with nonnegative-only coordinates in an $(n+1)$–basis. However it is also possible to code this space with nonnegative coordinates in any basis with more than $(n + 1)$ elements. We will derive a method for automatically finding a basis with the least number of basis vectors in which the data set can be expressed with nonnegative coordinates. If the data set has inherent dimensionality n, this basis will have $n + 1$ elements. It will be overcomplete but minimally so.

5.2.10 The Minimum Overcomplete Basis

It is well known [13] that additive noise may be used to introduce a regularization term into a neural network. The addition of noise into the negative feedback network acts in a different manner, as will become clear in the next Sections.

Additive Noise

Nonlinear PCA was derived as an approximation to the minimisation [101] of $J = E_x(||\mathbf{x} - W\mathbf{y}||^2) = E_x(||\mathbf{x} - W\mathbf{f}(\mathbf{a})||^2)$. Now we add noise to the outputs so $\mathbf{y} = \mathbf{f}(\mathbf{a}) + \mu$ where μ is a vector of independently drawn noise from a zero mean distribution and define $\mathbf{f} = f(\mathbf{a})$ so that,

$$\begin{aligned} J' &= E_{x,\mu}(||\mathbf{x} - W\mathbf{y}||^2) \\ &= E_{x,\mu}(||\mathbf{x} - W(\mathbf{f} + \mu)||^2) \\ &= E_{x,\mu}(\mathbf{x}^T\mathbf{x} - (\mathbf{f} + \mu)^T W^T \mathbf{x} - \mathbf{x}^T W(\mathbf{f} + \mu) + (\mathbf{f} + \mu)^T W^T W(\mathbf{f} + \mu)) \\ &= E_x(\mathbf{x}^T\mathbf{x} - \mathbf{f}^T W^T \mathbf{x} - \mathbf{x}^T W\mathbf{f} + \mathbf{f}^T W^T W\mathbf{f}) + E_\mu(\mu^T W^T W\mu) \\ &= J + E_\mu(\mu^T W^T W\mu) \end{aligned}$$

We have removed terms containing single expectations of μ with respect to μ from the equation as they are drawn from zero mean noise.

Now the Subspace Algorithm (Chapter 2) results in an orthogonal matrix as does the nonlinear PCA algorithm under certain constraints and so J' may be written as

$$J' = J + \sum_i ||\mathbf{w}_i||^2 \sigma_i^2 \qquad (5.22)$$

Intuitively, it can be seen that when the added noise has low magnitude then the first term of (5.22) dominates and so (nonlinear) PCA is performed in the normal manner. If the noise variance is increased, then the learning is moderated by this additional weighted noise term, which has the effect of forcing some weight vectors to have only zero weight values (the degenerate solution).

So the addition of noise to the outputs has the effect of introducing a second pressure into the learning rule of the nonlinear PCA algorithm. This is a natural way in which to introduce a sparsification term on to the weights. Also as the noise is simply added on to the outputs after the application of the nonlinearity there is less computational expense than adding a specific weight decay term. Additionally, it is interesting to note that real neurons operate in a noisy environment.

The Minimum Overcomplete Basis

In Section 5.2.9, we discussed how, if any data space may be represented in n dimensions, then it may also be represented by nonnegative-only coordinates in a basis with $n + 1$ basis vectors. It is desirable to predict the minimum number of outputs that are required to represent the causes in the data so that we may form a more efficient coding of the data. This is difficult as any nonnegatively constrained coordinates determined by more than n basis vectors will adequately account for the data.

The addition of noise on the outputs enables the network to find this minimum overcomplete basis (MOB) that will account for the data, as was illustrated in the previous section. In the standard bars example, this is a network in which only 16 outputs respond to the data regardless of how many outputs there are in the network in total. Each output then identifies an individual bar or is zero, and partial bars are not shared across the outputs. This happens because the coding cost in the network increases if the network learns partial bars. For example, if one bar is learned by two outputs, each output learning half a bar each, then more noise is carried back on the feedback weights, thus creating a larger reconstruction error. Each weight vector in the network must have a vector length of 1 so the weighted noise error fed back through the network when a bar is learned in two halves is $2*4*\sqrt{\frac{1}{4}}*noise = 4*noise$. The weighted noise error fed back through the network when a bar is learned as a whole is $8 * \sqrt{\frac{1}{8}} * noise \approx 2.8 * noise$, which is a considerably lower coding cost.

The major advantage of additive noise is that it reduces the number of basis terms used to code an input: we are finding the minimum overcomplete basis.

Additive noise has a further advantage in that it can be added in a number of ways. For example, if it is added uniformly on to all outputs, then all of the weight vectors are penalised equally and only features that are strong enough to dominate the noise are learned. So for the standard bar data, if we have 20 outputs in the network trying to learn 16 individual bars, then only 16 outputs will respond, the weight vectors connected to the other 4 all having zero weight values.

5.2.11 Simulations

Random Mixes of Horizontal and Vertical Bars

In this section a variety of aspects of the network are illustrated with the benchmark "bars data" [46]. Unless otherwise stated, the networks are trained over 50 000 presentations of the data with a learning rate of 0.05 which is annealed linearly to zero over the training period. The squares in the following figures are individual weight vectors each connecting to one output, arranged two-dimensionally so that we can conveniently view the results. The diameter of circles within the squares represent the individual weight values where black is a positive weight and white is a negative weight.

With the standard bars data set, but with a network with 24 outputs, all of the bars are found but some of the bars are shared by two or three outputs. The threshold implementation of the network is more successful at identifying the bars in this case: Fig. 5.6 shows the converged weights when using this implementation of the network on the data. Therefore, when more outputs than bars are used in the network, then all of the individual bars are identified, and redundant weight vectors simply contain noisy values. We use the *soft threshold* function with the artificial data here as it is very flexible to work with and forgiving of nonoptimal network parameters. The other threshold functions from this family once optimised, however, yield results that are virtually identical.

We have found that the soft threshold nonlinearity is more effective than the plain rectification of the outputs. With exactly as many outputs as bars then both networks identify all of the bars easily, whereas when there are more outputs in the network than bars that make up the data set, then all of the bars are identified, but in the case of the rectified network some bars are identified more than once, and junctions of of bars or combinations of bars are also found.

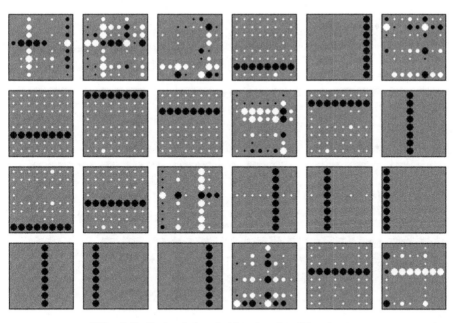

Fig. 5.6. Trained threshold network – 24 outputs.

Additive Noise

Additive noise, as can be seen from the results presented below (zero mean, Gaussian noise, of standard deviation 0.01, is added to every output) is beneficial in all of these networks when added to the outputs after the application of the nonlinearity. Fig. 5.7 shows that the additive noise enables each of these networks to identify all of the individual bars. That is, only as many outputs are used as are required in the coding; the weights connected to the other outputs each learn values that are close to zero.

As stated earlier we can add noise in a graduated way across the outputs so that the first output has zero mean Gaussian noise of standard deviation 0.001 added to it and every subsequent output having double the amount of noise as the previous output. In a network with 20 outputs this has the effect of forcing the first 16 outputs to learn all of the bars, but the last four to learn nothing. In this way we can control the location on the outputs where factors may be learned.

Illusory Causes

The human ability to find structure out of combinations of local and global information is illustrated in Figure 5.8. It is difficult to avoid seeing the white triangle, yet it is not there. If you are asked to give a compact description of the image, you would invariably mention this triangle. We illustrate the

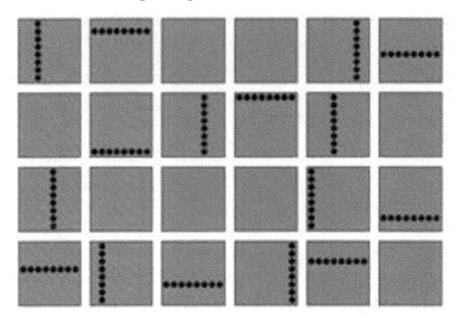

Fig. 5.7. Trained soft threshold network with additive noise – 24 outputs. Note that some weight vectors have no significant weight values.

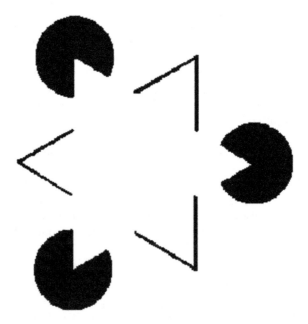

Fig. 5.8. The Kanizsa triangle illustrates the human ability to identify patterns which do not exist.

negative feedback network performing in the same way in this section. In the situation where bar patterns are nonsparse (the bars appear with a random probability of 7/8 each) then networks with built–in sparse priors (see Appendix B) cannot be expected to identify the individual bars. With the threshold network the results (Figure 5.9) confirm that the network converges to a very sparse representation, i.e. the illusory bars between the actual bars patterns. As the network operates so as to find a sparse response with the minimum descriptive length and because the individual bars are appearing in dense patterns together, then the network cannot learn the individual bars. Instead the network learns the spaces between bars patterns, which is the appropriate sparse response (illusory bars). Note that, in this experiment that the horizontal and vertical bars are not mixed so as to allow the weights to learn a more visually interesting response.

Although most of the weight vectors have prominent negative values, one of the weight vectors has small positive values on all of its weights. This weight vector is used to ensure that the output values respond with a significant positive value (normally a magnitude of over 3) to the illusory bars.

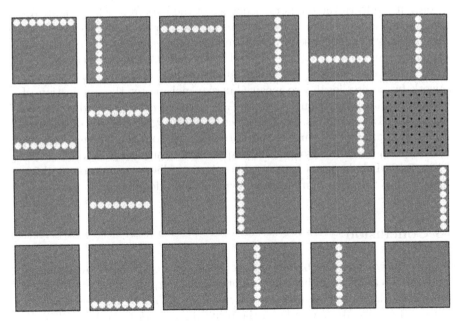

Fig. 5.9. The noisy soft threshold network discovers illusory bars – 24 outputs.

Using Noise to Modularise the Network Response

If the bars appear as part of horizontally or vertically moving sequences (each sequence beginning with a probability of 1/6), then by adding lateral connections on to the outputs of the network to include temporal context and adding modular noise to the network, it is possible to order the outputs so that temporally close features are coded spatially close at the output neurons. For example, this form of the network can be used to force vertical and horizontal bars to be learned in different modules of the output space.

The lateral connections come into play after the feedforward stage of the algorithm to give a but before the application of the nonlinearity $y = f(a)$ (see [26] for details). If the bars appear as part of horizontally or vertically moving sequences, then this network now can order the outputs so that temporally close features are coded at the output neurons in a manner that is spatially close.

Values are given to the lateral weights proportional to the distance between any particular pair of outputs and so an output's activation is increased in proportion to the previous values of the other outputs weighted by the lateral connections, as shown above. The method of setting the lateral connections here is simply to fix values of the weights to an output's four nearest neighbours i.e. an output is only connected to its nearest two neighbours on either side. An asymmetry can be put into these lateral weight values to encourage the outputs to learn the temporal sequence of the bars from left to right. Although symmetrical lateral weight values also work well, sometimes the bars are not exactly coded in sequence. Lateral weight values to the right of an output are -0.2 and -0.7 (nearest) and to the left 0.2 and 0.7 (nearest).

We create two "wells of attraction" at the outputs of the lateral connected network by adding local zero mean Gaussian noise after the application of the nonlinearity proportional to $\mid \cos \frac{2*\pi*i}{M} \mid$, where i is the identifier of the output, $i = 1, ..., M$. This has the effect of encouraging one set of features to be coded around the output that is one–third from the left and one set of features to be coded around the output that is one–third from the right of Fig. 5.10. This happens because the noise is of lower magnitude at these points and so the error minimisation term of (5.22) dominates.

5.3 Conclusion

We have shown how the basic PCA network can be amended by having a simple rectification on either the weights or the outputs which results in a network which can self-organise in order to identify individual causes from a mixture of causes. We have related the resulting network to both (nonlinear) Principal Component Analysis and to Factor Analysis and shown its capabilities mainly on artificial data sets. Experiments with other data sets and extensions of the basic network can be found in the theses of Darryl Charles [24] and Donald MacDonald [121]. Also in [66], we have shown that the non-negativity constraints may be applied to both the bigradient algorithm [180] and a class of algorithms derived from the generalised eigenproblem [188] in

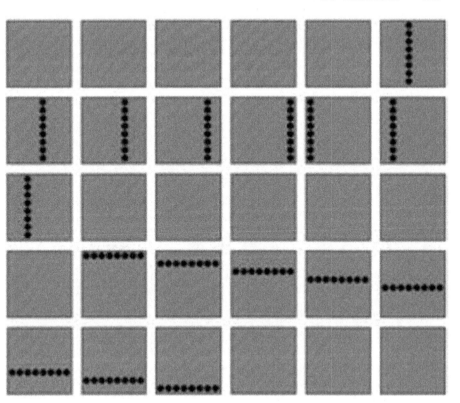

Fig. 5.10. Trained weights of noisy soft threshold network using noise to modularize the response of the network.

order to create a network which also identifies independent factors in a data set.

6

Exploratory Data Analysis

Cross-fertilisation between the fields of artificial neural networks and statistics has recently proved fruitful. In unsupervised learning, the realisation that simple neural network architectures are capable of performing classical statistical analysis has allowed insight into the operation of simple Hebbian neural networks and allowed the results of neural networks to be related to human psychophysical performance. Principal Component networks have been the major outcomes of this research. Here we use the same neural network architecture as in previous Chapters and show that it has other important statistical properties.

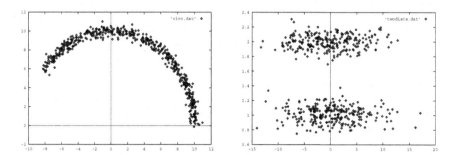

Fig. 6.1. Two distributions where Principal Component Analysis does not find the structure in the data: in the first distribution, the first PC would be an almost horizontal chord of the arc; in the second, it would lie on the diagonal of the rectangle. So projecting onto either principal component axis would hide the structure in the data.

Principal Component Analysis (PCA) has proved to be a powerful tool for the investigation and analysis of large data sets. However, some structure in data sets is not identifiable by means of the linear associations (correlations) among the variables; such effects as clustering or definition of edges of data sets are easily identified using the human eye on low–dimension projections of data but are not achievable by using the tools of classical multivariate analysis. For example, Fig. 6.1 (right) shows two ellipsoids representing the shapes of two data clusters; the first Principal Component is diagonal (either from top left to bottom right or bottom left to top right, depending on the actual samples used) yet the structure in the data – the two clusters – is not visible in the projection onto this direction. This problem increases in severity as the dimensionality of the data increases. The success of PCA has, in part, been because those directions which contain most of the variance in a data set will tend to contain most of the structure in the data set. However, this relationship is not logically necessary.

Exploratory Projection Pursuit (EPP) defines a recent form of exploratory data analysis methods which attempt to find "interesting" directions in high dimensional data (for reviews see [84, 94]). We introduce a nonlinearity to our PCA network and show that it is capable of performing an EPP.

6.1 Exploratory Projection Pursuit

The group of methods based on Projection Pursuit is based on one central idea: rather than solving the difficult problem of identifying structure in high–dimensional data, project the data onto a low–dimensional subspace and look for structure in the projection. However not all projections will reveal the data's structure equally well. Therefore we define an index that measures how "interesting" a given projection is, and then represent the data in terms of the projections that maximise the index and are therefore maximally "interesting". We will initially restrict our attention to one–dimensional subspaces i.e. we will identify an index for each line in the space and attempt to maximise the index in order to make projections of the raw data onto the line as interesting as possible.

Clearly the choice of index is the crucial factor in Projection Pursuit, and the index is specified by our desire to identify interesting directions. Therefore we must define what we mean by "interesting directions".

6.1.1 Interesting Directions

Friedman [50] notes that what constitutes an interesting direction is more difficult to define than what constitutes an uninteresting direction. The idea of "interestingness" is usually defined in relation to the oft-quoted observation of Diaconis and Freedman [39] that most projections of high-dimensional data onto arbitrary lines through most multidimensional data give almost Gaussian distributions. This would suggest that if we wish to identify "interesting" features in data, we should look for those directions $\mathbf{w}$, projections onto which are as non-Gaussian as possible. Thus, we will look for an $I(\mathbf{w})$, an index function of the direction **w**, which is maximum when the projection of the distribution onto **w** is furthest from Gaussian.

Two common measures of deviation from a Gaussian distribution are based on the higher–order moments of the distribution (see Fig. 2.4). Skewness is based on the normalised third moment of the distribution and basically measures if the distribution is symmetrical. Kurtosis is based on the normalised fourth moment of the distribution and measures the heaviness of the tails of a distribution. A bimodal distribution will often also have a negative kurtosis and therefore kurtosis can signal that a particular distribution shows evidence of clustering. Whilst these measures have there drawbacks as measures of deviation from normality (particularly their sensitivity to outliers), their simplicity makes them ideal for explanatory purposes.

In passing, we note that if we know what type of interesting structure we expect to find in the data set, instead of moving away from the uninteresting Gaussian distribution, we could move towards the interesting direction.

6.2 The Data and Sphering

Because a Gaussian distribution with mean a and variance x is no more or less interesting than a Gaussian distribution with mean b and variance y – indeed this second–order structure can obscure higher–order and more interesting structure – we remove such information from the data. This is known as "sphering". That is, the raw data is zeroed, projected onto the principal component directions and multiplied by the inverse of the square root of its eigenvalue to give data in all directions which has mean zero and is of unit variance. This removes all potential differences due to first– and second–order statistics from the data. To do this, the eigenvalue–eigenvector decomposition of the covariance matrix[1] is performed i.e. for input data X, we find the covariance matrix

$$\Sigma = E(X - E(X))(X - E(X))^T = UDU^T \tag{6.1}$$

where U is the eigenvector matrix, D is the diagonal matrix of eigenvalues and the T denotes the transpose of the matrix. New samples drawn from the distribution are then transformed to the principal component axes to give variables **y** where

$$\mathbf{y}_i = \frac{1}{\sqrt{D_i}} \sum_{j=1}^{n} U_{ij}(X_i - E(X_i)), \text{ for } 1 \leq i \leq m, \tag{6.2}$$

where n is the dimensionality of the input data and $m(\leq n)$ is the dimensionality of the sphered data. Typically $m \ll n$ and so this operation makes

---

[1] In practise, we make no distinction between statistics generated by samples from the distribution and those of the distribution itself.

high-dimensional data more manageable. It is important to note that any linear combinations of the y-values also retains these properties of the mean and variance e.g. see [125], Corollary 3.2.1.3.

This is the data in which we wish to find interesting directions.

6.3 The Projection Pursuit Network

The network is the same negative feedback network which we have used throughout the book. The sole difference is that a function of the output activations is calculated after the feedback stage but before the stage of changing the weights, and this function is used in the simple Hebbian learning procedure. We have for N–dimensional input data and M output neurons

$$s_i = \sum_{j=1}^{N} w_{ij} x_j \tag{6.3}$$

$$e_j = x_j - \sum_{k=1}^{M} w_{kj} s_k \tag{6.4}$$

$$r_i = f(s_i) = f\left(\sum_{j=1}^{N} w_{ij} x_j\right) \tag{6.5}$$

$$\Delta w_{ij} = \eta_t r_i e_j \tag{6.6}$$

$$= \eta_t f\left(\sum_{k=1}^{N} w_{ik} x_k\right) \left\{ x_j - \sum_{l=1}^{M} w_{lj} \sum_{p=1}^{N} w_{lp} x_p \right\} \tag{6.7}$$

where r_i is the value of the function $f()$ on the ith output neuron. Thus (6.7) may be written in matrix form as

$$\Delta W(t) = \eta(t)[I - W(t)W^T(t)]\mathbf{x}(t) f(\mathbf{x}^T(t) W(t)) \tag{6.8}$$

where t is an index of time and I is the identity matrix.

The set of network rules described above is a generalisation of those for the negative feedback network which performs PCA. Note also the difference between this network and the nonlinear generalisation of PCA discussed in Chapter 5: in that chapter, the nonlinear function was calculated before the feedback and was derived as an error minimisation; in this chapter, the nonlinear function is calculated after the (linear) feedback and will be derived as a constrained maximisation in the next section.

6.3.1 Extending PCA

Following [101], we can derive (6.8) as an approximation to the maximisation of a function, J, of the weights $J(W) = \sum_{i=1}^{M} E(g[\mathbf{x}^T \mathbf{w}_i]|\mathbf{w}_i)$.

We must ensure that the optimal solution is kept bounded; otherwise there is nothing to stop the weights from growing without bound. Formally,

$$\text{Let } J(W) = \sum_{i=1}^{M} E(g[\mathbf{x}^T\mathbf{w}_i]|\mathbf{w}_i) + \frac{1}{2}\sum_{i=1}^{M}\sum_{j=1}^{M}\lambda_{ij}[\mathbf{w}_i^T\mathbf{w}_j - a_{ij}] \tag{6.9}$$

where the last term enforces the constraints $\mathbf{w}_i^T\mathbf{w}_j - a_{ij}$ using the Lagrange multipliers λ_{ij}. As usual, we differentiate this equation with respect to the weights and with respect to the Lagrange multipliers. This yields respectively, at a stationary point,

$$\frac{\partial J(W)}{\partial W} = E(\mathbf{x}g'(\mathbf{x}^T W)|W) + W\Lambda = 0 \tag{6.10}$$

$$\text{and } W^T W = A \tag{6.11}$$

where $g'(\mathbf{x}^T W)$ is the elementwise derivative of $g(\mathbf{x}^T W)$, A is the matrix of parameters a_{ij} (often the identity matrix) and Λ is the matrix of Lagrange multipliers. Equations (6.10) and (6.11) define the optimal points of the process. Premultiplying (6.10) by W^T and inserting (6.11), we get

$$\Lambda = -A^{-1}W^T E(\mathbf{x}g'(\mathbf{x}^T W)|W)$$

and using this value and reinserting this optimal value of Λ into (6.10) yields the equation,

$$\frac{\partial J(W)}{\partial W} = [I - WA^{-1}W^T]E(\mathbf{x}g'(\mathbf{x}^T W)|W) \tag{6.12}$$

We will use an instantaneous version of this in a gradient ascent algorithm

$$\Delta W \propto \frac{\partial J(W)}{\partial W}$$

to yield

$$\Delta W = \mu[I - WA^{-1}W^T]\mathbf{x}g'(\mathbf{x}^T W) \tag{6.13}$$

We will be interested in the special case where the W values form an orthonormal basis of the data space and so $A = I$, the identity matrix. Therefore, we can equate (6.13) with (6.8).

Karhunen and Joutsensalo point out that the algorithm is approximative since the expression for Λ is derived from the optimum solution and used from the beginning of the algorithm. As we shall see in Section 6.3.4, the implications of the approximation are profound: to be used in a gradient ascent algorithm, $\frac{\partial J}{\partial W}$ must be continuous and with positive slope in the iteration intervals. We shall see that these constraints can only be justified, in general, on the set of points where the second constraint in (6.9) is satisfied a priori.

6.3.2 The Projection Pursuit Indices

Now for projection pursuit, we wish to maximise a specific index. But note that from the derivation in the last section, when we wish to maximise an index function we must use its derivative in the learning algorithm: the function $f()$ in (6.7) is equivalent to the function $g'()$ in (6.13). Thus to maximise a projection pursuit index e.g. for skewness, we could use a learning process like that described in (6.13) noting that to maximise the skewness index we must use the derivative of the index in the learning process.

We wish to emphasise the properties of the negative feedback network rather than those of specific indices. Thus we choose to report on the network's self-organisation with the simplest possible indices. The indices which we investigate in this section are either directly based on the higher moments of the input data or are functions of them (see Fig. 2.4):

- To measure skewness in a Normal distribution, $N(\mu, \sigma)$ we use

$$g(s) = \frac{E\{(s - \mu)^3\}}{\sigma^3}$$

  where s is a random variable drawn from the distribution with mean μ and standard deviation σ. Now our data distributions have all been sphered i.e. $E(x) = 0; E[(x - E(x))^2] = 1$ and our weights, $\mathbf{w}_i$, are normalised and therefore every direction $\mathbf{s}_i$ has the same first and second moments. Thus $g(\mathbf{s}) = \mathbf{s}^3$ is a measure of the skewness of the distribution. Thus, in the algorithm (6.8), we use

$$f(s) = k * s^2 \propto \frac{d}{ds} s^3$$

  Now in all Normal directions, this measure will be zero, but in a direction with a skewed distribution, there will be a nonzero skew value.
- Similarly, kurtosis[2] is measured by

$$g(s) = \frac{E\{(s - \mu)^4\}}{\sigma^4}$$

  Therefore as above, to measure a kurtotic deviation, we could use

$$f(s) = k * s^3 \propto \frac{d}{ds} s^4$$

- We can also use functions (see Section 6.3.6) whose expansions are dominated by either odd or even powers of s to measure kurtosis or skewness respectively.

---

[2] Typically, 3 is subtracted from this measure in order to make the kurtosis of a truly Normal distribution 0. However, since we use its derivative, we have simply used the stated measure.

The first two are the simplest possible measures of departure from Gaussianity yet are generally not used because of their susceptibility to outliers. Thus we have derived the third set of measures. We will use the naive sample-based versions of the measures making no adjustments for any potential differences between sample and distribution moments (see e.g. [125] for a discussion of such differences). We further treat each test as measuring only one facet, although we are aware that tests for skewness and kurtosis are distributionally dependent (see e.g. the discussion in [80]).

Traditional statistical methods require a computationally intensive recalculation of the distribution's moments from a reasonable sample of data points from the distribution each time a measure must be recalculated. However, it will be shown that a Hebbian learning rule for neural networks based on a measure of the instantaneous moments does in fact find that direction of maximum interest in the sense of Section 6.1.1.

6.3.3 Principal Component Analysis

The negative feedback network introduced here is identical to that used previously as a Principal Component network. The transfer of activation is exactly the same as described in this chapter; however, there was previously no nonlinear activation function at the output neurons. This is equivalent to a network with $f(x) = x$, the identity transformation. Now since f is the derivative of the function we wish to maximise, we can see that the PCA network is maximising the second moment of the distribution i.e. as we know, PCA is finding that direction with greatest variance. In fact, in the simulations described below, we use the above network twice – the first time to project the data onto the eigenvectors corresponding to the Principal Components and the second time to carry out Exploratory Projection Pursuit. The fact that the same network structure is capable of performing a PCA as well as EPP is unsurprising since Huber [84] has shown that PCA may be viewed as a particular case of Projection Puruit. Thus, for the PCA network, we are choosing $f(x) = x \propto \frac{d}{dx}(x^2)$ and so the original network is seen to be maximising the second–order statistics of the distribution i.e. finding the eigenvectors corresponding to maximal eigenvalues.

This suggests that Oja's Subspace Algorithm can be derived in terms of a gradient ascent procedure. However, Baldi and Hornik [4] have shown that this algorithm is not derivable from such a procedure. The reason for this apparent contradiction is found in the approximation assumptions used in the derivation of the algorithm and will be discussed in the next section.

6.3.4 Convergence of the Algorithm

The derivation of the algorithm was based on gradient ascent using (6.9). Therefore, this equation must define a function of W which is twice differentiable with respect to W.

Consider first the convergence of the algorithm on the set of points restricted to the surface $\|\mathbf{w}\| = c$, where $\|.\|$ denotes the Euclidean norm and c is a constant. On this set, the second term of the equation,

$$\frac{1}{2} \sum_{i=1}^{M} \sum_{j=1}^{M} \lambda_{ij} [\mathbf{w}_i^T \mathbf{w}_j - \delta_{ij}], \tag{6.14}$$

is a constant and we may then return to the original maximisation, which we will denote $J'(W) = \sum_{i=1}^{M} E(g[\mathbf{x}^T \mathbf{w}_i] | \mathbf{w}_i)$ on this set. Note first that each of the functions used in this chapter is twice differentiable.

Used as an instantaneous algorithm, we have, for the presentation of a single pattern $\mathbf{x}$:

- For kurtosis, $J'(W) = \sum_{i=1}^{M} (\mathbf{x}^T \mathbf{w}_i)^4$. Then $\frac{\partial^2 J'}{\partial \mathbf{w}_{ij}^2} = 12(\mathbf{w}_i.\mathbf{x})^2 x_j^2 \geq 0$.

  Thus the function $f(s) = ks^3, k > 0$ will converge to that direction with maximum kurtosis when the convergence takes place on the set of all points which satisfy $\|\mathbf{w}\| = c$. Similarly the function, $f(s) = ks^3, k < 0$ will always cause convergence to those directions with minimum kurtosis. Therefore to test kurtosis in a situation where the form of the data is unknown, we can (in parallel) test for both positive and negative kurtotic distributions e.g. with $f(s) = k_1 s^3, k_1 > 0$ and $f(s) = k_2 s^3, k_2 < 0$.

- For skewness, $J'(W) = \sum_{i=1}^{M} (\mathbf{x}^T \mathbf{w}_i)^3$. Now $\frac{\partial^2 J'}{\partial \mathbf{w}_{ij}^2} = 6(\mathbf{w}_i.\mathbf{x}) x_j^2$. Thus if $\Delta W \propto \frac{\delta J'(W)}{\delta W}$, we have a gradient ascent rule if $E(\mathbf{w}_i.\mathbf{x})$ is greater than 0, i.e. we will ascend till we converge to the direction with greatest positive skewness. If $E(\mathbf{w}_i.\mathbf{x})$ is less than 0, we will descend till we converge to the direction with most negative skewness. Thus one function can be used to test for both positive and negative skewness. It is important to recall that this is an exploratory data investigation tool: we do not care if the structure has positive or negative skewness – only that it is deviating from a Gaussian distribution.

  However, this does leave open the possibility that there exists a stage in the convergence when skewness in two directions reaches a stable point of convergence which is a mixture of two optimal states, though we have never seen this situation experimentally.

Therefore the algorithm may be viewed as gradient ascent on the hypersphere satisfying $\|\mathbf{w}\| = c$. Now we must consider the convergence of the algorithm in general; consider (6.10) with respect to a particular vector of weights into output neuron, i, for a function $g(s) = s^k$. Then we have

$$\frac{\partial J}{\partial \mathbf{w}_i} = k\mathbf{x}(\mathbf{w_i}.\mathbf{x})^{k-1} + W\lambda_i \tag{6.15}$$

$$\frac{\partial^2 J}{\partial \mathbf{w}_i^2} = k(k-1)\text{Diag}\{\mathbf{x}\mathbf{x}^T\}(\mathbf{w}_i.\mathbf{x})^{k-2} + \lambda_{ii}I \tag{6.16}$$

where Diag{.} is an operator which sets all off-diagonal entries to 0 and λ_i is the vector of Lagrange coefficients for the direction $\mathbf{w}_i$. Now since the data is sphered, $E(x_j^2) = 1$. Thus we only have a positive gradient in those directions, $\mathbf{w}_i$, which satisfy

$$(\mathbf{w}_i.\mathbf{x})^{k-2} > \frac{-\lambda_{ii}}{k(k-1)} \tag{6.17}$$

Recalling that λ_{ij} determines the relative weight accorded to the function J' and the constraint $[\mathbf{w}_i^T \mathbf{w}_j - \delta_{ij}]$ and we can see that the use of the final converged value of λ_{ij} in the converging algorithm causes a more serious problem than merely being an approximation. The algorithm is not guaranteed to converge.

In practice, this has not been found to be a problem. One possible heuristic would be to start the weights normalised and then converge across the surface. However there is the possibility that the convergence process will be slower using this. Empically, little difference has been found between starting with small (near 0) random weights and starting with normalised vectors.

6.3.5 Experimental Results

We have shown that this algorithm causes the weights to converge to identify the higher–order structure in a data set. Since we later wish to compare various algorithms, we create artificial data sets in which we know the nature of the structure in the data set and for which we know where the structure lies. Thus we can measure how quickly the networks converge to the correct solutions.

We therefore create five dimensional data such that four dimensions contain data drawn independently and identically distributed from zero mean, Gaussian distributions while the fifth dimension contains kurtotic data: we draw this also from a Gaussian distribution but randomly (in 20% of cases) substitute the sample with a small random number drawn from a uniform distribution between -0.005 and $+0.005$. We use the above algorithm with $f(y) = y^3$ and repeat the experiment 100 times. The cubic nature of this function makes learning inherently unstable and in 23 cases we found overflow. The convergence of the other 77 cases are shown in the left part of Fig. 6.2 in which we show the mean value of the cosine of the angle between the network weights and the optimal value at each iteration (central line). We also show in this figure one standard deviation above and below the mean. In three experiments, the network did not approach convergence after 80 000 iterations. The results of the remaining 74 cases are shown in the right part of Fig. 6.2.

More Than One Interesting Direction

Since the projection pursuit method is designed to find interesting directions worthy of human investigation, and since humans can visually investigate

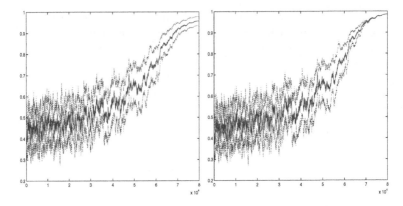

Fig. 6.2. Left: convergence of 77 experiments towards the optimal (kurtotic) direction with $f(y) = y^3$. The graph shows the mean value of the cosine of the angle between the network weights and the optimal weight at each iteration and one standard deviation on either side of the mean. Right: convergence of the best 74 experiments.

functions over a plane, we are often interested in finding two independent directions in a data set which contain interest. We first consider the situation when each interesting direction has the same type of interest. Experiments have shown that, in such situations, the network usually finds one direction more interesting than the others; the weights will converge to that direction on which the projection of the data has the largest deviation from the statistics of a normal distribution, e.g. when the kurtosis index is 3.67 in dimension 4 and 3.66 in dimension 7, the kurtosis index function invariably causes convergence to dimension 4.

However, it may be appropriate to find all interesting filters. This situation sometimes [50] is dealt with by "structure removal" using a transformation of the interesting filter to create a Normal distribution in that filter. This method has the disadvantage that such transformations may affect the Normality of other solution projections. However we note that the learning process used here not only finds but removes the interesting projections i.e. the residuals at the inputs consist of the original data minus the projections onto the learned interesting filters. Thus we suggest running the network until one interesting projection is found; then set these weights and restart learning with a new output neuron. This has been found to be very effective.

When the data contains projections which are interesting in different ways, we can investigate the data in different ways simultaneously. We can, for example, construct a network as before but with M output neurons each of which is searching for different characteristics in the sphered input data. So our final algorithm is:

1. Sphere the input data.

2. Create a set of output neurons with M different indices and train this network. For example, for the indices we have so far considered we may simply have two output neurons which use skew and kurtosis indices, respectively.
3. Visually examine (either individually, as lines, or in pairs, as planes) the projections found in order to identify humanly interesting projections.
4. Remove those neurons whose weights have not converged to interesting projections.
5. Repeat Steps 3–5 training the new neurons on the residuals after all current neurons have removed their projections.

6.3.6 Using Hyperbolic Functions

As an example of using hyperbolic functions we perform an experiment similar to the last one but report the convergence of $f(s) = \tanh(s)$. Since $f(s) = \tanh(s)$ has an expansion of $s - \frac{s^3}{3} + \frac{2s^5}{15} - \cdots$, it is an odd function. It can then be used to measure the kurtotic deviation from the normal distribution. In detail, using $\tanh(s)$ as the learning function, $f()$, in (6.8) maximises the integral of that function; thus, using $f(s) = \tanh(s)$ in the stochastic algorithm maximises

$$E\left(\int \tanh(s)ds\right) = E\left(\int \left(s - \frac{s^3}{3} + \frac{2s^5}{15} - \cdots\right)ds\right) \tag{6.18}$$

$$= E\left(s^2 - \frac{s^4}{12} + \frac{2s^6}{90} - \cdots\right) \tag{6.19}$$

$$= E(s^2) - E\left(\frac{s^4}{12}\right) + E\left(\frac{2s^6}{90}\right) - \cdots \tag{6.20}$$

$$= 1 - E\left(\frac{s^4}{12}\right) + E\left(\frac{2s^6}{90}\right) - \cdots \tag{6.21}$$

since s is a linear combination of sphered data.

Now for small s, the most important parts of the series are the first few terms; also, subsequent terms alternate between reinforcing the effect of the second term (being negative) and detracting from its effect. $\frac{s^4}{12} > \frac{2s^6}{90}$ in the interval $(-1.93, 1.93)$ which contains almost exactly 95% of the sphered data in Normal directions. This proportion will be slightly different for a non-normal direction, but the major conclusion must be that for the overwhelming majority of the data points the driving force of the learning is the cubed term in the expansion of $\tanh(s)$. Therefore if we use $\tanh(s)$ in our algorithm, we are minimising s^4 i.e. finding projections with least kurtosis.

Similarly we can use $f(s) = \text{sech}^2(s)$ which is the derivative of $\tanh(s)$ and is an even function, to find deviations in skewness from the normal distribution. We can, of course, search for both types of structure in parallel. In fact, not only does the convergence of the two indices not interfere with each other, the convergence of each may actually help the other if we use as input data for the second output neuron the residuals at **x** after the first neuron has subtracted its projection (a deflationary network, c.f. [160]). This has the effect of decreasing the dimension of the space of input data in which the second neuron must search for interesting projections. Note also that these are by no means the only functions which can be used. For example, $\tan^{-1}(s)$ can also be used for searching for kurtotic distributions as its expansion is also odd.

6.3.7 Simulations

In Fig. 6.3, we show the convergence of 100 simulations on the same type of data as before (left figure) and then the results of the best 90 experiments in the right figure when we use $f(y) = y - \tanh(y)$, a function which we will discuss in Section 6.6. We see that we not only improve stability but also improve the speed of convergence with this function. (Compare with Fig. 6.2).

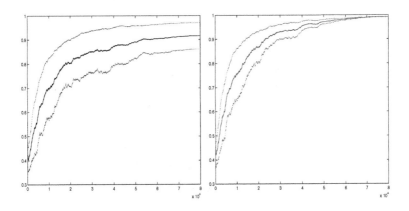

Fig. 6.3. Left: convergence of 100 experiments towards the optimal (kurtotic) direction. The graph shows the mean value of the cosine of the angle between the network weights and the optimal weight at each iteration and one standard deviation on either side of the mean. Right: convergence of the best 90 experiments. These graphs show results when using $f(y) = y - \tanh(y)$.

Similarly, we may create data which has one platykurtotic dimension by sampling from a zero mean Gaussian distribution and then randomly adding 5 to the sample or subtracting 5 from the sample. This gives a bimodal distribution which the function $f(y) = \tanh(y)$ can be used to find. Thus in Fig. 6.4, we show the results of 100 experiments searching for this dimension in the left figure and the best 99 (only one had not converged by iteration 80 000) in the right of the figure.

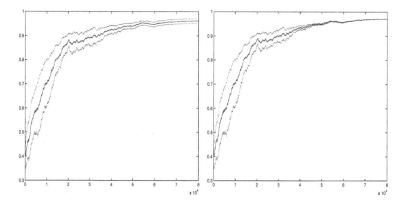

Fig. 6.4. Left: convergence of 100 experiments to optimal bimodal dimension when using tanh() in the learning rule. Right: convergence of the best 99 simulations.

6.4 Other Indices

With the method and network for Projection Pursuit now established, we can investigate other possible indices. We will investigate indices based on Information Theory and two specific indices: Friedman's index and Intrator's index. These have aroused a great deal of interest in their respective communities and are thought to be substantially the best currently available (though we must add the caveat that such assessments are usually made with respect to specific data sets). The most frequently referenced indices from the statistics community are those from Friedman [50] and Hall [65]. Both indices use polynomial approximations to analytically deduced indices of interestingness; such polynomials are usually introduced for reasons of computational efficiency. We chose to investigate Friedman's rather than Hall's since the former index is thought to be more generally effective than the latter (e.g. see [175] for a recent comparison).

In the neural network community, the series of articles written by Intrator (e.g. [87, 89, 90]) are given as example implementations of the Projection Pursuit methodologies. Other articles (e.g. [77]) do not specifically mention Projection Pursuit though often appearing to use a PP methodology. An interesting implementation of PP methodologies using radial basis function nets is given in [189].

6.4.1 Indices Based on Information Theory

As noted by Marriot (in Discussion of [94]), " a moment criterion, or any criterion dominated by third and fourth cumulants, will miss clustered projections that happen to be roughly symmetrical and nearly mesokurtic"; this has led to a search for alternative measures of non-normality.

Since the data is sphered, we may use the fact that for a constant variance, the distribution which exhibits maximum entropy is the Gaussian. Thus we may hope that the distribution which is furthest from a Gaussian has least entropy. Thus the entropy

$$H = \int p(x) \log p(x) dx \tag{6.22}$$

could be used as a measure of departure from normality. However, it is not possible to make an instantaneous calculation of $p(x)$ given the single input x. A measure using entropy would require to have a memory of previous inputs in order to calculate the relative frequency approximation to the entropy.

An alternative is to use the difference between the normal distribution and the actual distribution which can be quantified using the relative entropy (or Kullback–Leibler divergence [31], page 18). Thus we can measure the distance between our current distribution, defined by the probability distribution function $p(x)$, and the normal distribution with function $\phi(x)$, as

$$D(p||\phi) = \sum_x p(x) \log \frac{p(x)}{\phi(x)} \tag{6.23}$$

This gives a measure of the error involved in assuming the distribution of the x values is determined by $\phi(x)$ when, in fact, the x values are drawn from a distribution with pdf $p(x)$.

It is interesting in this case to begin with the information measure which we will use for $f(x)$ and derive the function which we will be maximising.

Consider a neuron which is expecting inputs from a normal $N(0, 1)$ distribution. In each uninteresting projection, the actual probability distribution which it sees will accord with its expectation that it is receiving data from the $N(0, 1)$ distribution; however, in an interesting projection, it will quantify its inputs' values based on the prior belief that these inputs are coming from a Normal distribution when in fact they are coming from a non-Normal distribution.

Thus we consider the measure $f(x) = -\log \phi(x)$, where $\phi(x)$ is the probability that it would have received input x had input x come from a Normal distribution. Thus the expected information which it believes that it is receiving from N samples of input data from the distribution is

$$
\begin{aligned}
D &= \frac{1}{N} \sum_{n=1}^{N} f(x_n) \\
&\approx \sum_{m=1}^{M} p(x_m) f(x_m) \\
&= -\sum_{m=1}^{M} p(x_m) \log \phi(x_m)
\end{aligned}
$$

where the input data has been binned into M boxes and the mean value of x in the mth box is x_m. The value $p(x_m)$ is the relative frequency of the sample in box m and is taken as an estimate of the probability of having an input x in box m. Now

$$D = - \sum_{m=1}^{M} p(x_m) \log \phi(x_m)$$

$$= \sum_{m=1}^{M} p(x_m) \log \frac{p(x_m)}{\phi(x_m)} - \sum_{m=1}^{M} p(x_m) \log p(x_m)$$

$$= D(p(x)||\phi(x)) - H_p \qquad (6.24)$$

where $D(a||b)$ is the relative entropy between the distributions a and b and H_a is the Shannon entropy of the distribution a. Note that since we are using a discrete (binned) version of $p(x)$, we may assume that $H_p \geq 0$.

Therefore, using the index $g(x) = - \log \phi(x)$, we are finding the distribution which maximises the difference between the distance from the Gaussian distribution and its own entropy. Note the double effect – a distribution which is furthest from Gaussian will have the least information in it.

6.4.2 Friedman's Index

Friedman [50] has developed an index which has attracted wide interest within the community of statistics users. Sun [175] has performed a useful and informative comparison of Friedman's and Hall's indices and has concluded that Friedman's is generally better. Hall [65] agrees. Briefly, Friedman's index is based on the transformation of the projection of the sphered data, $\mathbf{x}^T \mathbf{w} \rightarrow R = 2\Phi(\mathbf{x}^T \mathbf{w}) - 1$ where $\Phi(X)$ is the standard normal cumulative density function

$$\Phi(X) = \frac{1}{\sqrt{2\pi}} \int_{-\infty}^{X} \exp\left(-\frac{1}{2t^2}\right) dt \qquad (6.25)$$

If X follows a standard normal distribution, then R will be uniformly distributed in the interval $[-1,1]$. Therefore we take as a measure of the distance of $\mathbf{x}^T \mathbf{w}$ from the Normal distribution, the integral squared distance of the variable R from the uniform distribution,

$$\int_{-1}^{+1} \left[p(R) - \frac{1}{2}\right]^2 dR = \int_{-1}^{+1} p^2(R) dR - \frac{1}{2} \qquad (6.26)$$

Friedman expands R in Legendre polynomials so that

$$\int_{-1}^{+1} p^2(R) dR - \frac{1}{2} = \int_{-1}^{+1} \left[\sum_{j=0}^{\infty} a_j P_j(R)\right] p(R) dR - \frac{1}{2} \qquad (6.27)$$

where the Legendre polynomials are

$$P_0(R) = 1$$
$$P_1(R) = R$$
$$P_j(R) = \frac{(2j-1)RP_{j-1}(R) - (j-1)P_{j-2}(R)}{j}, \forall j > 1$$

The coefficients a_j are given by

$$a_j = \frac{2j+1}{2} \int_{-1}^{+1} P_j(R)p(R)dR = \frac{2j+1}{2}E[P_j(R)] \tag{6.28}$$

From this, we have an easily calculated projection index

$$g(\mathbf{w}) = \frac{1}{2} \sum_{j=1}^{J} (2j+1)(E[P_j(R)])^2 \tag{6.29}$$

for any direction $\mathbf{w}$, which is applied as

$$\hat{g}(\mathbf{w}) = \frac{1}{2} \sum_{j=1}^{J} (2j+1) \left[\frac{1}{N} \sum_{i=0}^{N} P_j(2\Phi(\mathbf{x}^T\mathbf{w}) - 1) \right]^2 \tag{6.30}$$

to give the sample version of the index. Note that this must be maximised under the constraint that $WW^T = 1$ to ensure a finite solution.

However, we require the derivative of $g(\mathbf{w})$ for our instantaneous measure, $f(\mathbf{s})$, and so we use

$$f(\mathbf{s}) \propto \sum_{j=1}^{J} (2j+1)E[P_j(R)]E\left[P'_j(R) \exp -(\mathbf{x}^T\mathbf{w})^2\right]$$

Now $P'_j(R)$ is also easily calculated via the recursion relation

$$P'_1(R) = 1$$
$$P'_j(R) = RP'_{j-1}(R) + jP_{j-1}(R)$$

Thus the instantaneous version of Friedman's index used in a neural network implementation is

$$f(\mathbf{s}) = \sum_{j=1}^{D} (2j+1)P_j P'_j \exp(-(\mathbf{x}^T\mathbf{w})^2)$$

Simulations on data such as used to test the polynomial indices have shown that such an index finds projections with either skew or kurtotic deviations from Normality with great reliability and accuracy.

6.4.3 Intrator's Index

Intrator [86, 87, 88, 89, 90, 91] has constructed a model for Exploratory Projection Pursuit derived from the Bienenstock, Cooper, Monro [12](BCM) model of cortical plasticity. This model has a learning function of

$$\frac{dw}{dt} = \mu(t)(\mathbf{x}.\mathbf{w})\left(\mathbf{x}.\mathbf{w} - \frac{4}{3}\theta_w\right)\mathbf{x} \qquad (6.31)$$

where $\theta_w = E[(\mathbf{x}.\mathbf{w})^2]$ provides a moving threshold which yields the dynamic flexibility necessary for stability. While this is not immediately transferable into the format (6.8), an approximate version of the index function

$$g(\mathbf{w}) = -\mu\left\{\frac{1}{3}E[(\mathbf{w}.\mathbf{x})^3] - \frac{1}{4}E[(\mathbf{w}.\mathbf{x})^2]^2\right\} \qquad (6.32)$$

yields a function $f(\mathbf{s}) = a * (\mathbf{w}.\mathbf{x})^2 + b * (\mathbf{w}.\mathbf{x})$ which has been found (unsurprisingly) to have almost identical convergence properties to the simple index for skewness since $E(\mathbf{w}.\mathbf{x}) = 0$.

A comparative study [53] of the network of this chapter and the BCM neuron found that the negative feedback model of this chapter performed better. To be fair, however, the comparative study was performed by the current author who may be thought to favor this model a priori.

6.5 Using Exploratory Projection Pursuit

To illustrate the use of EPP, is it demonstrated on a small database of bank customers, containing 1000 records, with 12 fields. (A few records are shown in Fig. 6.7). Information stored in the database includes a unique identifier, age, sex, salary, type of area in which they live, whether married or not, number of children and then several fields of financial information such as type of bank account, whether they own a Personal Equity Plan, etc.

Fig. 6.5 shows that the network has clearly identified 4 clusters in the data set using EPP with a tanh() non-linearity i.e. we are searching for negative kurtosis, an indicator of clusters. Manual investigation of the clusters readily reveals that the clusters are forming on the place of residence field – each cluster is specific to one of RURAL, TOWN, INNER CITY and URBAN sites.

Within each cluster, there is a smooth transition within the cluster with clearly identifiable types of customers in each group. For example, the projection of each cluster onto the first EPP direction also shows structure: in each case, the male customers are in the left strip while the female customers are in the right strip. In addition, the second projection may be crudely associated with a gradual change from young poor customers to old rich customers.

Data Results

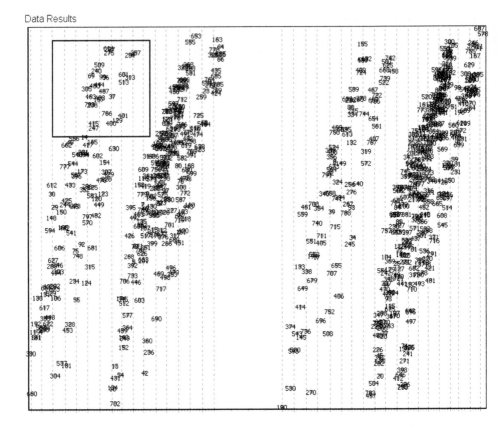

Fig. 6.5. The EPP network with a tanh() projection index finds four distinct clusters. The projection of each cluster onto the first EPP direction also shows structure: in each case, the male customers are in a left strip while the female customers are in a right strip. In addition, the second projection may be crudely associated with a gradual change from young poor customers to old rich customers.

Thus, for example, zooming in on the top left hand corner of Fig. 6.5 gives Fig. 6.6, the records in the square of which are shown in Fig. 6.7.

Not only are the records all from a rural district, but they also represent young, low–income customers. Looking at the records at the bottom end of the left cluster, gives very different records (see Fig. 6.8) from those at the top where we show older and richer customers.

6.5.1 Hierarchical Exploratory Projection Pursuit

EPP can only provide a linear projection of the data set. There may well be cases in which the structure of the data is not captured by a single linear projection of the data. In such cases, a hierarchical scheme may be beneficial.

Data Results

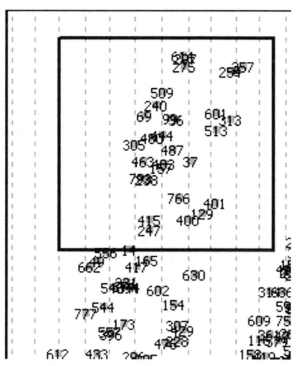

Fig. 6.6. Records found by zooming in on the top left-hand corner of Fig. 6.5. This is a enlarged image of Fig. 6.5, centering on the boxed area in the top left corner of Fig. 6.5. Some records in the square box are shown in Fig. 6.7.

ID12258	24	MALE	RURAL	16473.3	YES	2	YES	YES	YES	NO	NO
ID12545	21	FEMALE	RURAL	15797.1	YES	0	NO	YES	YES	NO	YES
ID12610	23	FEMALE	RURAL	11215.3	YES	2	YES	YES	YES	NO	YES
ID12584	23	MALE	RURAL	16403.8	YES	0	YES	YES	NO	NO	NO
ID12200	18	FEMALE	RURAL	15348.9	YES	0	YES	NO	YES	NO	NO
ID12588	19	MALE	RURAL	16625.9	YES	1	NO	NO	YES	NO	YES
ID12137	19	MALE	RURAL	13896.0	YES	3	YES	YES	YES	NO	NO
ID12867	40	FEMALE	RURAL	15233.1	NO	0	YES	YES	YES	YES	YES

Fig. 6.7. Records corresponding to those in the square in Fig. 6.6. They are from a rural area, with low income, and predominately of a young age.

ID12220	64	FEMALE	RURAL	49284.0	YES	3	YES	YES	YES	NO	YES
ID12293	64	FEMALE	RURAL	47562.9	YES	3	NO	YES	YES	NO	YES
ID12222	62	FEMALE	RURAL	51018.8	YES	1	YES	YES	YES	NO	YES
ID12442	66	MALE	RURAL	48346.1	YES	1	YES	YES	NO	NO	YES
ID12105	57	FEMALE	RURAL	51753.3	YES	0	NO	YES	NO	NO	NO
ID12663	57	FEMALE	RURAL	50897.6	YES	0	NO	YES	YES	NO	NO

Fig. 6.8. Records corresponding to the other end of the same cluster from those records in Fig. 6.7. Here the customers are older, with a higher income, but still from a rural area.

Thus we have developed software which allows the user to interactively investigate data sets: the user performs an EPP on the whole data set and then dynamically selects (by pointing and clicking a mouse) a subset of the data on which to perform a second EPP. Typically a user will choose to search for clusters in the data and then subsequently select one or more clusters in which to search for subsequent structure. This may be done for as many levels as seems suitable to the user for the current data set.

Note that we are doing more than zooming in when we use this hierarchical investigation. When the user selects a cluster, we perform an EPP on the records of that cluster which results in a reprojection of the cluster which is optimal for finding structure in the subset of records which make up the cluster. Fig. 6.9 shows an example. Again manual inspection revealed that the EPP has found a second cluster inside the first which contains the people without a current account. This shows that the Hierarchical Exploratory Projection Pursuit (HEPP) network will extract more information from the data set than is possible with an EPP network. The HEPP network can also be used to look for different characteristics in the second level of its exploration by changing the nonlinear function used, thus changing the projection of interest for which the network is looking.

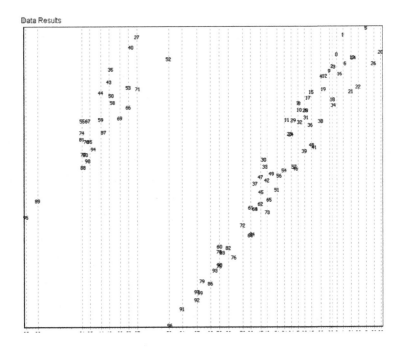

Fig. 6.9. HEPP projection of the left cluster of Fig. 6.5. This shows there are outliers in the left cluster. It transpires that records 89 and 95 are of people younger than the people in the rest of the leftmost cluster.

6.5.2 World Bank Data

The world bank data set consists of entries for 124 countries, with fields for the countries' name, gross national product (GNP), percentage growth, GNP per capita plus the percentage growth and finally GNP PPP (productivity per person) and percentage growth of GNP PPP. We first used the EPP network on the data set, and plot the results in Fig. 6.10.

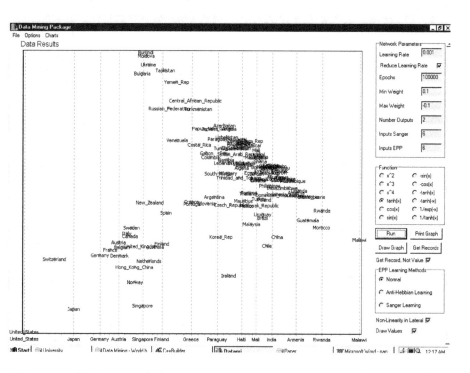

Fig. 6.10. Projections of the world bank data clearly separate the first world countries from the remainder of the countries.

These results cluster third world countries in the upper right half of Fig. 6.10, second world countries are left and lower; in the far bottom left lies the richest country in the world, the USA, with Switzerland, and Japan close by. The countries are shown based on their ranking in the three areas GNP, GNP per capita and GNP PPP in Table 6.1.

Fig. 6.11 shows the results from the HEPP network using tanh() to find clusters, when we use it on the cluster of countries in the upper right hand corner of Fig. 6.10. The second world countries have been separated from the third. The hierarchical use of the HEPP network has found more structure in the second cluster.

Table 6.1. The data of the countries selected as outliers by the EPP network.

Country	GNP	GNP per captia	GNP PPP
USA	1	8	1
Japan	2	2	5
Switzerland	17	1	3
Germany	3	6	12
France	4	10	10

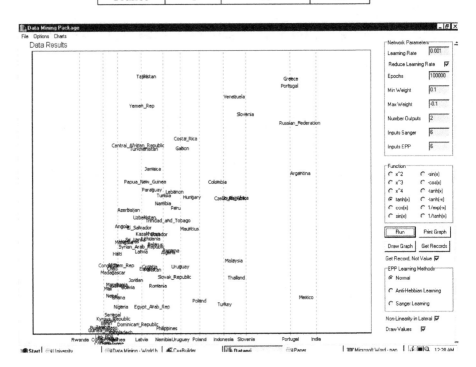

Fig. 6.11. A second level clustering of countries from the non-first world countries.

As can be seen from Fig. 6.11, the third world countries have clustered together, and the second world countries have moved out and towards the top right. Greece and Portugal are shown to be quite a distance from third world countries such as the Philippines, the Dominican Republic and Senegal. This is a good visual example of the power of HEPP to find higher–order structure in a data set: the second projection shows information that it is not possible to show with a single projection alone.

6.6 Independent Component Analysis

Girolami [58] has developed a slightly different form of the negative feedback EPP network described above and used it to perform ICA.

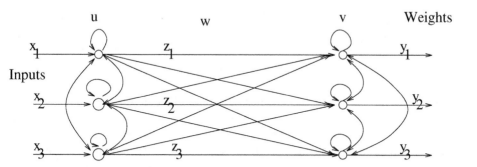

Fig. 6.12. The extended exploratory projection pursuit network. The first layer of weights, U, decorrelates the inputs; the interaction between the second layer of weights, W and the third, V eliminates statistical dependencies from the z values.

The first layer of weights which spheres the data is based on Foldiak's second model [46]; the equations are

$$z_i = x_i + \sum_{j=1}^{n} u_{ij} z_j$$

with learning rule

$$\Delta u_{ij} = -\alpha(1 - z_i z_j)$$

The net result is that the outputs, the components of $\mathbf{z}$, are decorrelated with about equal variance. Note that since, if the signals (such as voice signals) are not Gaussian, this does not lead to separation of the independent sources. The net result is the removal of the second-order statistics from the data – the covariance matrix of the z values should be diagonal.

Now $\mathbf{z}$ is fed forward through the W weights to the output neurons where there is a second layer of lateral inhibition. However, before the activation is passed through this layer it is passed back to the originating z values as inhibition and then a nonlinear function of the inputs is calculated:

$$y_i = \sum_{j} w_{ij} z_j$$
$$e_j \leftarrow z_j - w_{ij} y_i$$
$$s_i = y_i - \tanh(y_i)$$

Now we pass this output through the lateral inhibition to get the final output y values

$$y_i = s_i + \sum_j v_{ij} s_j \qquad (6.33)$$

and then the new weights are calculated

$$\Delta w_{ij} = \beta e_j y_i$$
$$\Delta v_{ij} = \gamma y_i y_j$$

The net result is the removal of any dependence from the output signals. The differences then from the network described above on artificial data are the following:

1. The method of sphering is rather different. There is some evidence that a sphering method which also equalises variance provides a better starting point for the exploratory investigation [9] known as ICA.
2. The index is $y_i - \tanh(y_i)$. Therefore, whereas previously we relied on the fact that $E(\tanh(y_i))$ was zero, this index is explicitly removing the first–order term each time. Thus, whereas previously our argument was true only in the expectation of the weight update, now the argument is true for each and every data presentation.

A Simulation Example

Five samples of five seconds of natural speech were recorded using the standard telecom sampling rate of 8 kHz. Two adult male and female voices were used along with that of a female child. The speakers each spoke their name and a six digit number. The samples were then linearly mixed using the 5×5 mixing matrix shown in Table 6.2 which is well conditioned with a determinant value of 1.42. The fourth–order statistics of the original signals are shown in Table 6.3.

Table 6.2. The mixing matrix used to create a babble of voices.

0.65	0.20	−0.43	0.60	0.467
−0.3	−0.49	0.7	−0.3	0.57
0.68	1.5	−0.8	0.41	1.34
−0.234	0.38	0.35	0.45	−0.76
0.85	−0.43	0.6	−0.7	0.4

The output signals played back are clear with no residual of the mixture as shown in Fig. 6.13. When we look at the converged weight matrices, we see that both U and V are diagonal and symmetric as would be expected. The magnitudes of the values in U indicate the large correlations in the incoming raw data, with the off-diagonal terms being typically within an order of

Table 6.3. Fourth-order cumulants of the individual voices.

Male 1	0.011347
Male 2	0.001373
Female 1	0.000288
Female 2	0.000368
Female 3	0.000191

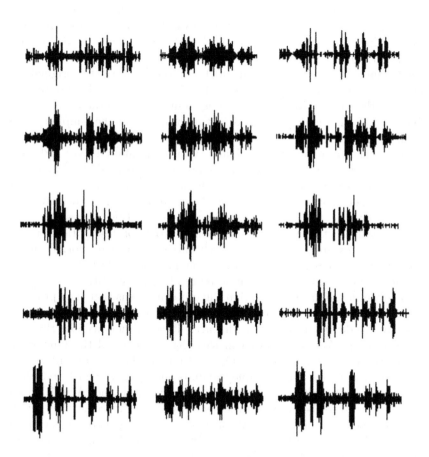

Fig. 6.13. Left column: the original signals. Centre: the mixture presented to the network. Right: the retrieved signals recovered by the network.

magnitude less than the diagonal terms. Compare this with the V weight matrix where the off-diagonal terms are all three orders of magnitude less than the diagonal terms, which is indicative of the whitened input to the layer of neurons.

6.7 Conclusion

We have introduced a neural network architecture which, using an extremely simple architecture and learning rule, has been shown to be capable of performing sophisticated statistical functions. The fact that the same network structure is capable of performing a PCA as well as EPP is unsurprising since Huber [84] has shown that PCA may be viewed as a particular case of Projection Pursuit. Thus, for the standard negative feedback network of Chapter 3, in performing a PCA we are choosing $g(x) = x \propto \frac{d}{dx}x^2$ and so the original network is seen to be maximising the second–order statistics of the distribution, i.e. finding the eigenvectors corresponding to maximal eigenvalues.

The initial PP indices discussed in this chapter are the simplest possible indices for the finding of non-Normal interesting directions; however, the method was shown to be equally valuable with Information Theory indices or with more sophisticated indices such as an instantaneous version of Friedman's index [50] or Intrator's index [90]. The important point to note, however, is that the method may be used with any function denoting a criterion which we wish to optimise. Long term it may be that information indices may prove to be the most effective indices in analysing a variety of distributions: we envisage further research into indices which maximise mutual information [7], which we will discuss in chapter 12, or maximise the effects of contextual information subject to externally imposed conditions e.g. [104].

The advantage of using Projection Pursuit concepts is that they provide a framework for understanding and integrating previous neural network models which have tended to introduce nonlinearity in an ad hoc fashion. However, we note that the format reviewed in this chapter required the function to be optimised to be differentiable; this need not be the case for the general neural network model. For example, Shapiro and Prügel-Bennet [169] have introduced a nonlinearity – a power law – into Oja's Subspace Algorithm but also used a threshold below which the neuron will not fire. Since they set the threshold to be zero, the analysis of convergence of a second–order network is understandable in PP terms, yet the fact that the threshold may be changed suggests a direction for future research of PP indices.

Koetsier [108] has twinned EPP networks so that a pair of networks jointly extract common or shared higher–order structure over a pair of data sets. This work will be discussed in Chapter 12.

7

Topology Preserving Maps

This chapter introduces three negative feedback artificial neural network architectures which perform a vector quantization. Vector quantization is used in signal processing applications to encode a high–dimensional signal in order to minimise processing/transmission costs. The basic aim is to associate with each group of vectors of the raw data a code which uniquely identifies that group. If the vectors of the group are sufficiently alike and the decoded code is sufficiently representative of the group, then the error when the code is used to represent a vector in the group can be made acceptably small. One further feature of the mapping which we desire is that it should retain an accurate representation of the topology of the data space. This is a rather complex feature to specify absolutely accurately so we shall initially content ourselves with a mapping in which nearby points in the data space are mapped to the same or nearby neurons in the coding space while ensuring that nearby neurons in the coding space are decoded to nearby points in data space.

7.1 Background

The most common type of artificial neural networks used to perform a topology preserving vector quantization is that developed by Kohonen. In the Kohonen network, when a data point is presented to the network, a competition takes place between the neurons and one neuron is declared the winner; its weights and those of its neighbours are moved towards the input pattern's values while those of the other neurons are moved further away. The net result is that if the same input pattern or one similar to it is presented again, the same neuron is most likely win again. It has been shown that, after a suitable training period, the neurons of the second layer form a map of the inputs which preserves some aspect of the topology of the input data. Such nets are known as self-organising nets as there is no teacher input to the net.

We will develop two negative feedback networks which quantise the data in a topology preserving manner but first we review competitive learning and Kohonen's algorithm.

7.1.1 Competitive Learning

The basic mechanism of simple competitive learning is to find a winning unit and update its weights to make it more likely to win in the future should a similar input be given to the network. We first have the activity transfer equation

$$y_i = \sum_j w_{ij} x_j, \forall i \qquad (7.1)$$

which is followed by a competition between the output neurons and then

$$\Delta w_{ij} = \eta(x_j - w_{ij}) \qquad (7.2)$$

for the winning neuron i. Note that the change in weights is a function of the *difference* between the weights and the input. This rule will move the weights of the winning neuron directly towards the input. If used over a distribution, the weights will tend to the mean value of the distribution since $\Delta w_{ij} \rightarrow 0 \iff w_{ij} \rightarrow E(x_j)$.

7.1.2 The Kohonen Feature Map

The interest in feature maps stems directly from their biological importance. A feature map uses the "physical layout" of the output neurons to model some feature of the input space. In particular, if two inputs $\mathbf{x}_1$ and $\mathbf{x}_2$ are close together with respect to some distance measure in the input space, then if they cause output neurons y_a and y_b to fire respectively, y_a and y_b must be close together in some layout of the output neurons. Further, we can state that the opposite should hold: if y_a and y_b are close together in the output layer, then those inputs which cause y_a and y_b to fire should be close together in the input space. When these two conditions hold, we have a feature map. Such maps are also called **topology preserving maps**.

Examples of such maps in biology include

- the retinotopic map, which takes input from the retina (at the eye) and maps it onto the visual cortex (back of the brain) in a two–dimensional map.
- the somatosensory map, which maps our touch centres on the skin to the somatosensory cortex.
- the tonotopic map, which maps the responses of our ears to the auditory cortex.

Each of these maps is believed to be determined genetically but refined by usage. e.g. the retinotopic map is very different if one eye is excluded from seeing during particular periods of development.

Hertz et al [73] distinguish between

- those maps which map continuous inputs from single (such as one ear) inputs or a small number of inputs to a map in which similar inputs cause firings on neighbouring outputs (left half of Fig. 7.1)
- with those maps which take in a broad array of inputs and map onto a second array of outputs (right half of Fig. 7.1).

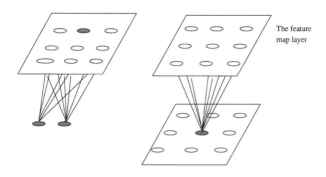

Fig. 7.1. Two types of feature maps: (a) a map from a small number of continuous inputs; (b) a map from one layer spatially arranged to another

Kohonen's algorithm [111] is exceedingly simple - the network is a simple 2-layer network and competition takes place between the output neurons; however, now not only are the weights into the winning neuron updated, but also the weights into its neighbours. Kohonen defined a neighbourhood function $f(i, i^*)$ of the winning neuron i^*. The neighbourhood function is a function of the distance between i and i^*. A typical function is the difference of Gaussians function (Figure 7.2); thus, if unit i is at point $\mathbf{r}_i$ in the output layer then

$$f(i, i^*) = a \exp\left(\frac{-|r_i - r_{i^*}|^2}{2\sigma^2}\right) - b \exp\left(\frac{-|r_i - r_{i^*}|^2}{2\sigma_1^2}\right) \qquad (7.3)$$

Notice that a winning neurons' chums – those neurons which are "close" to the winning neuron in the output space – are also dragged out to the input data while neurons further away are pushed slightly in the opposite direction.

The algorithm is:

1. Select at random an input data point.
2. There is a competition among the output neurons. That neuron whose weights are closest to the input data point wins the competition:

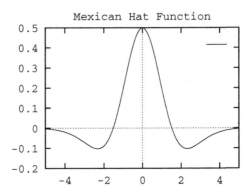

Fig. 7.2. The difference of Gaussian function.

$$\text{winning neuron, } i^* = \arg\min(\| \, \mathbf{x} - \mathbf{w}_i \, \|) \tag{7.4}$$

3. Now update all neurons' weights using

$$\Delta w_{ij} = \alpha(x_j - w_{ij}) * f(i, i^*) \tag{7.5}$$

where

$$f(i, i^*) = a \exp\left(\frac{-|r_i - r_{i^*}|^2}{2\sigma^2}\right) - b \exp\left(\frac{-|r_i - r_{i^*}|^2}{2\sigma_1^2}\right) \tag{7.6}$$

4. If not converged, go back to the start.

Kohonen typically keeps the learning rate constant for the first 1000 iterations or so and then slowly decreases it to zero over the remainder of the experiment (a simulation can take 100 000 iterations for self-organising maps). Two–dimensional maps can be created by imagining the output neurons laid out on a rectangular grid (we then require a two-dimensional neighbourhood function) or sometimes a hexagonal grid.

7.2 The Classification Network

Before we introduce a negative feedback topology-preserving mapping, we discuss the following very simple negative feedback competitive network based on nonlinear PCA discussed in Chapter 5:

$$s_i = \sum_{j=1}^{N} w_{ij} x_j \tag{7.7}$$

$$y_i = f(s_i) = f\left(\sum_{j=1}^{N} w_{ij}x_j\right) \tag{7.8}$$

$$e_j = x_j - \sum_{k=1}^{M} w_{kj}y_k \tag{7.9}$$

$$\Delta w_{ij} = \eta_t y_i e_j \tag{7.10}$$

$$= \eta_t f\left(\sum_{k=1}^{N} w_{ik}x_k\right)\left\{x_j - \sum_{l=1}^{M} w_{lj}\sum_{p=1}^{N} w_{lp}x_p\right\} \tag{7.11}$$

We noted that Karhunen and Joutsensalo [101] (in the context of feedforward networks) have shown that (7.11) is an approximation to the rule required to minimise the residuals at the inputs after the negative feedback is returned.

We will use the network described above to classify by using a function $f()$ defined by

$$f(s_i) = 1 \text{ if } i = \arg\max_j s_j \tag{7.12}$$

$$f(s_i) = 0 \text{ otherwise.} \tag{7.13}$$

Therefore we have an extremely simple network, which nevertheless will be shown to be capable of hierarchical classification. Clearly the learning rule is now equivalent to $\Delta w_{ij} = \eta(x_j - w_{ij})$, the standard competive learning rule for the winning neuron, so that

$$\Delta w_{ij} \to 0 \iff w_{ij} \to E_i(x) \tag{7.14}$$

where $E_i(x)$ is the average x for which neuron i is firing. Thus we can see that the weights will converge to the mean of the data to which the weight responds. Further, we can show that this network is stable, requiring no renormalisation or otherwise bounding of the network weights. As usual the–winner–take–all network can be modeled as a lateral inhibition network (c.f. Chapter 4) of the form used for the PCA network.

However, the most important property of this network is its ability to perform hierarchical decomposition of classification. Since the neuron's weights are converging to the mean of the distribution to which it is responding *and subtracting this mean* what remains is the difference between the mean and the distribution to which it is responding. This allows subsequent neurons which use exactly the same network learning to converge to subpatterns within the data.

7.2.1 Results

We describe an experiment using the network on artificial data. The data which we use is two dimensional and is shown diagrammatically in Fig. 7.3;

it comprises randomly chosen points from one of five equally probable two dimensional distributions. The distributions have centres (2,2), (2,−2), (−2,2), (−2,−2) and (−3,−3). The horizontal and vertical distances of individual samples from their respective distribution centres are independent and are drawn from a Gaussian distribution of standard deviation 1.

We use an initial network with two inputs – the x and y coordinates – and 10 outputs. Since we only have five classes, this gives us far more power than we need; however, we wish to simulate the situation in which the number of clusters is not known a priori. This task comprises for the network a simple problem – differentiating between the data clusters in the four quadrants – and a more difficult problem – that of differentiating between the two clusters in the 3rd (all negative) quadrant.

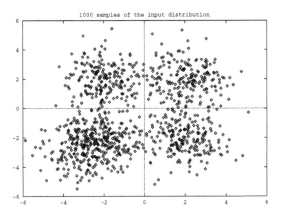

Fig. 7.3. The diagram represents the input data schematically: the data was taken from five independent Gaussian distributions with centres (2,2),(2,−2),(−2,2),(−2,−2) and (−3,−3), each distribution having a standard deviation of 1.

The weights for each neuron after convergence and its number of successes thereafter on 10 000 trials are shown in Table 7.1. These weights were learned in 1000 presentations of the input data (i.e. the network saw approximately 200 samples from each distribution); we have, however, trained the network on 200 000 iterations and found that the weights are stable at these points. The learning rate is annealed from 1 to 0 during the experiment.

Notice that during learning, neuron 9 has clearly been competing for the stewardship of the $(x_1 > 0, x_2 > 0)$ quadrant but has lost out to neuron 6. It is easy in such cases to monitor the success of such a neuron and delete it on testing the rate of success after learning.

Table 7.1. The filters to which the weights of the output neurons converged and the number of successes in 10 000 trials (after convergence).

Neuron	1	2	3	4	5	6	7	8	9	10
x_1-weight	0.00	-2.44	0.00	2.11	0.00	1.94	-1.86	0.00	1.60	0.00
x_2-weight	0.00	-2.47	0.00	-1.98	0.00	1.85	2.16	0.00	0.81	0.00
Successes	0	3992	0	2006	0	2043	1959	0	0	0

Clearly the network is very adept at differentiating between four groups, but, unsurprisingly, the two groups in the $(x_1 < 0, x_2 < 0)$ quadrant are treated as one.

Now the negative feedback network is not only finding the centres of each group, it is also subtracting out these centres because of the negative feedback. We can use this fact in finding subclasses of the sets which our network has already found by continuing the simulation with a second layer of output neurons which will learn on the distribution remaining *after* the first layer of output neurons has subtracted their activations. We could, in this situation, perform a top-down dichotomy of the data by creating layers containing only two output neurons each.

7.2.2 Stochastic Neurons

A network with similar properties can be created using stochastic neurons. We define a winner–take–all network using a roulette-wheel selection procedure: each neuron's probaility of firing is dependent on the sum of its weighted inputs by

$$P(\text{neuron}_j \text{ fires}) = \frac{\exp(s_j)}{\sum_k \exp(s_k)} \tag{7.15}$$

where initially the weights are small ($\approx 10^{-5}$) random numbers. Experiments have confirmed that this network also self-organises in a manner identical to that above to find increasingly refined hierarchies.

7.3 The Scale Invariant Map

We now extend this network to enable it to perform feature mapping by a method similar to that used by Kohonen to change simple competitive networks to feature maps.

Thus, in the learning rule above, we change not only the winning neuron's weights but also the weights of those neurons closest to it. We use a Gaussian mask[1] to model a differential effect: the weights of those closest to the winning

---

[1] A difference of Gaussian gives the Mexican hat field but was found to be unnecessary in this simulation and indeed more difficult to stabilise.

neuron are updated most highly. We decrease the diameter of the Gaussian during the simulation run. This, as usual [111], focuses the activation on a narrow set of neurons.

Consider a network with N dimensional input data and having M output neurons. Then the activation of the ith output neuron is given by

$$y_i = \sum_{j=1}^{N} w_{ij}x_j \qquad (7.16)$$

Now we invoke a competition between the output neurons. We will investigate networks using one of two possible criteria:

Type A: The neuron with greatest activation wins:

$$\text{winner } = \arg\max_i y_i \qquad (7.17)$$

Type B: The neuron closest to the input vector wins:

$$\text{winner } = \arg\min_i ||\mathbf{x} - \mathbf{w}_i|| \qquad (7.18)$$

In both cases, the winning neuron, the pth, is deemed to be maximally firing (=1) and all other output neurons are suppressed. Its firing is then fed back through the same weights to the input neurons as inhibition,

$$e_j = x_j - w_{pj}.1 \text{ for all } j \qquad (7.19)$$

where p is the winning neuron. Now the winning neuron excites those neurons close to it i.e. we have a neighbourhood function $\Lambda(p, j)$ which satisfies $\Lambda(p, j) \leq \Lambda(p, k)$ for all $j, k :|| p - j || \geq || p - k ||$ where $|| . ||$ is the Euclidean norm. In the simulations described in this chapter, we use a Gaussian whose radius is decreased during the course of the simulation. Then simple Hebbian learning gives

$$\Delta w_{ij} = \eta_t \Lambda(p, i).e_j \qquad (7.20)$$
$$= \eta_t \Lambda(p, i).(x_j - w_{pj}) \qquad (7.21)$$

For the pth winning neuron, the network is performing simple competitive learning but note the direct effect the pth output neuron's weight has on the learning of other neurons.

7.3.1 An Example

To illustrate convergence we use, as input data, a two–dimensional vector drawn randomly from the square $\{(x, y) : -1 < x \leq 1, -1 < y \leq 1\}$. With the rules used above we get results such as those shown in Fig. 7.4. The 25 output neurons' weights have organised in such a way that similar (and only similar) input values are mapped onto similar output neurons which is the usual definition of topographic mappings (e.g. [111]). The scale-invariance of the resultant mapping can be seen most clearly in Fig. 7.6 in which we have shown the points which were mapped onto three specific output neurons.

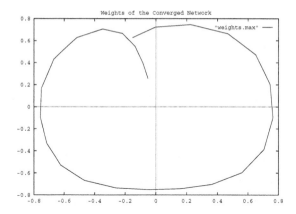

Fig. 7.4. The converged weights from the feature mapping network when the input data is drawn from $\{(x, y) : -1 < x \leq 1, -1 < y \leq 1\}$.

7.3.2 Comparison with Kohonen Feature Maps

It is well known (e.g. [111]) that, given similar input data to that used above, a one-dimensional Self-Organising Map (SOM) will self-organise to spread itself over the square to minimise the expected distance between the code points and the points of the square. We have called the current mapping a scale-invariant feature map since it ignores the magnitude of each input vector and responds solely to the relative proportion of the magnitude of the elements of the input vectors. The Kohonen feature map may be criticised on grounds of biological implausibility in that a single neuron takes responsibility for representing a set of inputs (the "grandmother" cell). However, if we increase the learning rate with the Scale Invariant feature map, an interesting effect comes into play: the mapping winds round upon itself so that each outer neuron (which is currently winning competitions) is backed up by a set of support neurons. The results of such an experiment are shown in Fig. 7.5.

The network learning rules can be extended to use a higher–dimensional neighbourhood function, though in dimensions higher than two, the resulting map is difficult to view. We have found that higher dimensional maps are more prone to twists such as those well known in the Kohonen SOM [73]. Such twists can take a very long time to untwist. Kohonen's strategy of beginning with a wide neighbourhood function and decreasing its width gradually is the most successful with this problem.

7.3.3 Discussion

As with the Kohonen SOM ([111], page VII), a full analysis of the current mapping is remarkably difficult. Our discussion here is descriptive rather than fully analytical.

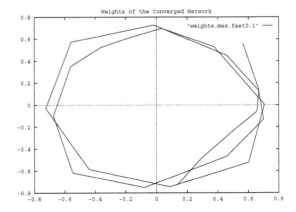

Fig. 7.5. The results when the learning rate is increased 100–fold. Notice that those weights which continue to occupy space within the outer ring of neurons, do not win any competitions. They can be thought of as backups for those neurons which are winning competitions: if one of the winners should fail, there exists a substitute which will step into its shoes.

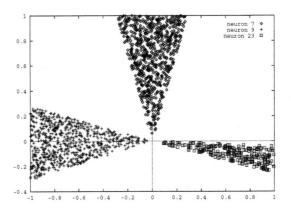

Fig. 7.6. Some points showing the winning areas for specific neurons in 10 000 trials of the fast learning network.

Consider an input distribution which is spherically symmetric and let us use a very simple (though still symmetric) mask in which the neighbourhood function in the weight update rule is given by

$$G(i_1) = 1 \text{ where } i_1 \text{ is the winning neuron}$$
$$G(i_1 + 1) = G(i_1 - 1) = g, \text{ for some g } > 0$$
$$G(j) = 0, \text{ when j } \neq i_1 - 1, i_1, i_1 + 1.$$

Consider three *consecutively numbered* neurons labelled A, B and C.

Now when neuron A wins the competition, neuron B's weights will also be changed according to:

$$\Delta \mathbf{w}_B = \eta_t . g . (\mathbf{x} - \mathbf{w}_A) \tag{7.22}$$

Note the crucial difference here between this form of learning and the usual competitive learning rules: here B's weights are being directly affected by the centre of gravity of the inputs to the winning neuron next to B. Thus, if the process is to converge,

$$\sum_{\mathbf{x} \in Set_A} g . (\mathbf{x} - \mathbf{w}_A) + \sum_{\mathbf{x} \in Set_B} (\mathbf{x} - \mathbf{w}_B) + \sum_{\mathbf{x} \in Set_C} g . (\mathbf{x} - \mathbf{w}_C) = 0 \tag{7.23}$$

where Set_L is the set of points sampled in sector L and $\mathbf{w}_L$ is the weight vector of the neuron winning in sector $L, L = A, B, C$. If we are drawing points randomly from the data set the limit of the above as the number of points tends to infinity is

$$\int_{Set_A} g(\mathbf{x} - \mathbf{w}_A) p(Set_A) d\mathbf{x} + \int_{Set_B} (\mathbf{x} - \mathbf{w}_B) p(Set_B) d\mathbf{x}$$
$$+ \int_{Set_C} g(\mathbf{x} - \mathbf{w}_C) p(Set_C) d\mathbf{x} = 0$$

where $p(Set_L)$ is the probability that a point drawn at random from the input distribution will be in sector L. A solution of this equation is the set of weights which converge to the centre of gravity of each sector where each sector is of equal probability (i.e. of equal angular spread) in each dimension. In this case the mapping created is a maximum entropy mapping. It should be noted that while this is found empirically to be true when the learning rate is small, a high learning rate (Fig. 7.6) disrupts this feature.

Notice also that, while individual input data points within slice A may have the effect of moving B's weights away from A, the mean effect of A's points on the weights into B will be to move them towards A's centre. Points in sector C will be having the opposite effect: if slices A and C are of equal width on either side of B, the effect of each will cancel out the other.

7.3.4 Self-Organisation on Voice Data

We illustrate the converged mapping of the Scale Invariant network on speech data. Typically such networks are trained on data which has been preprocessed by, for example, Fourier transforming the data to the frequency domain and often using the more computationally expensive cepstral coefficients (see e.g. [111]). We, however, wish to test the nework by using as crude voice data as possible as inputs.

The input data to the network then is raw voice data sampled at 8 kHz and subjected to no preprocessing. The network has been tested with 20, 32, 64 and 128 inputs where e.g. 64 inputs represents 64 consecutive inputs from the data stream (equal to 8 ms of data). Each presentation of the data consists of a randomly chosen starting point and the following 63 consecutive inputs. The results have been qualitatively similar for each size of network. We use 25 output neurons with a one–dimensional neighbourhood function and the maximum activation criterion to identify the winning neuron. Consider first one network which is trained on 8 speakers saying the word "far". The trained weights are shown in Fig. 7.7. The top diagram shows the weights into the first three neurons: we can see that:

- the neurons are extracting the frequency information from the input data
- the weights into the first neuron differ very little from those into the second and this in turn differs very slightly from those into the third neuron. This is a prerequisite for any network which is claiming to retain neighbourhood relations.

The second half of the diagram shows the weights into three neurons which are not neighbours. Clearly each neuron is extracting the same frequency information from the raw data but has learned to respond to different phases of this data. A full diagram with all 25 output neurons would show a complete coverage of all phases at this frequency.

7.3.5 Vowel Classification

In Fig. 7.8, we show the weights from converged networks which have been trained on different vowel sounds – "far", "pit" and "put". The fact that these are very different can be used to classify vowel sounds into their respective classes.

We now propose a network which will identify any vowel from its raw data samples. We use a network such as shown in Fig. 7.9. During the learning phase, each set of output neurons (Net A, Net B, ...) is trained on a different set of vowel data. Each learns the frequency information associated with that vowel, and individual neurons within that set will respond maximally at any particular time depending on the correspondence between the phase of the signal and the weights into the neuron. However during the vowel identification phase the network is fully connected – each input is connected to all output

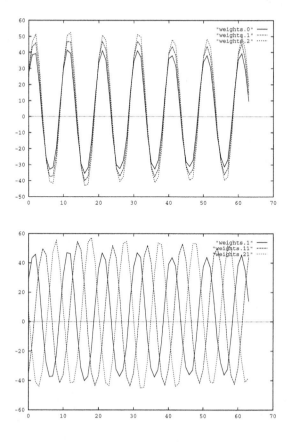

Fig. 7.7. Top diagram: the weights of the first three neurons trained on the vowel sound of "far". Bottom diagram: the weights of the first, eleventh and twenty first neurons on the same data.

neurons in all networks, Net A, Net B, ... When any vowel is presented to the network, the vowel can be identified by noting which of the output networks responds optimally since there is a ripple of activation across that network as particular neurons respond maximally to the particular phase of the inputs. Such a coherent ripple cannot be seen in the other output nets.

7.4 The Subspace Map

Recognition of patterns subject to transformations such as translation, rotation and scaling has been a difficult problem in artificial perception. The projection of objects on the retina is variable in position, size, orientation, luminance, etc., yet we are still able to recognise objects with great ease, re-

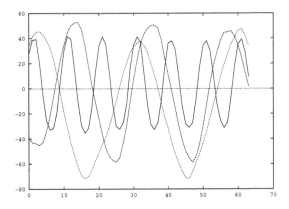

Fig. 7.8. A comparison of network weights when three different vowels are used for input data.

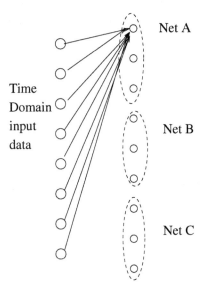

Fig. 7.9. Each group of output neurons comprises a single network trained on examples of a single vowel. Each output neuron responds maximally to a single type of input data.

gardless of such variations. A neural network solution to this problem has been proposed by Kohonen et al. [111] called the Adaptive Subspace Self-Organising Map (ASSOM). The motivation for the ASSOM is that if different samples can be derived from each other by means of a linear transformation, then they span a common linear subspace of the input space and projecting them onto this subspace can then filter out their differences. Using common linear subspaces as filters also motivates the method presented in this Section.

By constructing a SOM where each module defines a subspace, we can observe the organisation of clusters of different subspaces that each provides a different invariant filter. The neighbourhood relations in the map cause neighbouring modules to represent similar subspaces.

The map consists of an array of modules, normally in one or two dimensions, although higher-dimensional maps may be used. Each module of the map consists of the same arbitrary number of nodes which defines the dimensionality of the subspace represented by the module and a single output which is a quadratic function of the outputs of these nodes. (See Fig. 7.10).

Within each module, weights are updated using the Subspace Algorithm (either Oja's feedforward version (Chapter 2) or the negative feedback implementation (Chapter 3)), and each module learns the principal subspace of a subset of the data set. All of the nodes capture a local linear subspace but, collectively, they represent a nonlinear manifold. There is also a single output for each module which is quadratic sum of the activations of other nodes. The learning process requires the use of "episodes" of inputs, meaning that input vectors are presented to the network in batches and the weights are updated after a complete batch of inputs have been fed forward.

7.4.1 Summary of the Training Algorithm

The orthogonal projection of an input vector $\mathbf{x}$, onto a subspace $\mathcal{L}^{(i)}$ gives a projection $\hat{\mathbf{x}}^{(i)}$ and a residual $\tilde{\mathbf{x}}^{(i)}$ where $\mathbf{x} = \hat{\mathbf{x}}^{(i)} + \tilde{\mathbf{x}}^{(i)}$. Therefore, when data is fed forward in the network, each module, i, encodes the data as a vector $\mathbf{y}^{(i)}$ which is a lower–dimensional representation of the projected vector $\hat{\mathbf{x}}^{(i)}$. In order to select a module whose subspace most closely approximates a local part of the data set, it is necessary to consider several data points, and this leads to the use of episodes of data.

An episode $\mathcal{S}$ consists of a set of consecutive time instants and the set of inputs $\mathbf{x}(t_p)$ is taken from these sampling instants $t_p \in \mathcal{S}$. These vectors are projected onto the subspaces represented by the modules. For each episode of inputs, a representative winning node, c_r, is chosen and this is held constant for the duration of the episode. Since each node of the map represents a subspace $\mathcal{L}^{(i)}$ of the input space, the selection of the winning module, c_r, can be based on how closely the orientation of its subspace $\mathcal{L}^{(c)}$ approximates the orientation of the data in the episode. The representative winner is chosen to be the module with the maximum squared projections (7.24) which is equivalent to minimising the residuals since $\|\mathbf{x}\|^2 = \|\tilde{\mathbf{x}}\|^2 + \|\hat{\mathbf{x}}\|^2$:

inputs

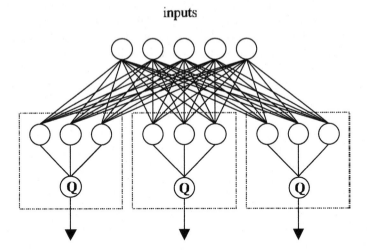

Fig. 7.10. The Subspace Map. Each module of the map (enclosed in dotted boxes) performs principal component analysis on a subset of data from the training set. There is also a single quadratic output in each module that is activated by the sum of squared outputs of the PCA nodes.

$$c_r = \arg\max_i \left\{ \sum_{t_p \in \mathcal{S}} \| \hat{x}^{(i)}(t_p) \|^2 \right\} \tag{7.24}$$

If the input space and subspaces have N and M dimensions, respectively, then an orthogonal projection onto a subspace gives a vector y of dimensionality M. The jth output of $\mathcal{L}^{(i)}$ is denoted by $y_j^{(i)}$ and is calculated by a weighted sum of its inputs (7.25). The activation of the quadratic output is then calculated by (7.26):

$$y_j^{(i)} = \sum_{k=1}^{N} w_{kj}^{(i)} x_k \tag{7.25}$$

$$Q^{(i)} = \sum_{j} = 1^M y_j^{(i)^2} \tag{7.26}$$

The weights of the winning module and its neighbouring modules are then updated using the Subspace Algorithm (7.27) with a radially decaying neighbourhood function to ensure that nodes that are close to the winner will be trained with a higher learning rate than those further away. The neighbourhood of a winning module is the set $\mathcal{N}^{(c_r)}$ of modules whose weights are updated. This set may include every module in the map; however, significant savings in computing time can be achieved if a threshold is added to the neighbourhood function such that modules are changed only if they are close to the winner. Those that are further away will have very low learning rates due to the neighbourhood function and it is acceptable to exclude them from training:

$$\Delta w_{jk}^{(i)} = h_{ci}(t)\alpha(t)\left(x_k y_j^{(i)} - y_j^{(i)}\sum_l w_{kl}^{(i)} y_l^{(i)}\right) \tag{7.27}$$

Therefore, for every module, $i \in \mathcal{N}^{(c_r)}$, the weights are updated to shift towards a principal subspace of $x(t_p \in \mathcal{S})$. This shift most strongly affects the winning module and the neighbourhood function determines the effect on its neighbouring modules. Competitive learning ensures that each module is trained on a subset of the training data and therefore, the weight vectors of each module form a principal subspace of a different class of input patterns (local PCA).

The neighbourhood function that we used $h_{ci}(t)$ is a Gaussian or difference of Gaussians (Mexican hat) with a width that is decreased in time; however, there are other neighbourhood functions that may be equally suitable. This function may be used with a narrowing threshold so that the number of modules in the neighbourhood of the winner, $\mathcal{N}^{(c_r)}$ is reduced.

7.4.2 Training and Results

We discuss two examples of using the map to create filters on data which may be of biological significance – sound data and artificial video data emulating moving objects.

Formation of Wavelet Filters from Raw Sound Data

The network was trained on raw speech data and formation of a mapping was observed that extracted phase and frequency information from voiced phonemes. In the experiments of Kohonen et al. [111], the ASSOM was used to generate wavelet filters from raw speech data and we now show that the subspace map presented will also form filters that are phase invariant.

To demonstrate the behaviour of this network, it has been trained on artificial sound data containing a range of frequencies between 200 Hz and 10000 Hz. This data was high-pass filtered by taking the differences between successive samples. The training vector, consisted of 64 consecutive samples of sound recorded at 8kHz giving an input window duration of 8.75 ms. The one-dimensional map consisted of 24 modules, each of which had a subspace dimensionality of two. Eight sets of inputs that were adjacent in time formed a single episode. Representative winners were selected using the maximum energy criterion.

Each input window was multiplied by a Gaussian weighting function with a full width half maximum, FWHM(t), of eight samples and this was increased to 20 during training. The initial learning rate was 0.008 and was reduced by 0.0005 after every 20 episodes.

Fig. 7.11 shows the weights of the converged map after 300 training episodes. At one end of the map the low frequencies have been learned and there is a smooth progression to the high frequencies at the opposite end, therefore the frequency of an input pattern corresponds to the position of the winning module. Each module has orthonormal basis vectors as a result of the principal subspace learning algorithm. Since the vectors are always orthogonal to the vectors, one can be considered to be a sine wavelet and the other a cosine wavelet. Sinusoidal oscillation is therefore eliminated from the square of the wavelet amplitude transform (A),

$$A^2 = F_c^2(t, \omega) + F_s^2(t, \omega) \tag{7.28}$$

and the organised subspaces are therefore approximately invariant to time shifts. When using the ASSOM, this property has to be forced by periodically re-orthonormalising the vectors.

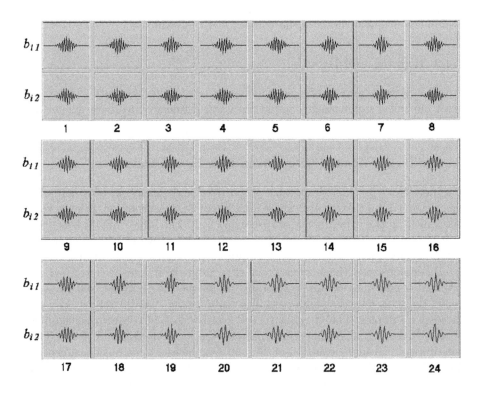

Fig. 7.11. The weights of the converged subspace map trained on sound data.

Invariant Image Filters

The network was trained on the bars data set (Chapter 5) but with two differences:

1. There is only a single bar in each image.
2. Eight different bar orientations were used (oriented at regular intervals between 0 and $\frac{7\pi}{8}$ radians from the vertical orientation.) In each episode, the orientation of the bar was held constant while its position was varied. A Gaussian weighting function was used at the inputs with a full width half maximum (FWHM) of sixteen pixels. This was to ensure that the orientation filters were centred on the sampling lattice.

A trained map is shown in Fig. 7.12. The diagram shows that each weight vector has positive and negative values in areas close to the centre of the sampling lattice and that weights of similar magnitudes are arranged along one direction. Pairs of weight vectors in the same module share the same direction. It is interesting to note that these vectors look similar to Gabor–type wavelets since they are localised and they exhibit orientation preferences. The Subspace Algorithm guarantees that trained weight vectors are orthogonal and therefore, within each module, the subspaces defined by the orthogonal Gabor–type wavelet pairs eliminate sinusoidal oscillations and are approximately invariant to translations. This is further discussed in the following section.

It is also important to point out that modules that are close together in the map have captured similar orientations and therefore the map has been smoothly ordered by the training process.

After training, the network was tested on the same data used for training but using only a single image in each feed-forward operation i.e. without using episodes of input vectors. The node with the highest activation in the quadratic output node was then selected as the winner (7.29) and the position of this node in the map shows the detected orientation of the bar in the inputs:

$$c_r = \arg\max_i Q^{(i)} = \arg\max_i \left\{ \sum_j y_j^{(i)^2} \right\} \qquad (7.29)$$

Fig. 7.13 shows the results from orientation selection using the trained network. In this example, the same module has the highest activation for every bar with the exception of two. These two correspond to lines that are far from the centre of the filter and reliability of the classification is therefore lower. (See Section 7.4.3.) No module has been activated significantly more than any other for these bars and they can be regarded as unclassified data rather than misclassified. The other bars clearly activated one group of modules significantly more than others and their orientations have therefore been successfully classified.

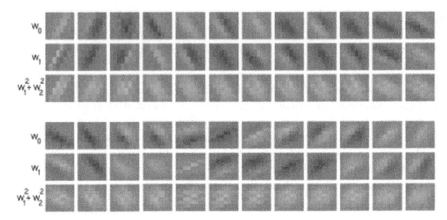

Fig. 7.12. The weights of a trained map. White pixels indicate positive weights and black pixels indicate negative. The weight vectors within each module show the same orientation and neighbouring modules have similar orientations. The vectors showing the sum of squared weights are approximately Gaussian and this confirms that the sinusoidal oscillations in the individual weight vectors can be eliminated (see Section 7.4.3).

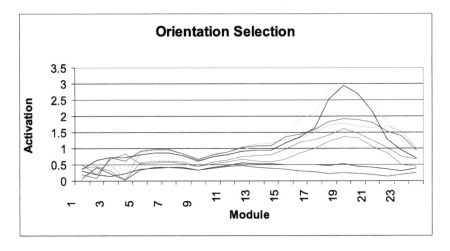

Fig. 7.13. Module activations plotted against the module number for several bars with the same orientation but with different positions. The winning module is in position 19 for every bar position except the outermost two.

7.4.3 Discussion

Assuming that the trained weight matrices are close approximations to orthogonal wavelet pairs, then we can simplify the analysis by considering one-dimensional cosine and sine wavelets respectively defined by:

$$\psi_c(t,\omega) = e^{-\frac{\omega^2 t^2}{2\sigma^2}}\cos(\omega t), \psi_s(t,\omega) = e^{-\frac{\omega^2 t^2}{2\sigma^2}}\sin(\omega t) \tag{7.30}$$

where t is time, ω is the angular frequency and σ is the variance of the Gaussian weighting component. An input signal, $\mathbf{x}$, is used and the output, Q, is given by :

$$Q = \sum_i (x_i\psi_c(i,\omega))^2 + (x_i\psi_s(i,\omega))^2 \tag{7.31}$$

$$= \sum_i \left(x_i e^{-\frac{\omega^2 i^2}{2\sigma^2}}\cos(\omega i)\right)^2 + \left(x_i e^{-\frac{\omega^2 i^2}{2\sigma^2}}\sin(\omega i)\right)^2 \tag{7.32}$$

$$= \sum_i x_i^2 e^{-\frac{\omega^2 i^2}{\sigma^2}}\left(\sin^2(\omega i) + \cos^2(\omega i)\right) \tag{7.33}$$

$$= \sum_i x_i^2 e^{-\frac{\omega^2 i^2}{\sigma^2}} \tag{7.34}$$

Therefore, sinusoidal oscillations are eliminated from (7.33) leaving only the Gaussian component, and the subspace defined by the orthogonal wavelets gives time–shift invariance. In the case of two dimensional complex Gabor wavelets, using x and y in place of the time variable t, Q is given by:

$$Q = \sum_x \sum_y e^{-\frac{\omega^2(x^2+y^2)}{\sigma^2}} \tag{7.35}$$

Consider a bar in the input data to be a plane, v, perpendicular to the $x - y$ plane with orientation θ_1. When this plane coincides with a Gabor function with orientation θ_2, the result is a curve. The integral of this curve indicates the difference between the orientations of the plane and the Gabor filter. This is effectively the same as calculating the weighted sum of an input vector where the input is a straight line in visual data and the weight matrix is similar to a complex Gabor function. The integral is described by (7.36):

$$\int e^{-\frac{\omega^2 v^2}{2\sigma^2}}\cos(v\sin(\theta_1 - \theta_2))dv \tag{7.36}$$

This integral is always at its highest absolute value when $\theta_1 = \theta_2$. For straight lines with orientation θ_2, some positions may give a zero value for the integral in (7.36) since the wavelet amplitude can be zero along that line. However, an orthogonal wavelet will coincide with a maximum absolute value along the same straight line and the sum of squares of these products eliminates any sinusoidal oscillations, as previously explained. Therefore, a pair of orthogonal Gabor wavelet filters will give the highest response to straight lines with orientation similar to that of the wavelet, regardless of the line position.

7.5 The Negative Feedback Coding Network

The various algorithms describing learning within the negative feedback network in previous chapters have been shown to extract the maximum information from sets of stochastic data. The next obvious question is to decide what a network should do with such information when it has been extracted. Some form of coding would be helpful in classifying such data.

The negative feedback coding network was developed in appreciation of the way in which Carlson [21] amended the basic network of Rubner and Schulten [158], a PCA network, in order to create a coding network.

While we have wished to emulate his success, we have the continuing design ethos based on the retention of as many of the attractive features of the basic negative feedback network as possible – those of simplicity, homogeneity, locality of information use and parallelism.

Our aim is to create a network which will take a set of raw data and code it so that different sections of the data are coded differently and such that data which have the greatest similarity are most alike in codes i.e. a topology-preserving network. A binary code is easiest to implement with a simple threshold. Since we require several bits for each codeword, we suggest that each input be connected to a set of coding output neurons. Raw data at the $\mathbf{x}$ input is converted to a binary coded vector $\mathbf{y}$ at the coding output neurons. There is only one major difference between this network and those investigated previously: each output neuron has a threshold above which its weighted inputs must sum in order to force a positive firing; if the threshold is not reached, the negative feedback will have a negative activation. For simplicity in exposition, we will consider only a scalar input, x. In detail the algorithm is:

- Set the residual at time 0 to be equal to the input, i.e. $e(0) = x$.
- For each output neuron,
  1. Calculate $w_i e(i)$ and set

  $$y_i = 1 \text{ if } w_i e > 0$$
  $$= -1 \text{ if } w_i e < 0$$

  2. Calculate the new residual

  $$e(i+1) = e(i) - v_i y_i$$

  3. Update weights w_i and v_i according to the same rule as previously, i.e. a simple Hebbian learning rule.

  $$w_i = w_i + \eta e y_i$$
  $$v_i = v_i + \eta e y_i$$

Note that the y_i values are either 1 or -1; changing the weights is the sole method of learning in the system to ensure that appropriate codes are found.

Each output neuron, in turn, receives a weighted sum (in this case of only 1) of the e values; however, each output neuron has a threshold above which its activation will have a positive value and below which the activation will be negative. We choose the threshold for *all* output neurons to be 0.

We wish to emphasise that we have not programmed a threshold nor any specific function which monitors the variance of the data and adjusts the network's response appropriately. We have retained an extremely simple network with only a single modification to the previous network.

7.5.1 Results

A typical set of results is shown in Table 7.2. These results are for a 6–output–neuron network which is learning from a set of x–values generated from a uniform distribution[2] between 2 and 4. The network used a learning rate of 0.01 and ran for 10 000 iterations.

Table 7.2. A section of coding for vectors produced by a 1–input 6–output neuron network for input data from a uniform distribution between 2 and 4. Learning rate = 0.01, number of trials = 10 000. We have replaced -1 with 0 to highlight the binary nature of the code.

Decimal		Decimal	
2.0	1 0 0 0 0 0	3.0	1 0 1 1 1 1
2.1	1 0 0 0 0 1	3.1	1 1 0 0 0 1
2.2	1 0 0 0 1 1	3.2	1 1 0 0 1 1
2.3	1 0 0 1 0 0	3.3	1 1 0 1 0 0
2.4	1 0 0 1 1 0	3.4	1 1 0 1 1 0
2.5	1 0 1 0 0 0	3.5	1 1 0 1 1 1
2.6	1 0 1 0 0 1	3.6	1 1 1 0 0 1
2.7	1 0 1 0 1 1	3.7	1 1 1 0 1 0
2.8	1 0 1 1 0 0	3.8	1 1 1 1 0 0
2.9	1 0 1 1 1 0	3.9	1 1 1 1 1 0

Table 7.3. The weights which the above network learned using only simple Hebbian learning

output neuron	1	2	3	4	5	6
Weight (w)	3.005	0.505	0.254	0.123	0.063	0.031

Several points are worth noting:

- First the coding seems fairly inefficient in that the first figure is always 1. This is due to our insistence that all means are zero. Thus the first code element is always 1 for inputs > 0 (see Section 7.5.2).

---

[2] We use a uniform distribution here to make it clear why each weight has converged to the actual value to which it has converged.

- If we wish a code where the first output neuron performs maximum discrimination, (i.e. in the above example, all inputs less than 3 would be coded as -1, all inputs > 3 would be coded as $+1$) we would use a threshold which will also learn; a rule such as

$$\theta_j = \theta_j + \alpha y_j$$

where θ_j is the threshold for the jth output neuron is an entirely local rule and easy to implement; however, in keeping with our design philosophy of maintaining simplicity, we have not implemented that here.
- The code is topology preserving – similar inputs have similar outputs (see Section 7.5.4).
- Experiments have shown that larger networks have no difficulty in providing more detailed codes and require only a slight increase in time as each element of the coding is done on the error remaining after the previous output neurons have performed their coding.
- A topological feature map using the negative feedback network has one major advantage over e.g. a Kohonen feature map: it can easily be re-implemented to show a hierarchy of subfeatures (see Section 7.5.6) by adding a new level of coding output neuron.
- Lower–valued digits are automatically coded more slowly and hence are less prone to an unusual input, an outlier, creating large changes.

7.5.2 Statistics and Weights

We will investigate what the weights are actually learning by considering their values at convergence. In general, we are using simple Hebbian learning, so

$$\Delta w_i = \eta e(i) y_i$$

where the subscript denote the ith output neuron and $e(i)$ denotes the value of e at time i. The proof that $v_i = w_i$ is similar to that shown in Chapter 3 and will not be repeated here. We investigate two interesting cases before looking at the general case:

1. A zero mean symmetric distribution.
   - Consider w_1, the weight to the first output neuron. Then,

$$\begin{aligned}
\Delta w_1 &= \eta e(1) y_1 \\
&= \eta (x - w_1 y_1) y_1 \\
&= \eta (x y_1 - w_1)
\end{aligned} \tag{7.37}$$

since $y_1^2 = 1$. So at convergence

$$E(\Delta w_1) = 0 \iff E(w_1) = E(x y_1) \tag{7.38}$$

With a zero mean symmetric distribution, $y_1 = -1$ when x is negative and $y_1 = +1$ when x is positive; therefore $y_1 x = |x|$ and so

$$w_1 = \overline{|x|}$$

at convergence i.e. w_i converges to the expected value of the absolute value of the input data, i.e. the mean absolute value. This has the effect of mapping the two halves of the distribution to a tighter (bipolar) distribution which is mostly contained within the interval $[-\overline{|x|}, \overline{|x|}]$.

2. A positive, compact distribution.

By compact, we mean a distribution which contains no holes.

• Consider w_1; as before, we can show that, at convergence,

$$E(\Delta w_1) = 0 \iff E(w_1) = E(xy_1) \qquad (7.39)$$

With this distribution, $y_1 = +1$ for all input values of x. So

$$w_1 = \overline{x}$$

i.e. the mean value of the input data

• Now consider w_2, the weight to the second output neuron. Then,

$$\begin{aligned}
\Delta w_2 &= \eta e(2) y_2 \\
&= \eta(e(1) - w_2 y_2) y_2 \\
&= \eta(e(1) y_2 - w_2) \qquad (7.40)
\end{aligned}$$

since $y_2^2 = 1$. But

$$\begin{aligned}
e(1) &= x - w_1 y_1 \\
&= x - \overline{x}
\end{aligned}$$

Now if $x > \overline{x}$, i.e. $x > w_1$, then $y_2 = 1$ while if $x < \overline{x}$, i.e. $x < w_1$, then $y_2 = -1$ Thus w_2 at convergence equals $\overline{|x - \overline{x}|}$. Therefore, at convergence, w_2 is bisecting the residual area of the distribution after w_1 has subtracted out the mean. So w_2 for this distribution is performing the task which w_1 performed for the symmetric distribution.

3. In general, for values of x drawn from any distribution

$$\begin{aligned}
\Delta w_i &= \eta e(i) y_i \\
&= \eta(e(i-1) - w_i y_i) y_i \\
&= \eta(e(i-1) y_i - w_i)
\end{aligned}$$

Therefore, $w_i \to E(e(i-1)y_i)$ at convergence

and if $e(i-1) > 0$ then $y_i = 1$
while if $e(i-1) < 0$ then $y_i = -1$.
Therefore, at convergence, $w_i = \overline{|e(i-1)|}$

In general, the y values correspond to the coding taking place while the e values represent the error after the coding has taken place.

The coding of a uniform distribution is shown in Figure 7.2. We use a uniform distribution this time to make it easy to corroborate that the network is performing an efficient coding; note from Table 7.3 that $w_1 = \overline{|x|} = \overline{x}$ and $w_2 = \overline{|x - \overline{x}|}$, etc..

7.5.3 Reconstruction Error

We show that, if we use the vectors produced by this method to reconstruct the original vectors, we can make the expected absolute reconstruction error from a final code (sometimes called the mean quantization error) arbitrarily small by simply adding new coding output neurons. We will devote a chapter (Chapter 8) to discussing optimising the learning rules of the negative feedback network with respect to the probability density functions of reconstruction errors or residuals. Let us assume that we cannot i.e that there exists an $\epsilon > 0$ such that all mean absolute reconstruction errors are greater than ϵ. We note from the above that this is equivalent to showing that the value of $w_i > \epsilon$ for all i.

Note that for a finite distribution, the maximum possible error after the first coding is

$$E_{Max} = \max(|x_{Max} - \overline{|x|}|, |x_{Min} - \overline{|x|}|, \overline{|x|})$$

where x_{Max}, x_{Min} is the largest (resp. smallest) possible member of the distribution. Thus E_{Max} is finite.

Consider a particular input x. From the results in the last section, because the system is creating the coding at y and subtracting the weighted coding at e, the value $e(i)$ is simply the error between the code found by the first i coding output neurons and the input x after i codings have taken place. Therefore $\overline{|e(i)|}$ is the mean absolute error after i codings of the input. Now,

$$e(i) = e(i - 1) - w_i y_i$$
$$= e(i - 1) - \overline{|e(i - 1)|} y_i$$

If $e(i - 1) < 0$, $y_i = -1$ and

$$e(i) = e(i - 1) + \overline{|e(i - 1)|}$$
$$= -|e(i - 1)| + \overline{|e(i - 1)|}$$

while if $e(i - 1) > 0$, $y_i = +1$ and

$$e(i) = e(i - 1) - \overline{|e(i - 1)|}$$
$$= |e(i - 1)| - \overline{|e(i - 1)|}$$

Therefore the amplitude of $e(i)$ is the difference between the absolute value of $e(i - 1)$ and the mean absolute value of $e(i - 1)$.

Thus

$$|e(i)| = |\{|e(i-1)| - \overline{|e(i-1)|}\}|$$
$$= |\{|e(i-1)| - w_i\}|$$
$$< ||e(i-1)| - \epsilon|$$

since all terms are positive. Therefore the absolute error at each stage is decreasing by more than ϵ. Therefore the mean absolute error is also decreasing by at least ϵ at each stage. Now the initial maximum error is E_{Max} which is finite and the absolute error is decreasing by a finite amount each time and so cannot remain above ϵ for all time.

Therefore, we can make the quantization error arbitrarily small by continuing the coding for a sufficient number of output neurons.

7.5.4 Topology Preservation

In stating that we have a topology preserving coding, we mean that similar inputs should be projected onto similar outputs and similar outputs should be the representations of similar inputs. This is only approximately true, in general, of feature maps using neural nets e.g. a Kohonen [110] map attempts to project the input space onto a network in such a way that the most essential neighbourhood relationships between data in the input space are preserved. Yet input data can be constructed which do not permit a 2-D (or 3-D) mapping to adequately represent all topological equivalences in the data. We give an intuitive notion of topology preservation here; a more formal proof is given in the next section.

Consider an n-unit coding of a set of input values. Let a particular point x be represented by $\mathbf{y}_i$ where $\mathbf{y}_i = \{y_{i1}, y_{i2}, ..., y_{in}\}$. The subscript i denotes an ordering of the $\mathbf{y}$ values (i.e. of the coding) such that $\mathbf{y}_i < \mathbf{y}_j$ for all $i < j$.

Then any input $x + \Delta x$ is represented by the vector $\mathbf{y}_i$ or by $\mathbf{y}_{i-1}$ or $\mathbf{y}_{i+1}$ for all Δx such that $|\Delta x| < w_n$, the nth weight. i.e. similar inputs are represented by similar outputs.

Now consider two distinct input values, x and u, which are represented by the same $\mathbf{y}_j$. Then

$$x = \sum_{i=1}^{n} w_i y_{ji} + \Delta x$$

$$u = \sum_{i=1}^{n} w_i y_{ji} + \Delta u$$

where Δx and Δu are the errors in the representations after the first n codings have taken place. Therefore,

$$x - u = \Delta x - \Delta u$$
$$\leq 2w_n$$

From the previous section, we know that the value of w_n can be made arbitrarily small and so a single code can be made to represent only similar values. Clearly a similar argument will show that if the values, x and u, are represented by contiguous codes, the values x and u can only be at most $3w_n$ apart. Thus, similar codes represent similar values.

7.5.5 Approximate Topological Equivalence

We will use the results from Section 7.5.3 in this proof: recall that for every $\epsilon > 0$, there exists an n such that $w_n < \epsilon$.

Let M be the metric space defined by that (sub)set of the real numbers defined by the probability distribution of the raw data and the metric, d, defined by the usual Euclidean distance metric.

Let M_1 be the metric space defined by the set of codes (of the real numbers in M) and the metric, d_1, defined by the usual Euclidean distance metric (now calculating in binary).

It is not possible to prove that there is a topology-preserving function which maps M to a particular value of M_1; however, it is possible to prove that there exists a mapping from the set M to a member of the family $\{M_1^1, M_1^2, M_1^3, ..., M_1^n, ...\}$ where M_1^n is the coding which has length n (i.e. formed by using n coding output neurons). In other words, we can make our mapping as close to a topology preserving mapping as possible by choosing n appropriately.

We shall create an ordering, $\{C_i\}$ of the codes in M_1^n based on the size of their binary values. Thus $C_i < C_j$ for $i < j$. Note that for this coding on M_1^n, if $f(x)$ is the function which codes the inputs i.e. $f : M \to M_1$, then $f^{-1}(C_i) - f^{-1}(C_{i-1}) = w_n$, the weight of the nth level of the coding. Note also that the greatest distance between values coded by the same code is also w_n.

- First consider the mapping $f : M \to M_1$.
  Then take any point $c \in M$. $\forall \epsilon > 0$, we require to prove that $\exists$ a $\delta > 0$ such that
  $$d_1(f(c), f(x)) < \epsilon$$
  $\forall x \in M : d(c, x) < \delta$. Choose the coding M_1^n such that $w_n < \frac{1}{2}\epsilon$. Let $\delta = w_n$. Let f map c to class C_i, the ith class of M_1^n. Then for all x such that $d(c, x) < \delta = w_n$, $f(x)$ is either in class C_i or in one of its neighbours C_{i-1} or C_{i+1}. So $f(x)$ is within $2w_n$ of $f(c)$ i.e. $f(x) \in (f(c) - \epsilon, f(c) + \epsilon)$, when $d(x, c) < 2w_n$, i.e.

  $$d_1(f(c), f(x)) < \epsilon \text{ when } d(c, x) < \delta$$

- Now consider the mapping $f : M_1 \to M$.
  Then take any point $C_i \in M_1$. Given any $\epsilon > 0$, we must prove that
  $$\exists \delta > 0 : d(f(C_i), f(x)) < \epsilon, \forall x \in M : d_1(C_i, x) < \delta$$

Choose n such that $w_n < \frac{1}{2}\epsilon$; this defines the actual representative of M_1 as M_1^n. We chose δ to be equal to 1. Then $\forall x \in (C_i - \delta, C_i + \delta)$ is equivalent to $x \in C_{i-1}, C_i$ or C_{i+1}.

Now the maximum distance between the values which code to C_{i-1} or C_{i+1} and those which map to C_i is $2w_n$ i.e. two times the remaining error after the nth coding. i.e. $d(f(x), f(c_1)) \leq 2w_n < \epsilon$ for all x in the declared interval.

Now any particular negative feedback coding network is not truly topology preserving; however, it can be made arbitrarily close to such a network by increasing the length of the coding. Therefore any particular coding is performing an approximate topology-preserving mapping.

We define an ϵ_0-*topology preserving network* as a negative feedback network in which for all points $c \in M$, $\forall \epsilon > \epsilon_0, \exists \delta > 0 : d_1(f(c), f(x)) < \epsilon, \forall x \in M : d(c, x) < \delta$ where $f(x), f(c)$ are the binary codes.

Note that this is equivalent to defining $w_n = \frac{1}{2}\epsilon_0$. Then each network in the sequence of negative feedback networks of increasing discrimination is an ϵ_0-topology preserving network with the value of ϵ_0 defined as $2w_n$ for the specific mapping M_1^n.

7.5.6 A Hierarchical Feature Map

We now propose the complete network composed of two separate negative feedback networks, rather like the EPP network in Chapter 6. The first half of the network is the basic negative feedback network described in Section 3.3 which will perform PCA; the right half of the network is the negative feedback coding network described in Section 7.5. The first section will extract the maximum information from the raw input data i.e. data will be projected onto those directions which contain maximum information; the second section will code the data along each dimension independently. All parts of the system use unsupervised learning. Our learning rule continues to be simple Hebbian learning with no weight decay or clipping of weights.

Since the network is topology preserving in each direction, it can be shown to be topology preserving in the space spanned by these directions. Therefore the construction can be viewed as forming a feature map which is topology preserving in the major information directions of the data. While it is possible to create continuous multidimensional maps which are topologically different from their projection onto this subspace, each such anomaly will tend to swing the principal components in the direction of the anomaly suggesting that such anomalies can at most be a minor part of the input data. Further, as will be shown, augmenting this type of feature map to take account of such features is a simple process.

The inherently modular nature of the network allows us to consider the effects of augmenting the network as a purely local process. This modular nature is a direct consequence of the Principal Component Analysis performed by the first section which leads to orthogonal input vectors for the second section.

Augmenting a Map

The desire to augment a map may be brought about by two circumstances:

1. The map is too crude since too little information has been extracted from the original input data.
   To extract more information from the raw data, we must find a new Principal component along which to project the data. Therefore, we must create a new data extraction output neuron i.e. in the central layer of the network. Now, by adding our new output neuron at the end of the learning process described in (3.22), (3.23) and (3.24), we are not disturbing the learning in any of the other directions which have already been found. Therefore the new direction can be found without disturbing the principal components already found; therefore the existing codes are not disturbed and the coding of the new dimension can be done independently of the existing codes.

2. The map is too crude since too little discrimination has taken place in a particular direction
   Note again that the modular nature of the map allows the discrimination in each direction to be modified independently of that in the other directions. Further in the coding within a particular direction, we may simply add a new coding output neuron into the network and, provided it learns after all the others have learned, it will simply learn to bisect the remaining information after the others have subtracted their activations. In other words, a new coding output neuron will simply provide increased discrimination within that direction and will affect neither the coding in other directions nor the existing coding in its direction.

Experimental results have confirmed this analysis.

Note that the potential for improving a map after it has been constructed is an improvement on the Kohonen map whose parameters must be specified in advance.

7.5.7 A Biological Implementation

It is well known that biological neurons are not accurately modelled by either simple linear summation neurons or by neurons which have step functions as activation functions. Instead a degree of nonlinearity of response has been found which is usually modeled by a sigmoidal function.

Using tanh() as an activation function for the output neurons allows a unified model of the above two types of output neurons to be created:

- tanh() is approximately linear in its middle section. We may adjust the range of middle section over which it is linear by using the parameter λ in $y_i = \tanh(\lambda \sum_j w_{ij} x_j)$. To get a large linear section requires a small value of λ. Experimental results have confirmed that $\lambda = 0.1$ is sufficiently small to approximate a linear function with which Principal Components can be found as before.
- tanh() may be made more dichotomous by adjusting the parameter, λ, upwards. A value of $O(10)$ is sufficient to create an output neuron which performs a smoothed coding of the input data (but see below). The larger the value of λ, the more step–like the function becomes.

Therefore a single type of output neuron with different λ parameters can perform both the information extraction and the information coding described above. Both sets of output neurons will have an activation function, $\tanh(\lambda \sum_j w_{ij} x_j)$. The sole difference is that output neurons in the first layer have the parameter $\lambda = 0.1$, while those in the second have the parameter $\lambda = 10$. The first value is, of course, dependent on the distribution of the input data and is based on distributions with single figure standard deviation and mean; the second depends on the amount of discrimination (length of the code) required.

Using such an activation function with the information extracting output neurons has several implications:

1. The output values, **y**, at these output neurons are all in the range $(-1,1)$
2. Thus the weights in the first part of the network will grow much larger than previously
3. The learning rate in this network can be made much larger than before as the **y** values are constrained to this small range. Previous networks used a learning rate of $O(0.0001)$; with a tanh() activation function a learning rate of $O(0.1)$ is possible.

However, there is a drawback to the use of this activation function in the coding layer: with the values stated, codes for the first three coding output neurons in each direction agree with those found with the step function network. However, for the later coding output neurons a lack of discrimination develops leading to very imprecise codes. By the 6th coding output neuron, we have lost the precision necessary for a topology-preserving network.

7.6 Conclusion

Three methods of creating topology-preserving feature maps have been presented: the first depends on an introduction of competition to the existing negative feedback network; the second used subspaces of clusters of the data; the third is a two stage process involving projection of the input data onto the directions of maximum information followed by a discrimination process within each direction.

The Scale Invariant network ignores the magnitude of the input vectors and quantises only on the directional information. This is a useful quantisation when only the relative proportion of each of several input data streams in important not the actual magnitude of the signal. Our original experiments with speech data were because humans can decode speech at a range of different volumes.

The Subspace map envisages a stream of data and was motivated by Kohonen's ASSOM. In this type of mapping, we wish to be able to decode a mapping while ignoring transformations which are not material to the mapping. Thus we may build in translation or rotation invariance into a code. The conjecture for both the ASSOM and the Subspace map of this chapter is that certain typical transformations of a data set define a low–dimensional manifold in which a coded object may be found. With these type of maps, we often find filters which perform wavelet–like operations.

In the coding network, the simple negative feedback network performs the projection while the negative feedback coding network performs the discrimination. Both use only simple Hebbian learning with no normalisation, weight decay or clipping of weights. The sole difference between the two sets of output neurons is that the coding output neurons have a threshold which must be achieved before their activation becomes positive. An attempt to produce a biologically feasible unification of the two types of output neurons was partially successful in that a single type of output neuron with different values of a parameter in its activation function has been shown to be capable of performing both tasks necessary to produce the feature map; the single addition of an activation function $\tanh(\lambda x)$ allowed us to dispense with the threshold in the coding output neurons.

We have, in other contexts [123], made a comparison of different types of topology-preserving networks and found, unsurprisingly, that the value of different coding networks depends on the use that will be made of the coding. Thus it is not possible to say that e.g. the Scale Invariant map is better than the Subspace map in a general sense. One can only be compared with another with respect to the specific use which will be made of the coding.

8

Maximum Likelihood Hebbian Learning

Throughout previous chapters, we have used the residuals after feedback as a means of enabling self-organisation to filters of the data sets which have been optimal for the particular effect for which we were striving. In this chapter, the focus is on the residuals themselves in that we consider how the network will learn in response to various types of residuals with different probability density functions (pdfs). We are, in effect, attempting to find optimal learning rules which will enable us to create residuals whose pdfs match pdfs which we state we will find interesting a priori. This use of the word "interesting" may evoke a memory of our use of the term in Chapter 6, and so it should. We will, in the second half of this chapter, sphere the data and then, by deciding in advance that it will be interesting to leave particular residuals after feedback, create filters which are rather similar to those which we created using the Exploratory Projection Pursuit network of Chapter 6. We will compare these two sets of rules and then combine them in a joint learning rule which attempts to get the best effects from both.

We begin the chapter, though, with a comparison of a specific rule which we relate to robust Principal Component Analysis (we will not sphere the data at this stage) and show that it is more robust than the standard rules in the presence of shot noise; because it is so important, we give this specific learning rule a specific name, ϵ-insensitive Hebbian learning, and go on to illustrate the effectiveness of the method in an anti-Hebbian rule and in a topology preserving network.

8.1 The Negative Feedback Network and Cost Functions

Throughout this book, we have used the residuals after feedback in the Hebbian learning rules

$$\Delta w_{ij} = \eta_t y_i e_j = \eta_t f(y_i) \left\{ x_j - \sum_{l=1}^{M} w_{lj} y_l \right\}$$

We can use the residuals after feedback to define a general cost function associated with this network as

$$J = f_1(\mathbf{e}) = f_1(\mathbf{x} - W\mathbf{y}) \tag{8.1}$$

where previously $f_1 = ||.||^2$, the (squared) Euclidean norm. It is well known (e.g. [13, 172]) that, with this choice of $f_1()$, the cost function is minimised with respect to any set of samples from the data set on the assumption of i.i.d. Gaussian noise on the samples.

It can be shown that, in general (e.g. [172]), the mimimisation of J is equivalent to minimising the negative log probability of the error or residual, $\mathbf{e}$, which may be thought of as the noise in the data set. Thus, if we know the probability density function of the residuals, we may use this knowledge to determine the optimal cost function.

We will shortly discuss data sets whose inherent noise is rather more kurtotic than Gaussian (Table 8.6 gives an example). An approximation to these density functions is the (one-dimensional) function

$$p(e) = \frac{1}{2 + \epsilon} \exp(-|e|_\epsilon) \tag{8.2}$$

where

$$|e|_\epsilon = \begin{cases} 0, & \forall |e| < \epsilon, \\ |e - \epsilon|, & \text{otherwise}, \end{cases} \tag{8.3}$$

with ϵ being a small scalar ≥ 0. Using this model of the noise, the optimal $f_1()$ function is the ϵ-insensitive cost function

$$f_1(e) = |e|_\epsilon \tag{8.4}$$

Therefore when we use this function in the (nonlinear) negative feedback network we get the learning rule

$$\Delta W \propto -\frac{\partial J}{\partial W} = -\frac{\partial f_1(e)}{\partial e} \frac{\partial e}{\partial W}$$

which gives us the learning rules

$$\Delta w_{ij} = \begin{cases} 0 & \text{if } |x_j - \sum_k w_{kj} y_k| < \epsilon \\ \eta y_i \text{sign}(x_j - \sum_k w_{kj} y_k) = \eta y \text{sign}(e) & \text{otherwise} \end{cases} \tag{8.5}$$

We see that this is a simplification of the usual Hebbian rule using only the sign of the residual rather than the residual itself in the learning rule. We will find that in the linear case allowing ϵ to be zero gives generally as accurate results as nonzero values, but the special data sets in subsequent sections require nonzero ϵ because of the nature of their innate noise.

8.1.1 Is This a Hebbian Rule?

The immediate question to be answered is "does this learning rule qualify as a Hebbian learning rule given that the term has a specific connotation in the Artificial Neural Networks literature?". We may consider e.g. covariance learning [119] to be a different form of Hebbian learning but at least it still has the familiar product of inputs and outputs as a central learning term. The first answer to the question is to compare what Hebb wrote with the equations above. We see that Hebb is quite open about whether the presynaptic or the postsynaptic neuron would change (or both) and how the mechanism would work. Indeed it appears that Hebb considered it unlikely that his conjecture could ever be verified (or falsified) since it was so indefinite [47]. Secondly, it is not intended that this method ousts traditional Hebbian learning but that it coexists as a second form of Hebbian learning. This is biologically plausible – *"Just as there are many ways of implementing a Hebbian learning algorithm theoretically, nature may have more than one way of designing a Hebbian synapse"*[20]. Indeed the suggestion that long term potentiation *"seems to involve a heterogeneous family of synaptic changes with dissociable time courses"* seems to favour the coexistance of multiple Hebbian learning mechanisms which learn at different speeds.

We now demonstrate that this new simplified Hebbian rule performs an approximation to Principal Component Analysis in the linear case and finds independent components when we have nonlinear activation functions.

8.2 ε-Insensitive Hebbian Learning

The ε-insensitive Hebbian learning rule is defined by

$$y_i = \sum_{j=1}^{N} w_{ij} x_j$$

$$e_j = x_j - \sum_{k=1}^{M} w_{kj} y_k$$

$$\Delta w_{ij} = \begin{cases} 0, & \text{if } |e_j| < \epsilon, \\ \eta.y\text{sign}(e_j), & \text{otherwise}, \end{cases} \tag{8.6}$$

We show that the network converges to the same values to which the more common PCA rules [138, 160] converge.

8.2.1 Principal Component Analysis

To demonstrate PCA, we use the ε-insensitive rules on artificial data. When we use the basic learning rules (8.6) on Gaussian data, we find an approximation to a PCA being performed. The weights shown in Table 8.1 are from an experiment in which the input data was, as before, chosen from zero mean Gaussians in which the first input has the smallest variance, the second the next smallest and so on. Therefore, the first Principal Component direction is a vector with zeros everywhere except in the last position which will be a 1 thus identifying the filter which minimises the mean square error. In this experiment, we have three outputs and five inputs; the weight vector has converged to an orthonormal basis of the principal subspace spanned by the first three principal components: all weights to the inputs with least variance are an order of magnitude smaller than those to the three inputs with most variance. This experiment used 5 000 presentations of samples from the data set, $\epsilon = 0.1$ and the learning rate was initially 0.01 and was annealed to 0 during these 5 000 iterations.

Table 8.1. The subspace spanned by the first three principal components is captured after only 5 000 iterations, with $\epsilon = 0.1$.

0.002	0.061	**−0.399**	**0.368**	**−0.839**
0.028	0.008	**0.855**	**0.477**	**−0.195**
0.005	−0.025	**0.327**	**−0.798**	**−0.507**

We have seen in Chapter 2 that Oja's Subspace Rule may be transformed into a PCA rule by using deflationary techniques [160].

$$e_j(k) = e_j(k-1) - w_{kj}y_k \qquad \forall j$$
$$\Delta w_{kj} = \eta_t y_k e_j(k) \qquad \forall j \tag{8.7}$$

for $k = 1, 2....$ Thus the feedforward rule is as before, but feedback and learning occur for each output neuron in turn. Similarly, we may find the actual Principal Components by using a deflationary rule with ϵ-insensitive Hebbian learning.

$$\Delta w_{kj} = \eta_t y_k.\mathrm{sign}(e_j(k)) \qquad \forall j \tag{8.8}$$

for $k = 1, 2....$

Table 8.2. The actual principal components are captured after only 5 000 iterations, with $\epsilon = 0.5$.

−0.018	0.033	0.026	0.049	**1.006**
−0.018	−0.020	0.021	**−1.003**	0.007
−0.004	0.036	**−0.991**	−0.036	0.066

Table 8.2 shows the results when five-dimensional data of the same type as before was used as input data; the learning rate was 0.1 decreasing to 0 and ϵ

was 0.1. These results were taken after only 5 000 iterations. The convergence is very fast: a typical set of results from the same data are shown in Table 8.3 where the simulation was run over only 1 000 presentations of the data ($\epsilon = 0$ in this case).

Table 8.3. The actual principal components are almost found after only 1 000 iterations, with $\epsilon = 0$.

−0.005	0.045	−0.002	0.013	**1.001**
−0.036	0.010	−0.133	**0.991**	0.002
−0.011	−0.008	**0.990**	0.141	−0.001

We may expect that since the learning rule is insensitive to the magnitude of the input vectors **x**, the rule is less sensitive to outliers than the usual rule based on mean square error. We can show [55] that the PCA properties of the deflationary ϵ-insensitive network (8.8) are relatively unaffected by the addition of noise from a uniform distribution in $[-10,10]$ to the last input in 30% of the presentations of the input data (Fig. 8.4). In comparison, the standard deflationary network (8.7) responds to this noise (as one would expect, Fig. 8.5).

We note that this need not be a good thing, however in the context of real biological neurons we may wish each individual neuron to ignore high magnitude shot noise and so the ϵ-insensitive rule may be optimal. Finally the insertion of a differentiated θ_i term in the calculation of the residual (as in Oja's Weighted Subspace Algorithm [141] which was discussed in Chapter 2) also causes convergence to the actual Principal Components but was found to be two orders of magnitude slower than the deflationary technique described above.

Table 8.4. The 30% outliers are ignored by the ϵ-insensitive rule.

0.025	−0.040	0.047	0.033	**0.995**
0.015	0.014	0.010	**−1.001**	0.033
−0.048	−0.065	**0.997**	0.022	−0.045

Table 8.5. The standard Sanger rule finds the noise irresistable.

0.065	−0.043	−0.035	0.060	**−0.997**
0.069	−0.015	0.018	**−0.996**	−0.071
−0.959	−0.002	−0.216	−0.095	−0.118

8.2.2 Anti-Hebbian Learning

Now the ϵ-insensitive rule was derived in the context of the minimisation of a specific function of the residual. It is perhaps of interest to enquire whether similar rules may be used in other forms of Hebbian learning. We investigate this using anti-Hebbian learning. Földiák [46] has suggested a neural net model which has anti-Hebbian connections between the output neurons.

The equations which define its dynamical behaviour are

$$y_i = x_i + \sum_{j=1}^{N} w_{ij} y_j$$

In matrix terms, we have

$$\mathbf{y} = \mathbf{x} + W\mathbf{y}$$
$$\text{and so, } \mathbf{y} = (I - W)^{-1}\mathbf{x}$$

He shows that, with the familiar anti–Hebbian rule,

$$\Delta w_{ij} = -\alpha y_i y_j, \text{ for } i \neq j$$

the outputs, $\mathbf{y}$, are decorrelated.

Now the matrix W must be symmetric and has only nonzero nondiagonal terms i.e. if we consider only a two input, two output net,

$$W = \begin{pmatrix} 0 & w \\ w & 0 \end{pmatrix} \tag{8.9}$$

However the ϵ-insensitive anti-Hebbian rule is nonsymmetrical and so if w_{ij} is the weight from y_i to y_j, we have

$$\Delta w_{ij} = -\eta y_j \text{sign}(y_i), \text{ if } |y_i| > \epsilon,$$
$$\Delta w_{ij} = 0, \qquad \text{otherwise.} \tag{8.10}$$

To test the method, we generated two-dimensional input vectors, $\mathbf{x}$, where each element is drawn independently from $N(0,1)$ and then added another independently drawn sample from $N(0,1)$ to both elements. This gives a data set with sample covariance matrix (10 000 samples) of

$$\begin{pmatrix} 1.9747 & 0.9948 \\ 0.9948 & 1.9806 \end{pmatrix} \tag{8.11}$$

The covariance matrix of the outputs, $\mathbf{y}$, (over the 10 000 samples) from the network trained using the ϵ-insensitive learning rule is

$$\begin{pmatrix} 1.8881 & -0.0795 \\ -0.0795 & 1.1798 \end{pmatrix} \tag{8.12}$$

We see that the outputs are almost decorrelated. It is interesting to note that

- the asymmetrical learning rules have resulted in nonequal variances on the outputs,
- but the covariance (off-diagonal) terms are equal, as one would expect.

It is our finding that the outputs are always decorrelated, but the final values on the diagonals (the variances) are impossible to predict and seem to depend on the actual values seen in training, the initial conditions, etc..

A feedforward decorrelating network, $\mathbf{y} = (I + W)\mathbf{x}$, may also be created with the ε-insensitive anti-Hebbian rule with similar results.

8.2.3 Other Negative Feedback Networks

The Scale Invariant Map

We may now report that if we use the ε-insensitive learning rule within the Scale Invariant Map of Chapter 7, we get the learning rules

$$\Delta w_{ij} = \begin{cases} 0, & \text{if } |e_j| < \epsilon, \\ \eta y \Lambda(p, i)\text{sign}(e_j), & \text{otherwise.} \end{cases}$$

and convergence to the same type of mapping is also achieved. As before, the resultant mapping is found more quickly and is more robust against shot noise.

The Factor Analysis Network

In the bars data set, noise is innate, caused by the presence of other bars. Consider a single horizontal bar composed of 8 pixels. If it is not present and a vertical bar *is* present, the pixel in the horizontal bar corresponding to the vertical bar will be blackened. Thus, there is a nonzero probability (Table 8.6) that some activation will be passed forward through the weights to the neuron responding to the horizontal bar even when the horizontal bar is not present.

Table 8.6. Approximate probabilities that the given number of pixels will be firing in a given bar. Probabilities for 4–7 pixels not shown are less than 0.05.

Number of pixels	0	1	2	3	8
Probability	0.3	0.34	0.17	0.05	0.125

We see from Table 8.6 that there is a significant probability that one or two pixels from a bar will be activated even when the bar is not present. This indeed was the problem which motivated ε-insensitive Hebbian learning: the probability density function

$$p(e) = \frac{1}{2 + \epsilon} \exp(-|e|_\epsilon) \tag{8.13}$$

is an approximation to the above densities. We showed in Chapter 5 that imposing a constraint of non-negativity on the weights (or outputs) of the negative feedback network with standard Hebbian learning changes the PCA network to a Principal Factor Analysis network capable of identifying the individual bars (something the PCA network cannot do). The rectification on the outputs is a special case of the nonlinear activation passing rule (5.10); we have similar results to those reported herein when we use continuous functions which approximate the rectification.

Now we consider the ϵ-insensitive learning rule with the non-negative weight constraint: whenever the rule has the effect of making a particular weight negative we set that weight to zero. This does not necessarily constrain its future growth. We have also used a network which rectifies the outputs i.e. in (5.10), $f(t) = t$ if $t > 0$, $f(t) = 0, t \leq 0$ with similar results.

The weights in a trained network on an 8*8 grid (i.e. 16 bars) are shown in Fig. 8.1. The bars have clearly been identified. The learning rate was 0.01, decreasing to 0 over the course of the simulation which took 50 000 iterations. The value of ϵ should be greater than 0 and best results are obtained with ϵ between 0.1 and 0.5.

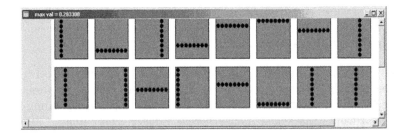

Fig. 8.1. Each row represents the weights into one output neuron. The eight horizontal and eight vertical lines have been found.

Similarly, we have [55] used the ϵ-insensitive rule with nonlinear PCA rules. However we wish to emphasise that this rule is one of a family which can be used as an exploratory data tool. We thus extend the method in the next section so that different types of structure can be identified.

8.3 The Maximum Likelihood EPP Algorithm

Let us now consider the residual after the feedback to have probability density function

$$p(\mathbf{e}) = \frac{1}{Z} \exp(-|\mathbf{e}|^p) \qquad (8.14)$$

Then we can denote a general cost function associated with this network as

$$J = -\log p(\mathbf{e}) = (\mathbf{e})^p + K \qquad (8.15)$$

where K is a constant, Therefore performing gradient descent on J we have

$$\Delta W \propto -\frac{\delta J}{\delta W} = -\frac{\delta J}{\delta \mathbf{e}} \frac{\delta \mathbf{e}}{\delta W} \approx \mathbf{y}(p(|\mathbf{e}|^{p-1}\text{sign}(\mathbf{e}))^T \qquad (8.16)$$

We would expect that for leptokurtotic residuals (i.e. more kurtotic than a Gaussian distribution), values of $p < 2$ would be appropriate, while for platykurtotic residuals (i.e. less kurtotic than a Gaussian), values of $p > 2$ would be appropriate. It is well known in the ICA community [85] that it is less important to get exactly the correct distribution when searching for a specific source than it is to get an approximately correct distribution i.e. all supergaussian signals can be retrieved using a generic leptokurtotic distribution and all subgaussian signals can be retrieved using a generic platykurtotic distribution.

Therefore the network operation is:

Feed forward: $y_i = \sum_{j=1}^{N} w_{ij} x_j, \forall i$

Feedback: $e_j = x_j - \sum_{i=1}^{M} w_{ij} y_i$

Weight change: $\Delta w_{ij} = \eta y_i \text{sign}(e_j)|e_j|^p$

Now the nature and quantification of the "interestingness" (Chapter 6) is in terms of how likely the residuals are under a particular model of the probability density function of the residual. As with standard EPP, we also sphere the data before applying the learning method to the sphered data.

8.3.1 Minimum Likelihood Hebbian Learning

Now, it is equally possible to perform gradient ascent on J. In this case we find a rule which is the opposite of the above rules in that it is attempting to minimise the likelihood of the residual under the current assumptions about the residual's pdf. The operation of the network is as before, but this time we have

Weight change: $\Delta w_{ij} = -\eta y_i sign(e_j)|e_j|^p$

This corresponds to the anti-Hebbian learning rule. One advantage of this formulation compared with the Maximum Likelihood Hebbian rule is that, in making the residuals as unlikely as possible, we are having the weights learn the structure corresponding to the pdf determined by the parameter p. With the Maximum rule, the weights learn to remove the projections of the data which are furthest from that determined by p. Thus, if we wish to search for clusters in our data (typified by a pdf with $p > 2$), we can use Maximum Likelihood learning with $p < 2$ [30], which would result in weights which remove any projections that make these residuals unlikely. Therefore the clusters would be found by projecting onto these weights. Alternatively we may use Minimum Likelihood learning with $p > 2$, which would perform the same job: the residuals have to be unlikely under this value of p, and so the weights converge to remove those projections of the data which exhibit clustering.

8.3.2 Experimental Results

Let us call the EPP algorithm of Chapter 6 the Output Functions algorithm and that of this chapter the Maximum Likelihood algorithm. Then we might reasonably wish to compare their properties.

We therefore create five-dimensional data as in Chapter 6 so that four dimensions contain data drawn independently and identically distributed from zero mean, Gaussian distributions, while the fifth dimension contains kurtotic data: we draw this also from a Gaussian distribution but randomly (in 20% of cases) substitute the sample with a small random number drawn from a uniform distribution between -0.005 and $+0.005$. We now use the Maximum Likelihood rules with $p = 3$ to search for leptokurtosis and $p = 1$ for the platykurtotic bimodal data set. We again performed 100 simulations and used the same values for the learning rate as before.

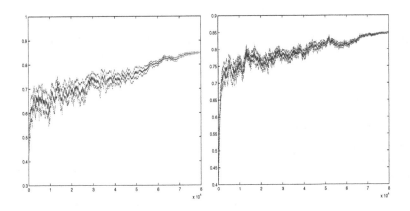

Fig. 8.2. The convergence of the Maximum Likelihood Learning rules. Left: using $p = 3$ on leptokurtotic data. Right: using $p = 1$ on the bimodal platykurtotic data.

Fig. 8.2 should be compared with Fig. 6.2 and Fig. 6.3. We see in this figure that convergence is not quite as accurate with the Maximum Likelihood method but that it seems more sure: we have no failures to report on this data set with this method. Also the initial convergence is much faster than before.

The Cetin Data Set

We first use a data set derived by H. Cetin [23] which consists of 16 sets of spectra in which each set contains 32×32 samples arranged in a chequer-board grid. It was used for comparative study of a variety of algorithms. The spectra themselves are samples taken from a spectral library.

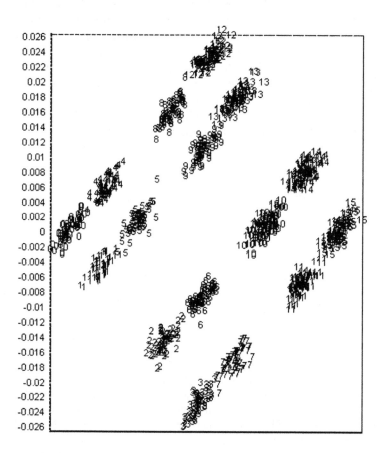

Fig. 8.3. Projection of the Cetin data set on the first two filters found by the Maximum Likelihood method with $p = 1$.

The projections of this data set onto the first two filters found by the Maximum Likelihood Hebbian rule with $p = 1$ is shown in Fig. 8.3. A comparison of this with that achieved by PCA reveals that the clusters are very much more compact and rather more isolated from one another in the Maximum Likelihood case. This result has been repeated with different initial conditions, learning rates, etc., and is rather close to the projections obtained when using the Exploratory Projection Pursuit of Chapter 6.

8.3.3 Skewness

Until now, we have discussed the maximum likelihood algorithm in terms of kurtosis: outliers are often signalled by positive kurtosis and clusters by negative kurtosis. However, this is not the only structure to be found in data sets: we showed that skewness could be found with the algorithm of Chapter 6 [54] with $f(y) = y^2$. More recently it has been suggested [122] that we can use $f(y) = \cos(y)$ in this algorithm to search for skewness. We first compare these methods on an artificial data set exhibiting skewness. The two output functions are compared in Fig. 8.4. One interesting conclusion we can draw from this study is that the cos() function, while perhaps more stable than simple squaring, does not converge so quickly or reliably.

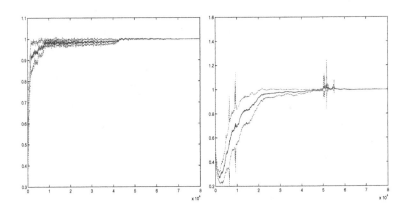

Fig. 8.4. The left diagram shows convergence of 81 out of 100 simulations on skewed data using y^2 as the output function; 18 simulations were unstable and 1 did not converge after 80 000 iterations. The right diagram shows convergence of 94 simulations which used the cos() function; 1 was unstable and another 5 did not converge after 80 000 iterations.

Skewness can also be identified by the Maximum Likelihood Algorithms. Consider a skewed (centred) distribution: it will typically have a large probability mass close to zero for positive (negative) values and much longer tails but no great lump of probability mass for negative (respectively positive) values. Therefore we suggest an algorithm so that

$$\Delta w_{ij} = \eta y_i \text{sign}(e_j), \text{ if } e_j < 0,$$
$$\Delta w_{ij} = \eta y_i \text{sign}(e_j)|e_j|^2, \text{ if } e_j > 0.$$

Thus we are modelling a skewed distribution with values of $p = 1$ on one side of the origin and $p = 3$ on the other. Of course, the skewness may be the other way round, in which case we reverse the equations. Results are shown in Fig. 8.5. We see that this algorithm is only partly successful: although, we achieve very fast convergence in all cases, we do not achieve more than 80% accuracy on this skewed data set.

As before, we attempted to combine the two algorithms; however, we were unable with the same learning rate to get convergence on this data set with a combined algorithm.

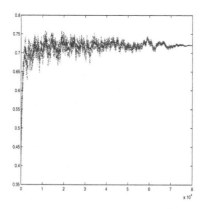

Fig. 8.5. Convergence of the algorithm using the new skew Maximum Likelihood method in 100 simulations.

8.4 A Combined Algorithm

The two EPP algorithms were derived from different perspectives and each used modifications to a Principal Component Analysis neural network *in different parts of the weight change algorithm*. This might suggest that we can get the best of both worlds by combining these algorithms. Fig. 8.6 shows the convergence of an algorithm which uses

Feed forward: $y_i = \sum_{j=1}^{N} w_{ij}x_j, \forall i$

Feedback: $e_j = x_j - \sum_{i=1}^{M} w_{ij}y_i$

Weight change: $\Delta w_{ij} = \eta f(y_i)sign(e_j)|e_j|^p$

where

- $f(.)$ is the tanh() function and $p = 1$ for the bimodal data.
- $f(y) = y - \tanh(y)$ and $p = 3$ for the leptokurtotic data.

We see (Fig. 8.6) that convergence in both cases was extremely reliable and extremely fast.

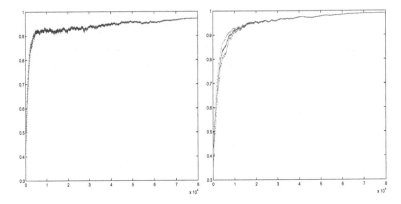

Fig. 8.6. Convergence of the algorithm using the combined learning rule. Left: bimodal data. Right: leptokurtotic data.

8.4.1 Astronomical Data

We now use a remote sensing data set, the 65-colour spectra of 115 asteroids used by [130]. The data set is composed of a mixture of the 52-colour survey by Bell et al. [11] together with the 8-colour survey conducted by Zellner et al. [187] providing a set of asteroid spectra spanning 0.3–2.5 μm. A more detailed description of the data set is given in [130]. When this extended data set was compared by [130] to the results of Tholen [176] it was found that the additional refinement to the spectra leads to more classes than the taxonomy produced by Tholen. The results of applying the higher-order EPP algorithm with the tanh() nonlinearity on this data set are shown in Fig. 8.7 (left); the results from the Maximum Likelihood EPP algorithm with $p = 1$ are shown in the right of this figure. Both projections have spread out the data well; we will leave it to the reader to decide if one is better than the other in some way.

Perhaps more interesting is the projection which we get when we use both nonlinearities in the learning rule. This is shown in Fig. 8.8.

8.4.2 Wine

A data set which may be retrieved from the UCI repository of machine learning databases is based on wine: it is a 13-dimensional data set of 178 samples from three different classes of wine. Fig. 8.9 shows the two-dimensional projections of the three EPP methods and (in the top left) the projection of the wine data onto the first two principal components. We see that there is very little difference between the three EPP projections and that all three are better than the PCA projection.

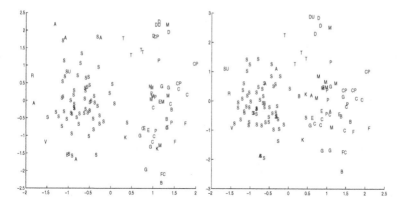

Fig. 8.7. The left figure shows the two dimensional projection onto the axes found by the original EPP algorithm while the right figure shows the projection onto the axes found by the Maximum Likelihood EPP algorithm.

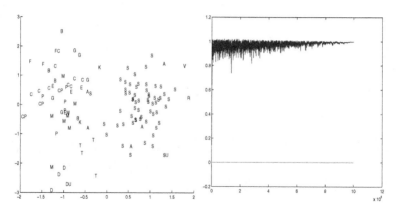

Fig. 8.8. The left figure shows the projection found by using both nonlinearities in the learning rule. The right figure shows how quickly convergence is achieved.

8.4.3 Independent Component Analysis

In Chapter 6, ICA was discussed and the connection to EPP was explained. As the maximum likelihood method is able to perform EPP, we can expect that it is also capable of performing ICA. We will show how to extract individual signals from mixtures of either supergaussian signals or mixtures of subgaussian signals, using the Maximum Likelihood network in a different way to that described in the previous sections. We do this by taking advantage of the central limit theorem, which states that mixtures of independent signals result in signals which are more Gaussian than the original signals. The more statistically independent signals we mix, the more Gaussian the resulting mixture becomes. This means that if we consider a mixture of kurtotic

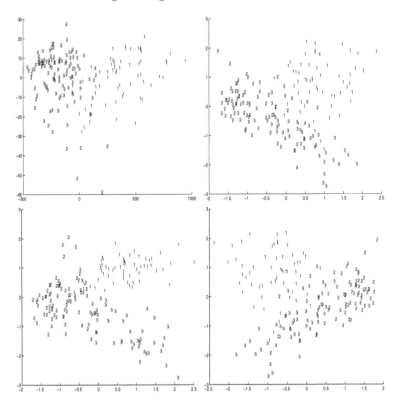

Fig. 8.9. The figure shows the projection on the first two filters found by (a) PCA, (b) Max. Likelihood EPP, (c) Output Functions EPP, (d) the combined EPP method, on the 13-dimensional inputs of the wine classification problem.

signals, the total kurtosis value of the residuals will be at a local minimum when there are as few kurtotic signals in the residuals as possible, i.e. if one of the signals has been completely eliminated from the mixture. If we have a mixture of supergaussian signals, completely removing one of these signals gives a residual which is more kurtotic than removing a little of each of the signals. Therefore, maximising the kurtosis of the residuals will identify one separate signal from a mixture of supergaussian sources and minimising the kurtosis of the residuals will pull out a signal from a mixture consisting of only subgaussian signals.

Extraction of a Signal from a Mixture

We begin with three mixed speech signals which we linearly mixed using mixing matrix A:

$$A = \begin{pmatrix} -0.3 & 0.4 & 0.2 \\ 0.6 & -0.9 & 0.4 \\ -0.5 & 0.5 & -0.3 \end{pmatrix} \quad (8.17)$$

Each of the signals comprised 40 000 samples of a speaker stating "Perhaps the most frequent use of ICA is in the extraction of independent voices from a mixture". The kurtosis of the individual signals is 6.8444, 7.7582 and 3.6833, respectively. We use a learning rate of 0.0001 and 100 000 iterations (randomly sampling with replacement from the 40 000 samples) to extract each signal in a deflationary manner. We use maximum likelihood learning with a value of p = 1 (a Laplacian distribution) which extracts one signal completely (see below) and then repeat the experiment with the residual mixture of the other two signals. The learning rule is

$$\Delta \mathbf{w}_i = \eta y_i * \text{sign}(\mathbf{e}) \quad (8.18)$$

where the sign() function acts on an elementwise basis. As with all the higher-order methods, we sphere the data, by multiplying by matrix $Q = \Sigma^{-\frac{1}{2}}$, where Σ is again the covariance matrix of the data.

As a measure of success, we use $W * Q * A$ i.e. the weights learned by the method times the sphering matrix times the mixing matrix:

$$W * Q * A = \begin{pmatrix} 0.0170 & -0.0050 & -\mathbf{1.0000} \\ \mathbf{1.0008} & -0.0093 & 0.0162 \\ 0.0122 & \mathbf{0.9995} & -0.0021 \end{pmatrix} \quad (8.19)$$

We see that the product matrix is very close to a permutation matrix showing that the signals have been extracted correctly.

Similarly, we have experimented with five sub Gaussian artificially generated signals randomly mixed. Their kurtosis values were −0.9845, −0.9638, −0.9769, −0.9795 and −0.9673, respectively. Notice that all sample kurtosis values are approximately equal. Again we used 40,000 samples, a learning rate of 0.0001 and no annealing of the learning rate. This time we used the rule

$$\Delta \mathbf{w}_i = \eta y_i * \text{sign}(\mathbf{e}) \otimes |\mathbf{e}|^3 \quad (8.20)$$

where again $|\mathbf{e}|^3$ is performed on an elementwise basis.

This somewhat more difficult problem required 500 000 iterations for each signal and the product matrix WVA is

$$W * Q * A = \begin{pmatrix} -0.033 & 0.021 & \mathbf{1.001} & -0.045 & 0.013 \\ -0.021 & -0.035 & 0.055 & \mathbf{0.997} & -0.028 \\ 0.038 & 0.022 & -0.001 & -0.012 & -\mathbf{0.971} \\ -0.006 & -\mathbf{0.996} & 0.027 & -0.039 & -0.099 \\ -\mathbf{0.998} & 0.012 & -0.036 & -0.004 & -0.021 \end{pmatrix} \quad (8.21)$$

Again we have almost a permutation matrix indicating that the sources have been recovered. Now the reason we needed to use 500 000 iterations is that the last signal is much the most difficult to extract (and we see that in the last column its accuracy is much worse). The reason for that lies in the fact that there is only one signal left in the "mixture" at this time: the network is attempting to structure the residuals to model the probability density function but if it is successful, there will be no residuals to model. This is somewhat of a conundrum. Of course, we can obviate this conundrum by simply noting that the fourth residual contains only the last signal, but this is somewhat unsatisfactory since we do not (in a truly blind problem) know a priori how many signals are in the mixture.

8.5 Conclusion

We have derived a slightly different form of Hebbian learning which we have shown capable of performing PCA-type learning in a linear network. However, more importantly in situations in which the noise is more kurtotic, we have shown that the network readily identifies the independent components of the signal.

We have also related the method of this chapter to Exploratory Projection Pursuit. We have therefore sphered the input data and shown that the resulting network can successfully search for structure of various types in the sphered data. We have applied the method to independent component analysis and shown that we can identify independent components from a mixture of signals.

Finally, we compared the method of Chapter 6, the Output Functions method, with that of this chapter, the Maximum Likelihood method, and showed that the former was more accurate, though the latter was faster and somewhat more reliable. Since each rule is a modification to the basic PCA rule at a different point in the learning rule, we are able to combine these modifications to achieve a rule which seems to be fast, reliable and accurate.

Part II

Dual Stream Networks

We now turn our attention to dual stream networks. Whereas in preceding chapters, we have considered the input data at each time instant as being a single-input vector, in experiments in subsequent chapters we will explicitly create two or more data streams and consider the input data at any one time as coming from these separate data streams. It is, perhaps, surprising that a technology such as artificial neural networks, which is based on an emulation of living organisms, has tended to neglect what seems to be one of the major features of sensory perception – the plurality of experience at each instant in time: not only do we have two each of our main sensors, but these sensors can simultaneously provide us with information. Thus, each of us is presented with the problem of integrating diverse information from different sensory modalities; it should be said that this is a problem which provides almost all of us with little or no problem, yet that fact disguises the inherent complexity of the task. As we have said, the study of artificial neural networks has been dominated by single stream networks but we hope that the work in Chapters 9 – 14 will excite interest in the field of dual stream networks.

Thus, typically we will have two sets of data presented at any one time; each of $\mathbf{x}_1$ and $\mathbf{x}_2$ can be considered as sensory inputs to our networks which arise because of some external events in the world. There is an implicit assumption that such external events are caused by a single underlying cause so that there is a relation between $\mathbf{x}_1$ and $\mathbf{x}_2$ which the dual stream artificial neural networks will be required to reveal. Typically we will have weights $\mathbf{w}_1$ and $\mathbf{w}_2$ through which the separate activations will be expected to pass independently of each other i.e. $y_1 = \mathbf{w}_1^T \mathbf{x}_1$ and $y_2 = \mathbf{w}_2^T \mathbf{x}_2$ are calculated separately[1] with no interaction taking place at this stage. However, under our implicit assumption that events which happen at the same time have a single underlying cause which we wish to uncover, we must let the learning of the weights $\mathbf{w}_1$ and $\mathbf{w}_2$ use information from the other data stream. Therefore typically

$$\Delta \mathbf{w}_1 = \eta f_1(\mathbf{x}_1, \mathbf{x}_2, y_1, y_2)$$
$$\Delta \mathbf{w}_2 = \eta f_2(\mathbf{x}_1, \mathbf{x}_2, y_1, y_2)$$

where the functions $f_1()$ and $f_2()$ are determined by the type of information which we believe is held in the two data streams.

Of course, events which appear at the same time are not the only type of events which are related: a single underlying cause may manifest itself at different times and so we must consider sets of events which are related in some way not necessarily temporally. Perhaps a concrete example which we will use later as a data set will clarify our approach: we discuss a data set

---

[1] Though we are considering a scalar output in this exposition, we can and do apply these methods to vector output neural networks.

of exam marks. Each of 88 students sit five exams. Two exams are given as closed-book exams while three exams are given as open-book exams. We make the implicit assumption that some students are better than others (at least at passing exams) and thus there is a single underlying cause for all five exam marks which is a measure of how good each student is at passing exams. Thus our x_1 data stream is two-dimensional and consists of the closed-book exams while our x_2 data stream is three-dimensional and consists of the open-book exams. We have 88 samples and at any one time we are presenting the same student's marks to both data streams simultaneously. We might subsequently hope to predict a new student's performance on the open-book exams given his closed-book results, or vice versa.

Therefore, in subsequent chapters, we discuss several methods that are able to find relationships between data sets. This is a useful approach for many engineering applications where multiple measurements are necessary in order to find a certain underlying signal. A process can be measured using different types of sensors and the information these sensors provide can be combined in order to get a clearer and more accurate measurement. Another approach is to use the same sensor, but at different times or at slightly different positions. For example, in feature analysis of images, neighbouring image patches can be used as input streams.

The development of ideas in these chapters is as follows:

- In Chapters 9 and 10, we develop several neural implementations of the statistical technique of Canonical Correlation Analysis which is designed to find projections which maximise the correlations between two data streams. The resulting networks are all shown to have negative feedback but, while, as with those of Part I, the feedback in each network comes from the associated output neuron, what directs learning is a positive Hebbian rule linking each input with the opposite output. We show in the summary of Chapter 10 that this seems to be a very general rule and we conjecture about its biological usefulness.
- We then consider extensions to canonical correlation analysis:
  1. We introduce nonlinearity into the process in two ways and find we can identify greater correlations than is possible with linear machines.
  2. We add penalty terms to the basic machines and show that these both make the convergence more robust and make the converged results more comprehensible.
  3. We search for shared higher-order stucture in two data streams simultaneously by twinning the methods developed in Chapters 6 and 8.
- Finally, we move away from neural methods by twinning the method of Principal Curves – we are searching for two curves which move through two data spaces searching for the greatest local correlations.

In Appendix D, we list some other dual stream approaches from the literature, though this is an emerging area and we do not claim to be exhaustive.

Two Neural Networks for Canonical Correlation Analysis

In this chapter, we derive two new methods for performing Canonical Correlation Analysis with Artificial Neural Networks. For the first network, we demonstrate the network's capabilities on artificial data and then compare its effectiveness with that of a standard statistical method on real data. We demonstrate the capabilities of the network in certain situations where standard statistical techniques are not effective, for example where we have correlations stretching over three data sets and where the maximum nonlinear correlation is greater than any linear correlation (see Chapter 11). The network is also applied to Becker's [8] random dot stereogram data and shown to be extremely effective at detecting shift information. We then derive the second network mathematically before comparing it to the first.

9.1 Statistical Canonical Correlation Analysis

Canonical Correlation Analysis [125] is used when we have two data sets which we believe have some underlying correlation. Consider two sets of input data, from which we draw iid samples to form a pair of input vectors, $\mathbf{x}_1$ and $\mathbf{x}_2$. Then in classical CCA, we attempt to find the linear combination of the variables which gives us maximum correlation between the combinations. Let

$$y_1 = \mathbf{w}_1^T \mathbf{x}_1 = \sum_j w_{1j} x_{1j} \tag{9.1}$$

$$y_2 = \mathbf{w}_2^T \mathbf{x}_2 = \sum_j w_{2j} x_{2j} \tag{9.2}$$

We wish to find those values of $\mathbf{w}_1$ and $\mathbf{w}_2$ which maximise the correlation between y_1 and y_2. Whereas Principal Components Analysis and Factor Analysis deals with the interrelationships within a set of variables, CCA deals with the relationships between two sets of variables. If the relation between y_1 and y_2 is believed to be causal, we may view the process as one of finding the best predictor of the set $\mathbf{x}_2$ by the set $\mathbf{x}_1$ and similarly of finding the most predictable criterion in the set $\mathbf{x}_2$ from the $\mathbf{x}_1$ data set. Thus, we later review a data set in which a set of exam results are split into those achieved by students when they had access to their books and those marks obtained when the students were denied their books during the exam. We might wish to use a student's open book exams to predict how well he/she might do in the closed book exams. One way to view canonical correlation analysis is as an extension of multiple regression (see [125], page 281). Recall that in multiple regression analysis the variables are partitioned into an $\mathbf{x}_1$-set containing q variables and a x_2-set containing p = 1 variable. The regression solution involves finding the linear combination of $\mathbf{x}_1$ which is most highly correlated with x_2.

Let $\mathbf{x}_1$ have mean μ_1 and $\mathbf{x}_2$ have mean μ_2. Then the standard statistical method (see [125]) lies in defining

$$\Sigma_{11} = E\{(\mathbf{x}_1 - \mu_1)(\mathbf{x}_1 - \mu_1)^T\} \tag{9.3}$$

$$\Sigma_{22} = E\{(\mathbf{x}_2 - \mu_2)(\mathbf{x}_2 - \mu_2)^T\} \tag{9.4}$$

$$\Sigma_{12} = E\{(\mathbf{x}_1 - \mu_1)(\mathbf{x}_2 - \mu_2)^T\} \tag{9.5}$$

$$\text{and } K = \Sigma_{11}^{-\frac{1}{2}} \Sigma_{12} \Sigma_{22}^{-\frac{1}{2}} \tag{9.6}$$

where T denotes the transpose of a vector. We then perform a singular value decomposition of K to get

$$K = (\alpha_1, \alpha_2, ..., \alpha_k) D(\beta_1, \beta_2, ..., \beta_k)^T \tag{9.7}$$

where α_i and β_i are the standardised eigenvectors of KK^T and $K^T K$, respectively, and D is the diagonal matrix of eigenvalues.

Then the first canonical correlation vectors (those which give greatest correlation) are given by

$$\mathbf{w}_1 = \Sigma_{11}^{-\frac{1}{2}} \alpha_1 \tag{9.8}$$

$$\mathbf{w}_2 = \Sigma_{22}^{-\frac{1}{2}} \beta_1 \tag{9.9}$$

with subsequent canonical correlation vectors defined in terms of the subsequent eigenvectors, α_i and β_i.

In most of the remainder of this book, we will make the simplifying assumption that the data has been centered so that $\mu_i = 0$. This makes the exposition simpler and is easily performed for any algorithm developed.

9.2 The First Canonical Correlation Network

The input data comprises two vectors $\mathbf{x}_1$ and $\mathbf{x}_2$. Activation is fed forward from each input to the corresponding output through the respective weights, $\mathbf{w}_1$ and $\mathbf{w}_2$ (see Fig. 9.1 and (9.1) and (9.2)) to give outputs y_1 and y_2.

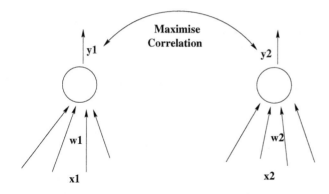

Fig. 9.1. The CCA network. By adjusting weights, $\mathbf{w}_1$ and $\mathbf{w}_2$, we maximise correlation between y_1 and y_2.

We wish to maximise the correlation $E(y_1 y_2)$ where $E()$ denotes the expectation which will be taken over the joint distribution of $\mathbf{x}_1$ and $\mathbf{x}_2$.

We may regard this problem as that of maximising the function $g_1(\mathbf{w}_1|\mathbf{w}_2)$ $= E(y_1 y_2)$ which is defined to be a function of the weights, $\mathbf{w}_1$, given the other set of parameters, $\mathbf{w}_2$. This is an unconstrained maximisation problem which has no finite solution, and so we must constrain the maximisation.

Typically in CCA, we add the constraint that $E(y_1^2 = 1)$ and similarly with y_2 when we maximise $g_2(\mathbf{w}_2|\mathbf{w}_1)$.

Using the method of Lagrange multipliers, this yields the constrained optimisation functions,

$$J_1 = E\left\{(y_1 y_2) + \frac{1}{2}\lambda_1(1 - y_1^2)\right\} \text{ and}$$

$$J_2 = E\left\{(y_1 y_2) + \frac{1}{2}\lambda_2(1 - y_2^2)\right\}$$

We may equivalently use

$$J = E\left\{(y_1 y_2) + \frac{1}{2}\lambda_1(1 - y_1^2) + \frac{1}{2}\lambda_2(1 - y_2^2)\right\}$$

but it will be more convenient in the following sections to regard these as separate criteria which can be optimised independently by implicitly assuming that $\mathbf{w}_1$ is constant when we are changing $\mathbf{w}_2$, and vice versa. We wish to find the optimal solution using gradient ascent and so we find the derivative of the instantaneous version of each of these functions with respect to both the weights, $\mathbf{w}_1$ and $\mathbf{w}_2$, and the Lagrange multipliers, λ_1 and λ_2. By changing the Lagrange multipliers in proportion to the derivates of J we are changing the relative strength of the constraint compared to the function we are optimising; this allows us to smoothly maximise that function in the region in which we are satisfying the constraint.

Noting that

$$\frac{\partial g_1(\mathbf{w}_1|\mathbf{w}_2)}{\partial \mathbf{w}_1} = \frac{\partial(y_1 y_2)}{\partial \mathbf{w}_1} = \frac{\partial(\mathbf{w}_1^T \mathbf{x}_1 y_2)}{\partial \mathbf{w}_1} = \mathbf{x}_1 y_2 \tag{9.10}$$

these yield, respectively,

$$\frac{\partial J_1}{\partial \mathbf{w}_1} = \mathbf{x}_1 y_2 - \lambda_1 y_1 \mathbf{x}_1 = \mathbf{x}_1 (y_2 - \lambda_1 y_1)$$

$$\frac{\partial J_1}{\partial \lambda_1} \propto (1 - y_1^2)$$

Similarly with the J_2 function, $\mathbf{w}_2$ and λ_2. This gives us a method of changing the weights and the Lagrange multipliers on an online basis. We use the joint learning rules

$$\Delta w_{1j} = \eta x_{1j}(y_2 - \lambda_1 y_1)$$
$$\Delta \lambda_1 = -\eta_0(1 - y_1^2)$$
$$\Delta w_{2j} = \eta x_{2j}(y_1 - \lambda_2 y_2)$$
$$\Delta \lambda_2 = -\eta_0(1 - y_2^2) \tag{9.11}$$

where w_{1j} is the jth element of weight vector, $\mathbf{w}_1$, etc. The actual weight update, $w_{ij} = w_{ij} + \Delta w_{ij}$, follows.

It has been found empirically that best results are achieved when $\eta_0 >> \eta$. However, just as a neural implementation of Principal Component Analysis may be very interesting but not a generally useful method of finding Principal Components, so a neural implementation of CCA may be only a curiosity. However, it has been shown that nonlinear extensions of PCA networks are able to search for the independent components of a data set (Chapters 5, 6 and 8) and that such extensions are therefore justifiable as engineering tools for investigating data sets. We therefore later extend our neural implementation of CCA by maximising the correlation between outputs when such outputs are a nonlinear function of the inputs. We investigate a particular case of maximisation of $E(y_1 y_2)$ when the values y_i are a nonlinear function of the inputs, $\mathbf{x}_i$.

9.3 Experimental Results

We report simulations on both real and artificial data sets of increasing complexity. We begin with data sets in which there is a linear correlation and we demonstrate the effectiveness of the network on Becker's [8] random dot stereogram data. We then extend the method in two ways not possible with standard statistical techniques:

1. We maximise correlations between more than two input data sets.
2. We consider maximising correlations where such correlations may be on nonlinear projections of the data.

9.3.1 Artificial Data

Our first experiment uses an artificial data set: $\mathbf{x}_1$ is a 4-dimensional vector, each of whose elements is drawn from the zero-mean Gaussian distribution, $N(0,1)$; $\mathbf{x}_2$ is a 3-dimensional vector, each of whose elements is also drawn from $N(0,1)$. In order to introduce correlations between the two vectors, $\mathbf{x}_1$ and $\mathbf{x}_2$, we generate an additional sample from $N(0,1)$ and add it to the first elements of each vector. In order to ensure that we are not simply responding to variance the data is then normalised so that the variance in each input is identical. Thus there is no correlation between the two vectors other than that existing between the first element of each.

Using an initial learning rate of 0.0001 which is decreased linearly to 0 over 50 000 iterations, the weights converge to the vectors (**0.679**, 0.023, −0.051, −0.006) and (**0.681**, 0.004, 0.005). This clearly illustrates the high correlation between the first elements of each of the vectors and also the fact that this is the only correlation between the vectors. We may compare the reported results with the optimal values of ($\sqrt{0.5}$,0,0,0) and ($\sqrt{0.5}$,0,0).

The effect of the constraint on the variance of the outputs is clearly seen when we change the distribution from which all samples are drawn to $N(0,5)$. The weight vectors converge to (**0.141**, 0.002, 0.003, 0.002) and (**0.141**, 0.002, −0.001) – the optimal results are ($\sqrt{0.02}$,0,0,0) and ($\sqrt{0.02}$,0,0). The differences in magnitude are due to the constraint, $E(y_i^2) = 1$ since

$$E(y_i^2) = 1 \Longleftrightarrow E(\mathbf{w}_i^T \mathbf{x}\mathbf{x}^T \mathbf{w}_i) = \mathbf{w}_i^T \Sigma_{xx} \mathbf{w}_i = 1 \tag{9.12}$$

where Σ_{xx} is the covariance matrix of the input data.

9.3.2 Real Data

Our second experiment uses a data set reported in [125], page 290; it comprises 88 students' marks on five module exams. The exam results can be partitioned into two data sets: two exams were given as close-book exams (C) while the other three were open-book exams (O). The exams were on the subjects of Mechanics(C), Vectors(C), Algebra(O), Analysis(O), and Statistics(O). We thus split the five variables (exam marks) into two sets – the closed-book exams (x_{11}, x_{12}) and the open-book exams (x_{21}, x_{22}, x_{23}). One possible quantity of interest here is how highly a student's ability on closed-book exams is correlated with his ability on open-book exams. Alternatively, one might try to use the open-book exam results to predict the closed-book results (or vice versa). The results shown in Table 9.1 were found using a learning rate of 0.0001 and 50 000 iterations. We have reported our results to four decimal places in this section to facilitate comparison with those reported in [125] which were found by standard statistical batch methods [125]. The $\mathbf{w}_1$ vector consists of the weights from the closed-book exam data to y_1, while the $\mathbf{w}_2$ vector consists of the weights from the open-book exam data to y_2. We note the excellent agreement between the methods. The highest correlations are given by a weighted average of x_{11} and x_{12} with the former receiving half the weight of the latter (since $\mathbf{w}_{11} \approx \frac{1}{2}\mathbf{w}_{12}$) and the average of x_{21}, x_{22} and x_{23} heavily weighted on x_{21} (since $\mathbf{w}_{21} >> \mathbf{w}_{22}, \mathbf{w}_{23}$).

Table 9.1. The converged weights from the neural network are compared with the values reported from a standard statistical technique [125].

Standard statistics maximum correlation	0.6630
$\mathbf{w}_1$	0.0260 0.0518
$\mathbf{w}_2$	0.0824 0.0081 0.0035
Neural network maximum correlation	0.6962
$\mathbf{w}_1$	0.0264 0.0526
$\mathbf{w}_2$	0.0829 0.0098 0.0041

9.3.3 Random Dot Stereograms

It has been suggested [8] that one of the goals of sensory information processing may be the extraction of common information between different sensors or across sensory modalities. One reason that this is possible is because of the coherence which exists in time and place in sensory input data. We may view the above network as a means of merging two different data streams, $\mathbf{x}_1$ and $\mathbf{x}_2$, which may be either representatives of two different modalities or as different representatives of the same modality where such representatives may be different in either time or place.

Becker [8] has developed this idea and experimented on a data set which is an abstraction of random dot stereograms: an example is shown graphically in Fig. 9.2. In Appendix D, we briefly discuss Becker's methods.

The central idea behind this is that two different neural units or neural network modules should learn to extract features that are coherent across their inputs.

If there is any feature in common across the two inputs, it should be discovered, while features which are independent across the two inputs will be ignored.

Each input vector consists of a one-dimensional random strip which corresponds to the left image and a shifted version of this which corresponds to the right image. The left image has components drawn with equal probability from the set $\{-1, 1\}$ and the right image is generated by choosing a randomly chosen global shift – either one pixel left or one pixel right – and applying it to the left image. We wish to find the maximum linear correlation between y_1 and y_2, which are themselves linear combinations of $\mathbf{x}_1$ and $\mathbf{x}_2$. Because the shifts are chosen with equal probability, there are two equal sets of correlations corresponding to left-shift and right-shift.

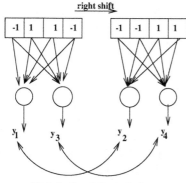

Fig. 9.2. The random dot stereogram data and network. The x_2 set is either a left-shifted or a right-shifted (as shown) version of the x_1 set. We find that w_1 and w_2 reliably find the left-shift and w_3 and w_4 the right-shift or vice versa.

In order to find these, we require two pairs of outputs and the corresponding pairs of weights (w_1, w_2) and (w_3, w_4). The learning rules for w_3 and w_4 in this experiment are analagous to those for w_1 and w_2; at each presentation of a sample of input data, a simple competition between the products $y_1 y_2$ and $y_3 y_4$ determines which weights will learn on the current input samples: if $y_1 y_2 > y_3 y_4$, then w_1, w_2 are updated, else w_3, w_4 are updated.

Using a learning rate of 0.001 and 100 000 iterations, the weights converge to the vectors shown in Table 9.2.

Table 9.2. The converged weights clearly show that the first pair of neurons has learned a right shift while the second pair has learned a left shift.

w_1	−0.002	**1.110**	0.007	−0.009
w_2	0.002	0.025	**0.973**	0.020
w_3	−0.014	0.026	**1.111**	−0.002
w_4	0.013	**0.984**	0.003	−0.007

The first pair of weights, w_1 and w_2, have identified the second element of x_1 and the third element of x_2 as having maximum correlation while other inputs are ignored (the weights from these are approximately 0). This corresponds to a right shift. This first pair of outputs has a (sample) correlation of 0.512. Similarly the second pair of weights has idenfied the third element of x_1 and the second element of x_2 as having maximum correlation while other inputs are ignored. The second pair has a (sample) correlation of 0.530 and corresponds to an identification of left shift.

Now, for some patterns there will be ambiguity since it is possible that by chance a right-shifted pattern will happen to match the weights $\mathbf{w}_3, \mathbf{w}_4$ and therefore a bank of pairs of pairs of neurons is required to perform as well as Becker's IMAX network [8] and Appendix D. For best results, each set of four neurons as above should see as input a slightly different part of the input data (though with a global left or right shift). We have shown [128] that a Factor Analysis network (Chapter 5) which takes as input the output of such a network can identify left and right shifts from this data set.

The table thus demonstrates that these high correlations come from one pair learning the shift-left transformation while the other learns the shift-right.

It should be noted at this point that [8] only uses a subset of 12 of the 16 possible patterns. Those which are ambiguous (such as $(-1,1,-1,1)$) are removed, whereas we are drawing our data from all 16 possible patterns. In addition, the method in [8] uses computationally expensive backpropagation of the derivatives of mutual information. We are able to find both correlations with a very simple network.

In Appendix D, we also review Kay and Phillips' [105] neural network method for integrating contextual and input data. We might consider e.g. from the first data set, $\mathbf{x}_1$ as the input data to the network and $\mathbf{x}_2$ as the contextual input. The results from the CCA network and the Kay and Phillips network are similar too, though Kay and Phillips use a probabilistic network with an nonlinear activation function designed to manage the effect of contextual information on the response of the network to input data.

Finally, the CCA network presented here may be criticised as a model of biological information processing in that it appears as a nonlocal implementation of Hebbian learning i.e. the $\mathbf{w}_1$ weights use the magnitude of y_2 as well as y_1 to self-organise. One possibility is to postulate non-learning connections which join y_2 to y_1, thus providing the information that $\mathbf{w}_1$ requires for learning. Alternatively we may describe the λ_1 parameter as a lateral weight from y_2 to y_1 and so the learning rules become

$$\Delta w_{1j} = (\eta \lambda_1) x_{1j} \left(\frac{y_2}{\lambda_1} - y_1 \right)$$
$$\Delta \lambda_1 = -\eta_0 (1 - y_1^2)$$

where we have had to incorporate a λ_1 term into the learning rate. Perhaps the second of these suggestions is more plausible than the first as a solution to the nonlocal feature of the previous learning rules since nonlearning connections hardwires some activation passing into the cortex.

9.3.4 Equal Correlations

We now create artificial data which contains two independent correlations of equal magnitude. We repeat the first experiment with the same artificial data but this time create correlations between x_{11} and x_{21} and correlations of equal magnitude between x_{12} and x_{22} by drawing two independent samples from $N(0, 1)$ and adding one to x_{11} and x_{21} and the other to x_{12} and x_{22}. The above network failed to converge to either of the correlations presumably because the correlations were of equal magnitude but independent of each other. However, by introducing asymmetry to the network (a tactic [51] often useful in neural networks, Chapter 4) in our constraints – we allow the outputs to have different power originally –

$$J_1 = E\left\{(y_1 y_2) + \frac{1}{2}\lambda_1(k_1 - y_1^2)\right\} \text{ and }$$

$$J_2 = E\left\{(y_1 y_2) + \frac{1}{2}\lambda_2(k_2 - y_2^2)\right\}$$

our weights converge to the CCA directions. We found that it is not necessary that $k_1 \neq k_2$ for all time, but merely ensure that during the first phase of convergence there is an inequality between these values. We also now report that the most accurate results (such as those reported in the first section on artificial data) are achieved when there is some asymmetry between the parameters k_1 and k_2 even when there is only one correlation between the data sets.

We will return to the significance of this feature when we develop specific neural networks from Becker's [7] models in Chapter 10.

9.3.5 More Than Two Data Sets

In integrating information from different sensory modalities, the cortex may be presented with the problem of integrating from more than two data sets simultaneously. Therefore, we extend the algorithm without introducing unnecessarily complex activation passing, which would become biologially implausible. We create an artificial data set which comprises three vectors each of whose first elements have correlations of equal magnitude: $\mathbf{x}_1$, $\mathbf{x}_2$ and $\mathbf{x}_3$ are each 3-dimensional vectors, each of whose elements is initially independently drawn from $N(0, 1)$. We now draw a sample from $N(0, 1)$ and add it to the first element of each of $\mathbf{x}_1$, $\mathbf{x}_2$ and $\mathbf{x}_3$ and attempt to maximise the correlation between y_1, y_2 and y_3. We opt to maximise three separate constrained objective functions:

$$J_1 = E\left\{(y_1 y_2) + \frac{1}{2}\lambda_1(1 - y_1^2)\right\} \text{ and }$$

$$J_2 = E\left\{(y_2 y_3) + \frac{1}{2}\lambda_2(1 - y_2^2)\right\} \text{ and }$$

$$J_3 = E\left\{(y_3 y_1) + \frac{1}{2}\lambda_3(1 - y_3^2)\right\}$$

We use gradient ascent on the instantaneous version of each of these functions with respect to both the weights, $\mathbf{w}_1$, $\mathbf{w}_2$ and $\mathbf{w}_3$, and the Lagrange multipliers, λ_1, λ_2 and λ_3. This gives us the learning rules

$$\Delta\mathbf{w}_1 \propto \frac{\partial J_1}{\partial \mathbf{w}_1} = \mathbf{x}_1 y_2 - \lambda_1 y_1 \mathbf{x}_1 = \mathbf{x}_1(y_2 - \lambda_1 y_1)$$

$$\Delta\mathbf{w}_2 \propto \frac{\partial J_2}{\partial \mathbf{w}_2} = \mathbf{x}_2 y_3 - \lambda_2 y_2 \mathbf{x}_2 = \mathbf{x}_2(y_3 - \lambda_2 y_2)$$

$$\Delta\mathbf{w}_3 \propto \frac{\partial J_3}{\partial \mathbf{w}_3} = \mathbf{x}_3 y_1 - \lambda_3 y_3 \mathbf{x}_3 = \mathbf{x}_3(y_1 - \lambda_3 y_3)$$

The derivates with respect to the Lagrange multipliers are similar to the previous rules (9.11) though we have found empirically that the best result are achieved when the λ's learning rate is again very greatly increased (now $\eta_0 \approx 200\eta$ to 1000η).

Using $\eta = 0.0001$, $\eta_0 = 0.05$ and 100 000 iterations, the weights converge to the values shown in Table 9.3. The three way correlation derived by this method is equal to three pairwise correlations.

Table 9.3. The weights of the converged three input vectors network. The network has clearly identified the correlation between the first element in each vector.

$\mathbf{w}_1$	**0.812**	0.013	0.027
$\mathbf{w}_2$	**0.777**	−0.014	0.030
$\mathbf{w}_3$	**0.637**	0.007	0.012

9.3.6 Many Correlations

However, the preceeding method only enables us to find one correlation in a data set. We can find more by deflationary methods [160], however, a better method is to create an objective function which contains the necessary criteria for finding more than one correlation. One obvious basis for this would be to insist that $\mathbf{y}_i \mathbf{y}_i^T = I$ where I is the $m * m$ identity matrix, with the $\mathbf{y}_i$, $i = 1,2$ vector of first (resp. second) outputs.

Thus, the criterion becomes to maximise

$$J = E\{(\mathbf{y}_1^T \mathbf{y}_2) + \Lambda_1(\mathbf{y}_1 \mathbf{y}_1^T - I) + \Lambda_2(\mathbf{y}_2 \mathbf{y}_2^T - I)\} \tag{9.13}$$

with $\Lambda_i, i = 1,2$ now a matrix of Lagrange multipliers. This gives us the learning rules

$$\Delta W_1 = \eta \mathbf{x}_1(\mathbf{y}_2 - \Lambda_1 \mathbf{y}_1)^T$$
$$\Delta \Lambda_1 = \eta_0(\mathbf{y}_1 \mathbf{y}_1^T - I)$$

where

$$\mathbf{y}_1 = W_1^T \mathbf{x}_1 \tag{9.14}$$

with W_1 the $m * n$ matrix of weights connecting $\mathbf{x}_1$ to $\mathbf{y}_1$. Similarly with W_2, Λ_2.

With this formulation we find a rotation of the canonical correlations. e.g. with artificial data in which $\mathbf{x}_1$ and $\mathbf{x}_2$ are both six-dimensional data vectors each of whose elements are randomly drawn from the zero mean Gaussian distribution $N(0,1)$ before an additional three samples are drawn from the same distribution and added to the first three elements of $\mathbf{x}_1$ and $\mathbf{x}_2$, we find the results in Table 9.4 (left). The fact that there are correlations in the first three elements has clearly been found and we see that there is only a very small magnitude in the other three positions of W_2. The weights, W_1 are similar.

Table 9.4. One set of weights, W_2. Each column represents the weights into one output. The weights from the last three outputs are approximately 0 in both cases. The right table shows the results when rectification of the weights is used.

−0.3848	0.2048	−0.1304	0.2764	0	0
−0.0647	−0.5468	−0.2438	0	0.2914	0
−0.1333	−0.0326	0.5218	0	0	0.3692
0.0052	0.0065	0.0099	0.0005	0.0032	0.0001
0.0025	−0.0063	−0.0009	0.0001	0.0005	0.0006
0.0047	0.0002	−0.0013	0.0007	0.0009	0.0076

However, the above network only finds the correlations spread out over the subspace spanned by the major correlations. We are constraining the expected value of $\mathbf{y}_1\mathbf{y}_1^T = I$ which means that

$$E(W_1^T \mathbf{x}_1 \mathbf{x}_1^T W_1) = I$$
$$\text{i.e.} \quad W_1^T \Sigma_{11} W_1 = I$$
$$\text{or} \quad W_1^T W_1 = I \text{ if } \Sigma_1 = I$$

where we have used Σ_{11} as the covariance matrix of the input vector $\mathbf{x}_1$. Now for our artificial data set this last condition holds and so any orthogonal matrix satisfies the constraints. The Λ_1 matrix learns the diagonal elements to ensure that $E(\mathbf{y}_{1i}^2) = 1$, but there is no interaction between the elements in $\mathbf{x}_1$ so that we do not ensure that $E(\mathbf{y}_{1i}\mathbf{y}_{1j}) = 0, i \neq j$, since $(\Lambda_1)_{ij} = 0$. The elements of the Λ matrix tend to be very heavily weighted on the main diagonal.

The trick ([25] and Chapter 5) which turns a PCA network into a Factor Analysis network was to insist that the weights remain positive. Recognising that the correlations we wish to find are positive, we may insist that the weights remain positive i.e. should the weight change mean that the weights become negative, we simply set that weight to 0. It subsequently has the opportunity to grow positive again. This is not too restrictive a practice since we may insist that another output neuron has all negative weights if we suspect an anticorrelation between x_1 and x_2. This change is also motivated by Dale's Law which states that a neuron may be inhibitory or excitatory but may not switch from one effect to the other. With this additional constraint we have the results shown in Table 9.4 (right): the actual correlation filters have been found. Of note, perhaps, is the fact that the off-diagonal elements of $\Lambda_i, i = 1, 2$ are very much larger, showing that we are now forcing interaction between the individual elements of the outputs.

We may use this method with the random dot stereogram method. With the above network the competition (between $y_1 y_2$ and $y_3 y_4$) is not necessary: one pair of neurons will learn a left shift while one pair learns a right shift.

9.4 A Second Neural Implementation of CCA

In Chapter 2, we discussed how Principal Component Analysis (PCA) can be performed by solving for $\mathbf{w}$ the eigenproblem

$$\Sigma \mathbf{w} = \lambda \mathbf{w} \tag{9.15}$$

where Σ is the covariance matrix of the data. Now any square matrix can have eigenvectors and eigenvalues and so the eigen problem for a general matrix A can be written

$$A\mathbf{w} = \lambda \mathbf{w} \tag{9.16}$$

We can consider a generalisation of this, which is to find eigenvectors $\mathbf{w}$ and eigenvalues λ solving the equation

$$A\mathbf{w} = \lambda B \mathbf{w} \tag{9.17}$$

which will be used in this section. This is equivalent to solving

$$B^{-1} A \mathbf{w} = \lambda \mathbf{w} \tag{9.18}$$

if B^{-1} exists. i.e. the eigenvector solution of (9.17) is the standard eigenvector solution of (9.18).

Now it may be shown [155] that an alternative method of finding the canonical correlation directions is to solve the generalised eigenvalue problem

$$\begin{bmatrix} 0 & \Sigma_{12} \\ \Sigma_{21} & 0 \end{bmatrix} \begin{bmatrix} \mathbf{w}_1 \\ \mathbf{w}_2 \end{bmatrix} = \rho \begin{bmatrix} \Sigma_{11} & 0 \\ 0 & \Sigma_{22} \end{bmatrix} \begin{bmatrix} \mathbf{w}_1 \\ \mathbf{w}_2 \end{bmatrix} \tag{9.19}$$

where ρ is the correlation coefficient. Intuitively, since $\Sigma_{ij} = E(\mathbf{x}_i \mathbf{x}_j)$ we are stating that $\mathbf{w}_2$ times the correlation between $\mathbf{x}_1$ and $\mathbf{x}_2$ is equal to the correlation coefficient times the weighted (by $\mathbf{w}_1$) variance of $\mathbf{x}_1$. Now this has multiple solutions since we are not constraining the variance of the outputs to be 1. If $\mathbf{w}_1^*$ and $\mathbf{w}_2^*$ are solutions to (9.19), then so are $a\mathbf{w}_1^*$ and $a\mathbf{w}_2^*$ for all real a. Thus this method will find the correct correlation vectors but not with a unique magnitude.

It has recently been shown [188] that solutions of the generalised eigenvalue problem

$$Aw = \lambda Bw \qquad (9.20)$$

can be found using gradient ascent of the form

$$\frac{d\mathbf{w}}{dt} = A\mathbf{w} - f(\mathbf{w})B\mathbf{w} \qquad (9.21)$$

where the function $f(\mathbf{w}) : R^n - \{0\} \rightarrow R$ satisfies

1. $f(\mathbf{w})$ is locally Lipschitz continuous.
2. $\exists M_1 > M_2 > 0 : f(\mathbf{w}) > \lambda_1, \forall \mathbf{w} :\| \mathbf{w} \|\geq M_1$ and $f(\mathbf{w}) < \lambda_n, \forall \mathbf{w} : 0 <\| \mathbf{w} \|\leq M_2$.
3. $\forall \mathbf{w} \in R^n - \{0\}, \exists N_1 > N_2 > 0 : f(\theta\mathbf{w}) > \lambda_1, \forall \theta : \theta \geq N_1$ and $f(\theta\mathbf{w}) < \lambda_n, \forall \theta : 0 \leq \theta \leq N_2$ and $f(\theta\mathbf{w})$ is a strictly monotonically increasing function of θ in $[N_1, N_2]$.

where λ_1 is the greatest generalised eigenvalue and λ_n is the least eigenvalue. Intuitively, what these criteria mean are the following:

1. The function is rather smooth.
2. It is always possible to find values of $\mathbf{w}_i, i = 1, 2$ large enough so that the functions of the weights exceed the greatest eigenvalue.
3. It is always possible to find values of $\mathbf{w}_i, i = 1, 2$ small enough so that the functions of the weights are smaller than the least eigenvalue.
4. For any particular value of $\mathbf{w}_i, i = 1, 2$, it is possible to multiply $\mathbf{w}_i, i = 1, 2$ by a scalar and apply the function to the result to get a value greater than the greatest eigenvalue.
5. Similarly, we can find another scalar so that, multiplying the $\mathbf{w}_i, i = 1, 2$, by this scalar and taking the function of the result gives us a value less than the smallest eigenvalue.
6. The function of this product is monotonically increasing between the scalars defined in 4 and 5.

Taking $\mathbf{w} = [\mathbf{w}_1^T \mathbf{w}_2^T]^T$, we find the canonical correlation directions $\mathbf{w}_1$ and $\mathbf{w}_2$ using

$$\frac{d\mathbf{w}_1}{dt} = \Sigma_{12}\mathbf{w}_2 - f(\mathbf{w}_1)\Sigma_{11}\mathbf{w}_1$$

$$\frac{d\mathbf{w}_2}{dt} = \Sigma_{21}\mathbf{w}_1 - f(\mathbf{w}_2)\Sigma_{22}\mathbf{w}_2$$

Using the facts that $\Sigma_{ij} = E(\mathbf{x}_i\mathbf{x}_j^T), i, j = 1, 2$, and that $y_i = \mathbf{w}_i.\mathbf{x}_i$, we may propose the instantaneous rules

$$\Delta \mathbf{w}_1 = \eta(\mathbf{x}_1 y_2 - f(\mathbf{w}_1)\mathbf{x}_1 y_1)$$
$$\Delta \mathbf{w}_2 = \eta(\mathbf{x}_2 y_1 - f(\mathbf{w}_2)\mathbf{x}_2 y_2)$$

For example, if we choose $f(\mathbf{w}) = \ln(\mathbf{w}^T(t)\mathbf{w}(t))$, we have:

$$\Delta w_{1j} = \eta x_{1j}(y_2 - \ln(\mathbf{w}_1^T \mathbf{w}_1)y_1)$$
$$\Delta w_{2j} = \eta x_{2j}(y_1 - \ln(\mathbf{w}_2^T \mathbf{w}_2)y_2) \qquad (9.22)$$

This algorithm is simpler than that used previously, in that we don't need to adjust the parameter λ any more. In a similar manner, we can use:

$$\Delta w_{1j} = \eta x_j(y_2 - \ln\left(\sum_j |w_{1j}|\right)y_1) \qquad (9.23)$$
$$\Delta w_{1j} = \eta x_j(y_2 - \ln(\max_{1 \leq j \leq n} |w_{1j}|)y_1) \qquad (9.24)$$
$$\Delta w_{1j} = \eta x_j(y_2 - (\mathbf{w}_1^T \mathbf{w}_1 - \phi)y_1) \qquad (9.25)$$
$$\Delta w_{1j} = \eta x_j(y_2 - \left(\sum_j |w_{1j}| - \phi)y_1\right) \qquad (9.26)$$
$$\Delta w_{1j} = \eta x_j(y_2 - (\max_{1 \leq j \leq n} |w_{1j}| - \phi)y_1) \qquad (9.27)$$

and similarly with the corresponding updates of $\mathbf{w}_2$. The functions (9.22)-(9.27) will be known as $f_1(), \cdots, f_6()$ in the following. These new algorithms only have a one-phase operation; there is no additional λ parameter to update.

It may appear that the rule (9.25) is equivalent to the update of the λ parameter in the previous section. This is only superficially true: firstly, the derivations are quite different; secondly, the ϕ parameter in (9.25) must satisfy the constraint [188] that it is greater than the greatest eigenvalue of the covariance matrix of the input data, whereas the previous rule (9.11) used the equivalent parameter to ensure that the variances of the outputs were bounded. The need for a larger value of ϕ in (9.25) has been verified experimentally.

9.5 Simulations

9.5.1 Artificial Data

To compare this new family of algorithms with the first neural algorithm, we use the artificial data discussed above. We generated an artificial data set to give two vectors $\mathbf{x}_1$ and $\mathbf{x}_2$. The vector $\mathbf{x}_1$ is a four-dimensional vector, each of whose elements is drawn from the zero-mean Gaussian distribution, $N(0,1)$; $\mathbf{x}_2$ is a three-dimensional vector, each of whose elements is also drawn from $N(0,1)$. In order to introduce correlations between the two vectors, $\mathbf{x}_1$ and $\mathbf{x}_2$, we generate an additional sample from $N(0,1)$ and add it to the first elements of each vector and then normalise the variances. Thus there is no correlation between the two vectors other than that existing between the first elements

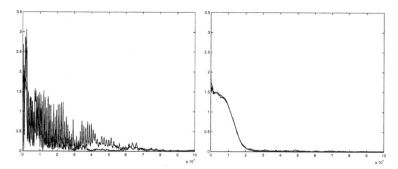

Fig. 9.3. Left: Convergence on artificial data using the first algorithm. Right: Convergence on artificial data using the second algorithm with $f_1()$.

of each. All simulations have used an initial learning rate of 0.001, which is decreased linearly to 0 over 100 000 iterations.

The left diagram in Fig. 9.3 shows the convergence of the weights with the first algorithm, while the right of Fig. 9.3 shows the convergence of the weights using the second method with function $f_1()$. The convergence is given as the angle between the weights at each iteration in our simulation and that of the optimal set of weights i.e.(1,0,0,0) for $\mathbf{x}_1$ and (1,0,0) for $\mathbf{x}_2$. In each figure, we graph two lines to express the convergence of $\mathbf{w}_1$ and $\mathbf{w}_2$. Comparing the diagrams in Figure 9.3, we can see the second algorithm converges much faster than the first algorithm and is very stable. All of the learning algorithms of this class have the same order of convergence speed and so the convergence speed does not depend on the specific form of $f(\mathbf{w})$. The learning algorithms are robust to implementation error on $f(\mathbf{w})$. The simple $f(\mathbf{w})$ reduces the implementation complexity.

9.5.2 Real Data

Again, in order to compare our family of methods with those reported above, we use the 88 students' marks on five module exams. Thus we have a two-dimensional $\mathbf{x}_1$ and a three-dimensional $\mathbf{x}_2$. One set of results is shown in Table 9.5. In our experiment, the learning rate was 0.0001 and the number of iterations was 50 000.

In Table 9.5, $\mathbf{w}_1$ vector consists of the weights from the closed-book exam data to y_1, while the $\mathbf{w}_2$ vector consists of the weights from the open-book exam data to y_2. We note the excellent agreement between the methods.

We will not repeat with the second neural CCA method the other experiments which we discussed with the first neural method, but simply state that the second method can be extended in exactly the same way as the first and will perform as well as the first in all cases [60].

Table 9.5. Correlation and weight values on the examination data.

Standard statistics: max. correlation	0.6630
$\mathbf{w}_1$	0.0260 0.0518
$\mathbf{w}_2$	0.0824 0.0008 0.0035
Second algorithm: max. correlation	0.6790
$\mathbf{w}_1$	0.0270 0.0512
$\mathbf{w}_2$	0.0810 0.0090 0.0040

9.6 Linear Discriminant Analysis

As a slight diversion from the main theme of this book, we note that we may use the above method to perform Linear Discriminate Analysis. Fisher's Linear Discriminant Analysis (LDA) is a method for seeking a direction or filter of a high-dimensional data set which is optimum for discrimination between two data sets. This may be compared with Principal Component Analysis which finds components useful for representing the data with minimal loss, but which need not find directions useful for discrimination.

Formally, let us have n_1 samples from a data set D_1 and n_2 samples from a data set D_2. Then LDA finds those vectors $\mathbf{w}$ which best differentiate between the samples from D_1 and D_2 when the samples are projected onto $\mathbf{w}$. If $\mathbf{x}_1$ is a sample from D_1 and $\mathbf{x}_2$ is a sample from D_2, then ideally $y_1 = \mathbf{w}.\mathbf{x}_1$ and $y_2 = \mathbf{w}.\mathbf{x}_2$ should be clearly differentiable for all $\mathbf{x}_1$ and $\mathbf{x}_2$. It can be shown (e.g. [42]) that this best line can be found by optimising the criterion function

$$J(\mathbf{w}) = \frac{\mathbf{w}^T S_B \mathbf{w}}{\mathbf{w}^T S_W \mathbf{w}} \tag{9.28}$$

where $S_B = (\mu_1 - \mu_2)(\mu_1 - \mu_2)^T$ is the between class covariance matrix and $S_W = \sum_{\mathbf{x} \in D_i} (\mathbf{x} - \mu_i)(\mathbf{x} - \mu_i)^T$ is the within class covariance matrix. This is known as the generalised Rayleigh Quotient and can be maximised by finding the eigenvector with maximum eigenvalue of the generalised eigenvalue problem

$$S_B \mathbf{w} = \lambda S_W \mathbf{w} \tag{9.29}$$

We thus create an artificial neural network which has a feedforward stage:

$$y_1 = \sum_i w_i x_{1i}$$

$$y_2 = \sum_i w_i x_{2i}$$

after which learning takes place

$$\Delta w_i = \eta(t_i - f(\mathbf{w})(y_1 x_{1i} + y_2 x_{2i})) \tag{9.30}$$

where $t_i = \sum_j A_{ij} \mathbf{w}_j$, with $A_{ij} = (\mu_{1i} - \mu_{2i}) * (\mu_{1j} - \mu_{2j})$ being the between sets covariance matrix. Exemplar results are shown in Fig. 9.4: the left diagram shows two uniform clusters which are linearly separable and the line which is found; it is optimal for the separation; the right diagram shows two Gaussian clusters which, though they are not separable, are best differentiated by projecting onto the line shown.

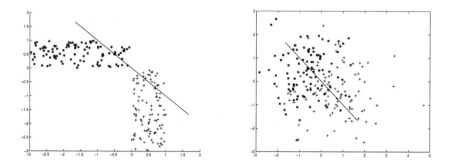

Fig. 9.4. Left: The two uniform distributions will be linearly separable by projecting onto the line found. Right: The two Gaussian clusters are most separated if projected onto the line.

9.7 Discussion

In this chapter, we have illustrated the performance of two linear canonical correlation neural networks on a variety of data sets, both real and artificial. We have been able to extend the CCA methods to deal with the case where the correlation extends to more than two input data sets, something which is not possible using the standard statistical methods of Canonical Correlation Analysis. The networks were also successfully operated on a data set designed to abstract the essential concepts of random dot stereograms and was shown to be capable of finding whether a particular pair of inputs exhibited a left or a right shift. The actual data set used contained all possible pairs of inputs, whether ambiguous or not, and identified the appropriate shift extremely reliably.

In Chapter 11, we will extend these methods so that they respond to nonlinear correlations. Since there is no simple closed-form solution to the maximisation of such correlations, an iterative method such as a neural network weight learning method is necessary for such problems.

We close this chapter by noting that the two methods are very similar in structure: both contain a Hebb-type learning term between each input and the opposite output and an anti-Hebb term between the input and the corresponding output. We again have negative feedback of activation though we have continued to use it specifically in the learning rule. Yet they were derived from very different starting points. We will see, in the next chapter, that similar rules can be derived from a variety of perspectives.

10

Alternative Derivations of CCA Networks

In the previous chapter, we discussed two new methods of performing Canonical Correlation Analysis (CCA) with artificial neural networks. In this chapter, we re-derive learning rules from a probabilistic perspective which then enables us, by use of a specific prior on the weights, to simplify the algorithm. We then derive CCA-type rules from Becker's models (see Appendix D), though with a very different methodology from that used in [7]. We finally derive a robust version of the above rules from probability theory and compare the convergence of all the various rules on artificial data sets.

10.1 A Probabilistic Perspective

We may also derive these learning rules from a probabilistic perspective. Let us assume that we have discovered the best linear combination of the elements of $\mathbf{x}_2$ which will match with greatest correlation a linear combination of $\mathbf{x}_1$. Then y_2 has the target distribution which we wish y_1 to match as closely as possible. Now if the distributions match exactly, there is a linear relationship between y_1 and y_2 i.e.

$$y_2 = \lambda_1 y_1 + \theta$$

For simplicity in the following, we will take $\theta = 0$. Now let us assume that the actual value of y_2 is Gaussian distributed with mean proportional to y_1. Then the probability of the output y_2 given the model and its parameters is

$$P(y_2|\mathbf{w}_1, \mathbf{x}_1, H) = \frac{1}{Z_n} \exp\{-\beta(\lambda_1 y_1 - y_2)^2\}$$

where Z_n is a normalising factor, H is the current model and β determines the spread of the distribution. We wish to find the most probable value of y_2. Then we may define J_1 to be the negative log of this probability to give an objective function which we can minimise with respect to the parameters $\mathbf{w}_1$ to get the learning rules

$$\Delta \mathbf{w}_1 \propto -\frac{\partial J_1}{\partial \mathbf{w}_1} = \beta \mathbf{x}_1 (y_2 - \lambda_1 y_1) \lambda_1 \qquad (10.1)$$

The learning rule for $\mathbf{w}_1$ is identical to that of the previous chapter with $\beta \lambda_1 = \eta$. This perspective also allows us to perform error descent on the unknown parameter λ_1 (ignoring for the moment difficulties with joint optimisations) or we may set λ_1 to a particular value determined by our prior beliefs about the data set, $\mathbf{x}_1$ and $\mathbf{x}_2$ and the parameters. We may consider the more general situation where

$$y_1 = f_1(\mathbf{w}_1^T \mathbf{x}_1)$$

for some nonlinear function $f_1()$. Then

$$\Delta \mathbf{w}_1 = -\frac{\partial J_1}{\partial \mathbf{w}_1} = \beta f_1' \mathbf{x}_1 (y_2 - \lambda_1 y_1) \lambda_1 \qquad (10.2)$$

where f_1' is the derivative of $f_1(t)$ with respect to t. We will see in the next chapter that this is a very similar rule to that derived from a constrained optimisation perspective. However it now allows us to experiment with rules optimal for different probabilities.

10.1.1 Putting Priors on the Probabilities

Let us now consider our prior beliefs about the weights $\mathbf{w}_1$. It is our prior belief that we require a nonzero value for y_1; in other words, we wish to ensure that our learning algorithm does not find values of $\mathbf{w}_1$ and $\mathbf{w}_2$ that give us the trivial solution $y_1 = y_2 = 0$. For reasons which will become obvious later, we will use the prior

$$P(\mathbf{w}_1 | \mathbf{x}_1, H) = \frac{1}{Z_w} \exp \left(\frac{1}{2} \gamma y_1^2 \right) \qquad (10.3)$$

where Z_w is a normalising constant. We may equally express this as a prior on y_1 since the relationship between $\mathbf{w}_1$ and y_1 is deterministic. Note that this implies that our prior belief in the joint magnitude of all weights into output y_1 is an exponential distribution. Now $\frac{1}{Z_w} \exp(\frac{1}{2} \gamma y_1^2) = \frac{1}{Z_w} \exp(\frac{1}{2} \gamma f_1^2(\mathbf{w}_1^T \mathbf{x}_1))$ and so with binary $\mathbf{x}_1$, we are at, or close to, the corners of a hypercube. Since most of the volume of a high-dimensional hypercube is near the corners, this prior displays our belief that some weights will have large values. Then the joint probability of weights and data can be written

$$P(y_2, \mathbf{w}_1 | \mathbf{x}_1, H) = P(y_2 | \mathbf{w}_1, \mathbf{x}_1, H).P(\mathbf{w}_1 | \mathbf{x}_1, H)$$

and so taking

$$J_2 = -\log(P(y_2|\mathbf{w}_1, \mathbf{x}_1, H).P(\mathbf{w}_1|\mathbf{x}_1, H)) \tag{10.4}$$

we have

$$\Delta\mathbf{w}_1 = y_2\mathbf{x}_1\beta\lambda_1 f_1' - \mathbf{x}_1 y_1(\beta\lambda_1^2 - \gamma)f_1'$$

If we use $y_1 = \tanh(\mathbf{w}_1^T\mathbf{x}_1)$

$$\Delta\mathbf{w}_1 = y_2\mathbf{x}_1\beta\lambda_1(1 - y_1^2) - \mathbf{x}_1 y_1(1 - y_1^2)(\beta\lambda_1^2 - \gamma)$$

We now set $\gamma = \beta\lambda_1^2$ to get a much simplified equation,

$$\Delta\mathbf{w}_1 = y_2\mathbf{x}_1\beta\lambda_1(1 - y_1^2)$$

which we use in the remainder of this chapter.

The prior above is an improper prior ($\int P(\mathbf{w}_1|\mathbf{x}_1, H) = \infty$), but this prior may be made proper if we bound the region over which the weights have non-zero prior probability,

$$P(\mathbf{w}_1|\mathbf{x}_1, H) = \begin{cases} \frac{1}{Z_w}\exp(\frac{1}{2}\gamma y_1^2), & \text{if } |\mathbf{w}_{1i}| < a, \forall i, \\ 0, & \text{otherwise.} \end{cases} \tag{10.5}$$

Note that the value of a must be sufficiently large that the nonzero part of the posterior distribution lies within the prior's support. If a is fixed, the width of the posterior of y_2 is proportional to the width of the least probable region round the origin for the prior on y_1.

However, we see that this method will not be successful in the linear case since the probability density function increases so strongly towards a that all weights will tend to a uniform vector all of whose elements are equal to a.

Finally, we note that using y_2 as the target density in this way is equivalent to the M step in the EM algorithm, while the adaption of the weights, $\mathbf{w}_2$ is equivalent to the E step. From the perspective of the y_2 neuron, these are reversed.

10.2 Robust CCA

We use a similar argument in this section to that used in Chapter 8. We use as a cost function for this network

$$J_1 = f_1(e) = f_1(y_1 - \lambda_2 y_2) \tag{10.6}$$

where, previously, $f_1 = ||.||^2$, the (squared) Euclidean norm. It is well known (e.g. [172, 13]) that with this definition of $f_1()$ the cost function is minimised with respect to any set of samples from the data set on the assumption of Gaussian noise on the samples.

In Chapter 8, we discussed how the mimimisation of J_1 is equivalent to minimising the negative log probability of the error, e, which is equivalent to the noise which y_1 assumes is on its data set. Thus we may determine the optimal cost function based on our knowledge of the noise in a data set.

If, for example, we were to believe that the noise followed a different distribution e.g. if $e \sim \frac{1}{Z} \exp(-|y_1 - \lambda_2 y_2|)$ then we may derive optimal learning rules from

$$\frac{\partial J_1}{\partial \mathbf{w}_1} = \mathbf{x}_1.\text{sign}(y_1 - \lambda_2 y_2)$$

$$\text{where sign}(t) = \begin{cases} -1, & \text{if } t < 0, \\ +1, & \text{if } t > 0. \end{cases}$$

In Chapter 8, we showed the importance of the (one-dimensional) probability density function

$$p(e) = \frac{1}{2 + \epsilon} \exp(-|e|_\epsilon) \tag{10.7}$$

where

$$|e|_\epsilon = \begin{cases} 0, & \forall |e| < \epsilon, \\ |e - \epsilon|, & \text{otherwise} . \end{cases} \tag{10.8}$$

Using this model of the noise, the optimal $f_1()$ function is the ϵ-insensitive cost function

$$f_1(e) = |e|_\epsilon \tag{10.9}$$

Therefore when we use this function in the twinned network we get the learning rule

$$\Delta w_{1i} = \eta x_{1i}\text{sign}(y_1 - \lambda_1 y_2) , \text{ if } |y_1 - \lambda_1 y_2| > \epsilon,$$
$$\Delta w_{1i} = 0, \qquad\qquad\qquad \text{otherwise.} \tag{10.10}$$

which is a simplification of the standard rule.

10.3 A Model Derived from Becker's Model 1

Perhaps the most influential researcher in the neural network field investigation problems such as understanding random dot stereograms has been Susanna Becker; a slightly more detailed discussion of her methods is given in Appendix D. For continuous distributions, Becker ([7], page 58) has three main models, two of which are used in practice. The first assumes that y_1 and y_2 are both zero mean Gaussian scalars. Then from y_1's perspective, y_1 is trying to predict y_2 which it assumes is telling the truth. i.e. y_1 is assuming that it is noisy version of the signal which y_2 gives perfectly. Then the information (Chapter 2) that y_2 provides about y_1 is

$$I_{y_1 y_2} = h(y_1) - h(y_1|y_2)$$
$$= \frac{1}{2}(\log \sigma_{y_1}^2 - \log \sigma_{(y_1|y_2)}^2) \tag{10.11}$$

Becker actually maximises the average of $h(y_1) - h(y_1|y_2)$ and $h(y_2) - h(y_2|y_1)$ even though both are equal to $I_{y_1 y_2}$. (This is done "for symmetry" [7], page 58). Now Becker uses backpropagation of the derivatives of I with respect to the weights. We shall simplify the process by making two additional assumptions:

1. The relationship between $E(y_1)$ and $E(y_2)$ is linear. i.e. $\exists \lambda_2 : E(y_1) = \lambda_2 E(y_2) + d$. Under the assumption that both are zero mean, $d = 0$.
2. We assume that $\sigma_{y_1}^2 = 1$, which accords with the standard CCA assumptions.

Under these assumptions, (10.11) becomes

$$I_{y_1 y_2} = -\frac{1}{2} \log\{\text{variance of } (y_1 - \lambda_2 y_2)\} \tag{10.12}$$

Using the assumptions of Gaussianity, maximising $I_{y_1 y_2}$ is then equivalent to maximising $E(y_1 - \lambda_2 y_2)^2$. If we opt for online learning, the best instantaneous estimator of this is $(y_1 - \lambda_2 y_2)^2$ and its derivative with respect to $\mathbf{w}_1$ gives the previous online learning rule. So we are now changing the weights to give a residual with minimum variance; we may formulate this as

$$J_1 = \frac{1}{2}(y_1 - \lambda_2 y_2)^2$$

$$\Delta \mathbf{w}_1 \propto -\frac{\partial J_1}{\partial \mathbf{w}_1} = -\mathbf{x}_1(y_1 - \lambda_2 y_2)$$

$$\Delta \lambda_2 \propto -\frac{\partial J_1}{\partial \lambda_2} = y_2(y_1 - \lambda_2 y_2) \tag{10.13}$$

Thus our learning rule for λ_i is very different from that in the previous section but we should ask "Is it necessary?". Under our assumptions, $\sigma_{y_1}^2 = \sigma_{y_2}^2 = 1$ and so we do not require the λ_i parameters which leads to

$$J_1 = \frac{1}{2}(y_1 - y_2)^2$$

$$\Delta \mathbf{w}_1 \propto -\frac{\partial J_1}{\partial \mathbf{w}_1} = -\mathbf{x}_1(y_1 - y_2) \tag{10.14}$$

In the comparison section of Lai's thesis [113], she calls (10.13) "BM1", and (10.14) "BM1, no Lambda" since they have been derived from starting points developed by Becker. We do, however, recognise that Becker's derived models are very different.

10.3.1 Who Is Telling the Truth?

Now, Becker points out that there is a contradition in her method: y_1 believes that it is adapting its weights to match the information given by y_2 i.e. y_2 is the signal and currently y_1 is a noisy version of this signal. Therefore,

y_1 should expect its variance to be greater than y_2's variance. Meanwhile y_2 believes just the opposite. "However, in practice this is not a problem, because each module is minimising $V(y_a - y_b)$, so our model becomes more accurate as learning proceeds." However we are introducing a new degree of freedom which can also be changed online so that we are managing the contradiction inherent in the weight update rules by allowing our belief in the correctness of y_i's assumptions to change in time. Similarly in the previous chapter, by using k_1, k_2 parameters with $k_1 > k_2$ we are actually making the prior assumption that y_1 is correct in its belief that it should have more variance than y_2 initially, and as the two sets of weights converge to maximise the correlations, we are changing our belief to one that gives equal confidence in both neurons' assumptions.

10.3.2 A Model Derived from Becker's Second Model

Becker also derives a second model based on the belief that both y_1 and y_2 are noisy versions of some underlying true signal. She suggests that a good approximation to the required mutual information is the mutual information between the average of y_1 and y_2 and the underlying signal. Therefore $y_i = sig + n_i, i = 1, 2$ where sig is the underlying signal and n_i is the noise on the ith output. Thus

$$I^* = I\left(\frac{y_1 + y_2}{2}; sig\right) = 0.5 \log \frac{V(sig + 0.5 * (n_1 + n_2))}{V(0.5 * (n_1 + n_2))}$$

$$= 0.5 \log \frac{V(sig + 0.5 * (n_1 + n_2))}{V(0.5 * (n_1 - n_2))}$$

$$= 0.5 * \log \frac{V(0.5 * (y_1 + y_2))}{V(0.5 * (y_1 - y_2))}$$

where $V(t)$ is the variance of t. Now we are interested in an instantaneous version of the algorithm and so we approximate this expectation by its instantaneous version (we may ignore the monotonic function 0.5*log()) to get

$$J_2 = \frac{y_1^2 + 2y_1y_2 + y_2^2}{y_1^2 - 2y_1y_2 + y_2^2} = \frac{1 + y_1y_2}{1 - y_1y_2} \tag{10.15}$$

assuming that the variances equal 1. This allows us to calculate the derivative of J_2 with respect to the weights, $\mathbf{w}_1$ and $\mathbf{w}_2$ to get

$$\Delta\mathbf{w}_1 = \frac{\eta\mathbf{x}_1y_2}{(1 - y_1y_2)^2} \tag{10.16}$$

In practice we use

$$\Delta\mathbf{w}_1 = \frac{\eta\mathbf{x}_1y_2}{(1 - y_1y_2)^2 + k} \tag{10.17}$$

to avoid problems with division by 0. Similarly for $\mathbf{w}_2$. A final model may be similarly derived as

$$\Delta \mathbf{w}_1 = \frac{\eta \lambda_1 \mathbf{x}_1 \lambda_2 y_2}{(1 - \lambda_1 y_1 \lambda_2 y_2)^2 + k} \tag{10.18}$$

where now the λ_i parameters may be adapted according to

$$\Delta \lambda_1 = \frac{\eta_0 y_1 \lambda_2 y_2}{(1 - \lambda_1 y_1 \lambda_2 y_2)^2 + k} \tag{10.19}$$

In the comparison section of [113], (10.17) is termed "BM2b" while (10.18) and (10.19) is termed "BM2". We should recognise that Becker derived very different models from the same starting point.

10.4 Discussion

In the thesis of Lai [113], in particular, there is an extensive investigation of the performance of the various algorithms (though not the second algorithm from the previous chapter) on a variety of artificial data sets; the point is very soundly made that it is "horses for courses": the nature of the data set and the noise within it interact to make one or another learning rule optimal for finding correlations in that particular data set. The theory and the simulations match very well. We will not discuss the experimental findings at all in this book, but merely point the interested reader to the thesis.

We wish to highlight the common theme underlying all of the above rules, regardless of their derivation: consider

1. The first CCA algorithm of Chapter 9:

$$\Delta w_{1j} = \eta x_{1j}(y_2 - \lambda_1 y_1)$$
$$\Delta \lambda_1 = \eta_0(1 - y_1^2) \tag{10.20}$$

2. The second set of CCA algorithms of Chapter 9:

$$\Delta \mathbf{w}_1 = \eta(\mathbf{x}_1 y_2 - f(\mathbf{w}_1)\mathbf{x}_1 y_1)$$
$$\Delta \mathbf{w}_2 = \eta(\mathbf{x}_2 y_1 - f(\mathbf{w}_2)\mathbf{x}_2 y_2)$$

3. The probabilistic rule:

$$\Delta \mathbf{w}_1 \propto -\frac{\partial J_1}{\partial \mathbf{w}_1} = \mathbf{x}_1(y_2 - \lambda_1 y_1)\lambda_1 \tag{10.21}$$

4. The rule derived from Becker's first model, BM1:

$$\Delta \lambda_2 \propto -\frac{\partial J_1}{\partial \lambda_2} = y_2(y_1 - \lambda_2 y_2) \tag{10.22}$$

5. The same rule but omitting λ :

$$\Delta \mathbf{w}_1 \propto -\frac{\partial J_1}{\partial \mathbf{w}_1} = \beta \mathbf{x}_1(y_2 - \lambda_1 y_1)\lambda_1 \tag{10.23}$$

6. The rule derived from Becker's second model, BM2:

$$\Delta\mathbf{w}_1 = \frac{\eta\lambda_1\mathbf{x}_1\lambda_2\mathbf{y}_2}{(1 - \lambda_1\mathbf{y}_1\lambda_2\mathbf{y}_2)^2 + k} \tag{10.24}$$

$$\Delta\lambda_1 = \frac{\eta_0 y_1\lambda_2 y_2}{(1 - \lambda_1 y_1\lambda_2 y_2)^2 + k} \tag{10.25}$$

7. The same rule but omitting λ:

$$\Delta\mathbf{w}_1 = \frac{\eta\mathbf{x}_1\mathbf{y}_2}{(1 - \mathbf{y}_1\mathbf{y}_2)^2 + k} \tag{10.26}$$

We see that each weight change rule begins with a Hebbian learning term and is often then followed by an anti-Hebbian weight change rule. In Appendix D, we discuss Stone's weight change rule [174] which also has Hebbian and anti-Hebbian elements in it, though it again was derived from very different principles. We conjecture that such an arrangement may have been very powerful in an evolving system which attempts to find the greatest correlations in elements of its environment.

11

Kernel and Nonlinear Correlations

In this chapter we consider three extensions to Canonical Correlation Analysis networks

- We derive a nonlinear CCA network for use where the highest correlations are found from nonlinear projections.
- Using the idea of kernel operations derived from Support Vector Machines, we derive two kernel CCA method and show that one is much to be preferred over the other.
- We show that a mixture of CCA networks can be used where a locally linear set of correlations varies in time or space.

11.1 Nonlinear Correlations

Now while the data set in Chapter 9 provides us with an abstraction of random dot stereograms, it cannot be a complete and accurate abstraction of how the cortex extracts depth information from surfaces: at any one time, we view not just single points but segments of a surface. Consider Fig. 11.1. We see that the relationship between the projection of the surface on the retinas is a function of the angle between the plane of the surface and the plane of the retinas. We do not wish to determine the precise relationship in any specific case since we wish to create a general-purpose depth analyser which does not depend on, e.g. both the retinas and the surface being flat, the pupils having a precise relationship with the limits of the viewed surfaces etc.

Therefore, first we investigate the general problem of maximising correlations between two data sets when there may be an underlying nonlinear relationship between the data sets.

11.1.1 Experiment Results

We generate data according to the prescription:

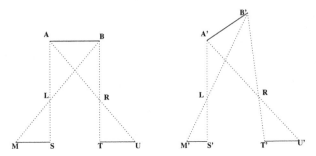

Fig. 11.1. The left figure represents visual information from a surface AB which is passed through pupils R and L to flat "retinas" MS and TU. The right figure represents the same scene when the external surface A'B' is not parallel to the plane of the retinas.

$$x_{11} = 1 - \sin\theta + \mu_1 \tag{11.1}$$

$$x_{12} = \cos\theta + \mu_2 \tag{11.2}$$

$$x_{21} = \theta - \pi + \mu_3 \tag{11.3}$$

$$x_{22} = \theta - \pi + \mu_4 \tag{11.4}$$

where θ is drawn from a uniform distribution in $[0, 2\pi]$ and $\mu_i, i = 1, ..., 4$ are drawn from the zero mean Gaussian distribution $N(0, 0.1)$. Equations (11.1) and (11.2) define a circular manifold in the two-dimensional input space while (11.3) and (11.4) define a linear manifold within the input space, where each manifold is only approximate due to the presence of noise ($\mu_i, i = 1, ..., 4$). The subtraction of π in the linear manifold equations is merely to centre the data.

Thus $\mathbf{x}_1 = \{x_{11}, x_{12}\}$ lies on or near the circular manifold while $\mathbf{x}_2 = \{x_{21}, x_{22}\}$ lies on or near the line.

We wish to test whether the network can find nonlinear correlations between the two data sets, $\mathbf{x}_1$ and $\mathbf{x}_2$, and test whether such correlations are greater than the maximum linear correlations. To do this we train the first network of Chapter 9 (9.11) on this data set but this time calculate y_3 and y_4 using

$$y_3 = \sum_j w_{3j} \tanh(v_{3j}x_{1j}) = \mathbf{w}_3\mathbf{f}_3 \text{ and}$$
$$y_4 = \sum_j w_{4j} \tanh(v_{4j}x_{2j}) = \mathbf{w}_4\mathbf{f}_4$$

The correlation between y_1 and y_2 neurons was maximised using the previous linear operation while that between y_3 and y_4 used the functions

$$J_3 = E\left\{(y_3y_4) + \frac{1}{2}\lambda_3(1 - y_3^2)\right\} \text{ and}$$

$$J_4 = E\left\{(y_3y_4) + \frac{1}{2}\lambda_4(1 - y_4^2)\right\}$$

whose derivatives give us

$$\frac{\partial J_3}{\partial \mathbf{w}_3} = \mathbf{f}_3 y_4 - \lambda_3 y_3 \mathbf{f}_3 = \mathbf{f}_3 (y_4 - \lambda_3 y_3)$$

$$\frac{\partial J_3}{\partial \mathbf{v}_3} = \mathbf{w}_3 y_4 (1 - \mathbf{f}_3^2) \mathbf{x}_1 - \lambda_3 \mathbf{w}_3 (1 - \mathbf{f}_3^2) \mathbf{x}_1 y_3$$

$$= \mathbf{w}_3 (1 - \mathbf{f}_3^2) \mathbf{x}_1 (y_4 - \lambda_3 y_3)$$

Similarly with the J_4 function, $\mathbf{w}_4$, $\mathbf{v}_4$, and λ_4. This gives us a method of changing the weights and the Lagrange multipliers on an online basis. We therefore have the joint learning rules

$$\Delta \mathbf{w}_3 = \eta \mathbf{f}_3 (y_4 - \lambda_3 y_3)$$

$$\Delta \mathbf{w}_4 = \eta \mathbf{f}_4 (y_3 - \lambda_4 y_4)$$

$$\Delta \mathbf{v}_{3i} = \eta \mathbf{x}_{1i} \mathbf{w}_{3i} (y_4 - \lambda_3 y_3)(1 - \mathbf{f}_3^2)$$

$$\Delta \mathbf{v}_{4i} = \eta \mathbf{x}_{2i} \mathbf{w}_{4i} (y_3 - \lambda_4 y_4)(1 - \mathbf{f}_4^2) \tag{11.5}$$

For the first method we use a learning rate of 0.001 for all weights and learn over 100 000 iterations. The learning rates are decreased to 0 during the course of the simulation.

The trained network finds a linear correlation between the data sets of 0.623 (equal to the correlation between y_1 and y_2) while the nonlinear neurons, y_3 and y_4, have a correlation of 0.865. Clearly by putting the data sets through the nonlinear tanh() function we have created a relationship whose correlations are greater than the linear correlations of the original data set. We show a test of the outputs from both the linear and nonlinear networks in Fig. 11.2 in which we graph the output values of the trained network from each pair of neurons against inputs where the θ values in (11.1)–(11.4) range from -3 to 3 and, of course, we do not add noise. We see that the linear network is aligning the outputs as well as may be expected but the nonlinear network's outputs are very closely aligned with each other over much more of the data set. The linear network does not have enough degrees of freedom to make the line bend toward the curves. We have similar results matching data from a noisy surface of a sphere with data from a noisy plane. We can compare these results with the nonlinear version of the second CCA algorithm on the same problem. We noted the similarity between the two types of CCA network in Chapter 9 but qualified that by observing that they were derived from very different criteria. Nevertheless, based on this obvious similarity, we have experimented with the functions $f_1(), ..., f_6()$ in the nonlinear case equivalent to (11.5)

$$y_3 = \sum_j w_{3j} \tanh(v_{3j} x_{1j}) = \mathbf{w}_3 \mathbf{g}_3 \tag{11.6}$$

$$y_4 = \sum_j w_{4j} \tanh(v_{4j} x_{4j}) = \mathbf{w}_4 \mathbf{g}_4 \tag{11.7}$$

to get the nonlinear update equations w_3 and w_4 using

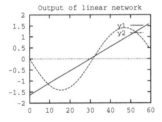
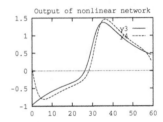

Fig. 11.2. The left diagram shows the outputs, y_1 and y_2, of the linear network, while the right diagram shows the outputs, y_3 and y_4, of the nonlinear network. Visual inspection would suggest that the outputs from the nonlinear network are more correlated. The actual sample correlation values achieved were 0.623 (linear) and 0.865 (nonlinear).

$$\Delta w_3 = \eta g_3(y_4 - f(w_3)y_3)$$
$$\Delta v_{3i} = \eta x_{1i} w_{3i}(y_4 - f(w_3)y_3)(1 - g_3^2)$$
$$\Delta w_4 = \eta g_4(y_3 - f(w_4)y_4)$$
$$\Delta v_{4i} = \eta x_{2i} w_{4i}(y_3 - f(w_4)y_4)(1 - g_4^2) \tag{11.8}$$

- we train one pair of weights $\mathbf{w}_1$ and $\mathbf{w}_2$ using rules (9.22);
- we train a second pair of weights, $\mathbf{w}_3$ and $\mathbf{w}_4$ using (11.8).

We use a learning rate of 0.001 for all weights and learn over 100 000 iterations. We did not attempt to optimize any parameters for either algorithm. In the nonlinear case, the second family of networks find a greater correlation (see Table 11.1). Note that the homogeneity of the family of functions has been broken in the nonlinear case.

Table 11.1. Nonlinear correlations on the artificial data set from the second CCA network of Chapter 9.

Function	Correlation
$f_1()$	0.859
$f_2()$	0.859
$f_3()$	0.859
$f_4()$	0.863
$f_5()$	0.831
$f_6()$	0.814

11.2 The Search for Independence

The objective of this section is to present neural networks for separation of mixed independent source signals. Let there be N independent non-Gaussian signals $(s_1, s_2, ..., s_N)$ which are mixed using a (square) mixing matrix A to get N vectors, x_i, each of which is an unknown mixture of the independent signals,

$$\mathbf{x} = A\mathbf{s} \tag{11.9}$$

Then the aim is to use an artificial neural network to retrieve the original input signals when the only information presented to the network is the unknown mixture of the signals. Note that the outputs, $\mathbf{y}$, are to be the elements of the original signal in some order i.e. we are not insisting that the first output of our neural network is equal to the first signal, the second equal to the second signal and so on. We merely insist that neuron i's output is solely one of the N original signals and is not mixed with any of the other signals.

There are two possibilities that we have considered:

1. We may attempt to extract pairs of maximal correlations and ensure that all pairs are orthogonal to each other. Thus we may use our linear or nonlinear CCA methods to maximise correlations between the two different input vectors.
2. We may alternatively attempt to extract one signal from the mixture and minimise correlation between this output and the other output, thus hoping to extract another signal (implicitly uncorrelated with the first) from the mixture.

If we begin with the first of these possibilities, the criterion we use is to maximise

$$J = E\left\{(\mathbf{y}_1^T\mathbf{y}_2) + \frac{1}{2}\Lambda_1(I - \mathbf{y}_1\mathbf{y}_1^T) + \frac{1}{2}\Lambda_2(I - \mathbf{y}_2\mathbf{y}_2^T)\right\} \tag{11.10}$$

with $\Lambda_i, i = 1, 2$ now a matrix of Lagrange multipliers. This gives us the learning rules

$$\Delta\mathbf{w}_1 = \eta\mathbf{f}_1(\mathbf{y}_2 - \Lambda_1\mathbf{y}_1)^T$$
$$\Delta\mathbf{v}_1 = \eta\mathbf{w}_1(\mathbf{y}_2 - \Lambda_1\mathbf{y}_1)\mathbf{x}_1(1 - \mathbf{f}_1^2)$$
$$\Delta\Lambda_1 = \eta_0(I - \mathbf{y}_1\mathbf{y}_1^T)$$

where

$$\mathbf{y}_1 = \mathbf{w}_1^T\mathbf{f}(\mathbf{v}_1.\mathbf{x}_1) \tag{11.11}$$

with $\mathbf{w}_1$ the $m * n$ matrix of weights connecting $\mathbf{x}_1$ to $\mathbf{y}_1$. Similarly with $\mathbf{w}_2$, Λ_2.

This model may be viewed as an abstraction of two data streams (e.g. sight and sound) identifying an entity in the environment by identifying the maximal correlations between the data streams. An alternative use of CCA is to attempt to identify independent components of data streams by minimising the correlations between data sets.

11.2.1 Using Minimum Correlation to Extract Independent Sources

Just as a Principal Component Analysis network can be changed into a Minor Component Analysis network by changing the sign of the learning rate, we can change our CCA network into a network which searches for the linear combination of the data set which has the minimum mutual correlation by changing the sign of the learning. However, one difficulty with this is that, performing descent on the correlations, it is possible to create negative correlations. We therefore wish to add an additional constraint that ensures that $E(y_1 y_2) \geq 0$. For the nonlinear neurons, this gives us the learning rule

$$J = E\left\{(\mathbf{y}_1^T \mathbf{y}_2) + \frac{1}{2}\Lambda_1(I - \mathbf{y}_1 \mathbf{y}_1^T) + \frac{1}{2}\Lambda_2(I - \mathbf{y}_2 \mathbf{y}_2^T) + \Lambda_3(\mathbf{y}_1 \mathbf{y}_2^T)\right\}$$

$$\Delta \mathbf{w}_1 = -\eta \mathbf{f}_1((I + \Lambda_3)(\mathbf{y}_2 - \Lambda_1 \mathbf{y}_1))$$
$$\Delta \mathbf{v}_1 = -\eta \mathbf{w}_1(I + \Lambda_3)(\mathbf{y}_2 - \Lambda_1 \mathbf{y}_1)\mathbf{x}_1(1 - \mathbf{f}_1^2)$$
$$\Delta \Lambda_1 = -\eta_0(I - \mathbf{y}_1 \mathbf{y}_1^T)$$
$$\Delta \Lambda_3 = \eta_2 \mathbf{y}_2 * \mathbf{y}_1^T$$

where we have used the Λ_i to denote matrices of parameters. Similarly with $\mathbf{w}_2, \mathbf{v}_2$ and Λ_2. This method then ensures that we do not find weight pairs which give a negative nonlinear correlation.

We wish to emulate biological information processing in that e.g. auditory information cannot choose which pathway it uses. All early processing is done by the same neurons. In other words, these neurons see the information at each time instant. Denoting the signals at time t by $\mathbf{s}(t) = \{s_1(t), s_2(t)\}$, we create mixtures at time t by

$$\mathbf{m}_1(t) = f_1(\mathbf{s}(t))$$
$$\mathbf{m}_2(t) = f_2(\mathbf{s}(t))$$

for some functions $f_1()$ and $f_2()$. The data set presented to the network is then

$$\mathbf{x}_1 = \{m_{1,1}(t), m_{1,2}(t), \cdots, m_{1,1}(t+1), \cdots, m_{1,2}(t+P_1)\}$$
$$\mathbf{x}_2 = \{m_{2,1}(t), m_{2,2}(t), \cdots, m_{2,1}(t+1), \cdots, m_{2,2}(t+P_2)\}$$

Good extraction of single sinusoids from a mixture of sinusoids (e.g. for 2×2 square mixes $\mathbf{m}_1, \mathbf{m}_2$) have been obtained using only

$$\mathbf{x}_1 = \{m_{1,1}(t), m_{1,2}(t), m_{1,1}(t+P_1), m_{1,2}(t+P_1)\}$$
$$\mathbf{x}_2 = \{m_{2,1}(t), m_{2,2}(t), m_{2,1}(t+P_2), m_{2,2}(t+P_2)\}$$

where P_1 and P_2 are the time delays which need not be equal.

Both methods have been successfully used on extraction of sinusoids; we report some experimental work in the next section.

11.2.2 Experiments

We have embedded two sine waves on the surface of a sphere and on the plane (recall that we previously were able to maximise correlations when we selected random, but corresponding, points on the surface of the sphere and on the plane). We generate artificial data according to the prescription (see Fig. 11.3):

$$x_{11} = \sin(s_1) * \cos(s_2) \tag{11.12}$$
$$x_{12} = \sin(s_1) * \sin(s_2) \tag{11.13}$$
$$x_{13} = \cos(s_1) \tag{11.14}$$
$$x_{21} = s_1 \tag{11.15}$$
$$x_{22} = s_2 \tag{11.16}$$

where e.g. $s_1(t) = \sin(t/8) * \pi$, $s_2(t) = \sin\{(t/5 + 3) * \pi\}$ for each of $t = 1, 2, ..., 500$. The results from the nonlinear mixture, $\mathbf{x_1}$, and $\mathbf{x_2}$ are shown in Fig. 11.4 . We see that the sine wave has been recovered. Similar results have been achieved when

$$x_{21} = s_1 + s_2$$
$$x_{22} = s_1 - s_2$$

though the quality of the recovered signal deteriorates when both $\mathbf{x_1}$ and $\mathbf{x_2}$ are nonlinear mixtures.

11.2.3 Forcasting

In this section, we wish to compare the forecasting ability of a first linear CCA network and the first nonlinear CCA network.

Power loading is strongly related to the maximum and minimum daily temperatures and there also exits a strong relationship between the power generated one day and the power generated one, two and seven days before[27]. All of these factors are taken into account when codifying the data used to train the network. Table 11.2 shows the 12 values which are used as input to the neural network with one neuron as the output for the supervised methods: load forecasting is essentially a signal-processing problem. Examining and extrapolating past load behavior, taking account of other influencing factors such as weather information, time of day, season of year and holidays, will enable reasonably accurate forecasting.

However, to make an accurate forecast, the effects of all the influencing parameters need to be considered simultaneously, yet, in general, the inter-relationships between these factors are varying, complex and nonlinear.

In this experiment, we use a data set from the Taiwan Power Company; for the supervised networks two years (1992 and 1993) were used to train the neural network, the 1994 data was used for validating, and the 1995 data has

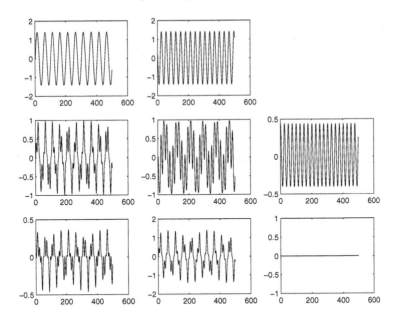

Fig. 11.3. The top line shows the underlying sines which we wish to recover. The second line shows the signals when they are embedded on the surface of a sphere, $\mathbf{x_1}$. The third line shows the signals as coordinates on the plane, $\mathbf{x_2}$.

Table 11.2. The output of the network is the expected power load peak P_i for day i. The inputs are shown below.

Neuron	Description
i_1	Peak of energy on the previous day
i_2	Max. temperature of the previous day
i_3	Min. temperature of the previous day
i_4	Peak of energy generated two days previously
i_5	Max. temperature of two days previously
i_6	Min. temperature of the day previously
i_7	Peak of energy for seven days previously
i_8	Max. temperature of seven days previously
i_9	Min. temperature of seven days previously
i_{10}	Max. forecast temperature
i_{11}	Min. forecast temperature
i_{12}	Day-of-the-week code

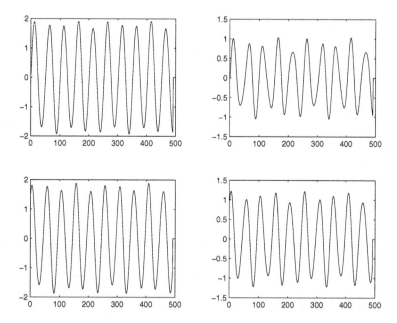

Fig. 11.4. One sine wave has been recovered from the nonlinear mixtures of the sines. The top line shows a pair of y_1 neurons, the bottom line the corresponding pair of y_2 neurons.

been used for the practical test. The results shown in Table 11.3 were found using a learning rate of 0.001 and 100 000 iterations (linear) and 500 000 iterations (nonlinear) for the CCA networks.

Table 11.3. The mean absolute percentage error(MAPE) from a linear and a nonlinear CCA network.

	Maximum correlation	MAPE
Linear correlation	0.960947	3.473950
Nonlinear correlation	0.963584	3.328755

11.3 Kernel Canonical Correlation Analysis

In this section, we extend nonlinear CCA by nonlinearly transforming the data to a feature space and then performing linear CCA in this feature space. We give comparative results on both artificial and real data sets. We also apply the method to Independent Component Analysis (ICA) to extract information from two data sets, each of which contains a mixture of signals.

11.3.1 Kernel Principal Correlation Analysis

Unsupervised kernel methods are a recent innovation based on the methods developed for Support Vector Machines [178]. Support vector regression for example, performs a nonlinear mapping of the data set into some high-dimensional feature space in which we may then perform linear operations. Since the original mapping was nonlinear, any linear operation in this feature space corresponds to a nonlinear operation in data space. This section is based very much on the analysis in [165, 166].

The Linear Kernel

To introduce the use of kernel methods, we review Kernel Principal Component Analysis [131, 157, 164, 165, 166, 167, 170, 171] which has been the most frequently reported linear operation involving unsupervised learning in feature space and is also a method which will underlie our Kernel Canonical Correlation Analysis (KCCA). We will review PCA and then show that it may be calculated in a somewhat different manner to that used normally. PCA finds the eigenvectors and corresponding eigenvalues of the covariance matrix of a data set. Let $\chi = \mathbf{x}_1, ..., \mathbf{x}_M$ be independent, identically distributed samples drawn from a data source. If each $\mathbf{x}_i$ is n-dimensional, $\exists$ at most n eigenvalues/eigenvectors. Let C be the covariance matrix of the data set; then C is $n * n$. Then the eigenvectors, $\mathbf{e}_i$ are n-dimensional vectors which are found by solving

$$C\mathbf{e}_i = \lambda\mathbf{e}_i \qquad (11.17)$$

where λ_i is the eigenvalue corresponding to $\mathbf{e}_i$. We will assume that the eigenvalues and eigenvectors are arranged in nondecreasing order of eigenvalues and each eigenvector is of length 1. We will use the sample covariance matrix as though it was the true covariance matrix and so

$$C \doteq \frac{1}{M} \sum_{j=1}^{M} \mathbf{x}_j \mathbf{x}_j^T \qquad (11.18)$$

Now each eigenvector lies in the span of χ; i.e. the set $\chi = \{\mathbf{x}_1, ..., \mathbf{x}_M\}$ forms a basis set (normally overcomplete since $M > n$) for the eigenvectors. So each $\mathbf{e}_i$ can be expressed as

$$\mathbf{e}_i = \sum_j \alpha_{ij} \mathbf{x}_j \qquad (11.19)$$

Now if we wish to find the principal components of a new data point, $\mathbf{x}$, we project it on the eigenvectors previously found: the first principal component is $\mathbf{x}.\mathbf{e}_1$, the second is $\mathbf{x}.\mathbf{e}_2$, etc. These are the coordinates of $\mathbf{x}$ in the eigenvector basis. There are only n eigenvectors (at most) and so there can only be n coordinates in the new system: we have merely rotated the data set. Now consider projecting one of the data points from χ on the eigenvector $\mathbf{e}_1$:

$$\mathbf{x}_k.\mathbf{e}_1 = \mathbf{x}_k. \sum_j \alpha_{1j}\mathbf{x}_j = \sum_j \alpha_{1j}\mathbf{x}_k.\mathbf{x}_j \qquad (11.20)$$

Now let $\mathbf{K}$ be the matrix of dot products. Then $K_{ij} = \mathbf{x}_i.\mathbf{x}_j$. Multiplying both sides of (11.17) by $\mathbf{x}_k$ we get

$$\mathbf{x}_k^T C \mathbf{e}_1 = \mathbf{x}_k^T \lambda \mathbf{e}_1 \qquad (11.21)$$

and using the expansion for $\mathbf{e}_1$ and the definition of the sample covariance matrix, C, gives

$$\frac{1}{M} K^2 \alpha_1 = \lambda_1 K \alpha_1 \qquad (11.22)$$

Now it may be shown [166] that all interesting solutions of this equation are also solutions of

$$K\alpha_1 = M\lambda_1\alpha_1 \qquad (11.23)$$

whose solution is that α_1 is the principal eigenvector of K and so we may find the eigenvectors $\mathbf{e}_i$ of C using (11.19).

Nonlinear Kernels

Now we preprocess the data using $\Phi : \chi \to F$. So F is now the space spanned by $\Phi(\mathbf{x}_1), ..., \Phi(\mathbf{x}_M)$. The above arguments all hold and the eigenvectors of the dot product matrix $K_{ij} = \Phi(\mathbf{x}_i).\Phi(\mathbf{x}_j)$ may be found similarly. But now the *kernel trick*: provided we can calculate K, we don't need the individual terms $\Phi(\mathbf{x}_i)$.

And the above argument shows that any operation which can be defined in terms of dot products can be kernelised. We thus kernelise CCA.

11.3.2 Kernel Canonical Correlation Analysis

Consider mapping the input data to a high-dimensional (perhaps infinite-dimensional) feature space, F. Now,

$$\Sigma_{11} = E\{(\Phi(\mathbf{x}_1) - \mu_1)(\Phi(\mathbf{x}_1) - \mu_1)^T\}$$
$$\Sigma_{22} = E\{(\Phi(\mathbf{x}_2) - \mu_2)(\Phi(\mathbf{x}_2) - \mu_2)^T\}$$
$$\Sigma_{12} = E\{(\Phi(\mathbf{x}_1) - \mu_1)(\Phi(\mathbf{x}_2) - \mu_2)^T\}$$

where now $\mu_i = E(\Phi(\mathbf{x}_i))$ for $i = 1, 2$. Let us assume for the moment that the data has been centred in feature space (we actually use the same trick as [163] to centre the data later (see [114] for details)). Then we define

$$\Sigma_{11} = E\{\Phi(\mathbf{x}_1)\Phi(\mathbf{x}_1)^T\}$$
$$\Sigma_{22} = E\{\Phi(\mathbf{x}_2)\Phi(\mathbf{x}_2)^T\}$$
$$\Sigma_{12} = E\{\Phi(\mathbf{x}_1)\Phi(\mathbf{x}_2)^T\}$$

and we wish to find those values $\mathbf{w}_1$ and $\mathbf{w}_2$ which will maximise $\mathbf{w}_1^T \Sigma_{12} \mathbf{w}_2$ subject to the constraints $\mathbf{w}_1^T \Sigma_{11} \mathbf{w}_1 = 1$ and $\mathbf{w}_2^T \Sigma_{22} \mathbf{w}_2 = 1$.

In practice, we will approximate Σ_{12} with $\frac{1}{M} \sum_i \Phi(\mathbf{x}_{1i}) \Phi^T(\mathbf{x}_{2i})$, the sample average. At this stage we can see the similarity with our nonlinear CCA: if we consider an instantaneous hill-climbing algorithm, we would derive precisely our NLCCA algorithm for the particular nonlinearity involved.

Now $\mathbf{w}_1$ and $\mathbf{w}_2$ exist in the feature space which is spanned by

$$\{\Phi(\mathbf{x}_{11}), \Phi(\mathbf{x}_{12}), ..., \Phi(\mathbf{x}_{1M}), \Phi(\mathbf{x}_{21}), ..., \Phi(\mathbf{x}_{2M})\}$$

and therefore can be expressed as

$$\mathbf{w}_1 = \sum_{i=1}^{M} \alpha_{1i} \Phi(\mathbf{x}_{1i}) + \sum_{i=1}^{M} \alpha_{2i} \Phi(\mathbf{x}_{2i})$$

$$\mathbf{w}_2 = \sum_{i=1}^{M} \beta_{1i} \Phi(\mathbf{x}_{1i}) + \sum_{i=1}^{M} \beta_{2i} \Phi(\mathbf{x}_{2i})$$

With some abuse of the notation we will use $\mathbf{x}_i$ to be the ith instance from the set of data i.e. from either the set of values of $\mathbf{x}_1$ or from those of $\mathbf{x}_2$ and write

$$\mathbf{w}_1 = \sum_{i=1}^{2M} \alpha_i \Phi(\mathbf{x}_i)$$

$$\mathbf{w}_2 = \sum_{i=1}^{2M} \beta_i \Phi(\mathbf{x}_i)$$

Therefore substituting this in the criteria we wish to optimise, we get

$$(\mathbf{w}_1^T \Sigma_{12} \mathbf{w}_2) = \frac{1}{M} \sum_{k,i} \alpha_k . \Phi^T(\mathbf{x}_k) \Phi(\mathbf{x}_{1i}) \sum_l \beta_l \Phi^T(\mathbf{x}_{2i}) \Phi(\mathbf{x}_l) \qquad (11.24)$$

where the sums over i are to find the sample means over the data set. Similarly with the constraints, and so

$$\mathbf{w}_1^T \Sigma_{11} \mathbf{w}_1 = \frac{1}{M} \sum_{k,i} \alpha_k . \Phi^T(\mathbf{x}_k) \Phi(\mathbf{x}_{1i}) . \sum_l \alpha_l \Phi^T(\mathbf{x}_{1i}) \Phi(\mathbf{x}_l)$$

$$\mathbf{w}_2^T \Sigma_{22} \mathbf{w}_2 = \frac{1}{M} \sum_{k,i} \beta_k . \Phi^T(\mathbf{x}_k) \Phi(\mathbf{x}_{2i}) . \sum_l \beta_l \Phi^T(\mathbf{x}_{2i}) \Phi(\mathbf{x}_l)$$

Using $(K_1)_{ij} = \Phi^T(\mathbf{x}_i) \Phi(\mathbf{x}_{1j})$ and $(K_2)_{ij} = \Phi^T(\mathbf{x}_i) \Phi(\mathbf{x}_{2j})$ we then have that we require to maximise $\alpha^T K_1 K_2^T \beta$ subject to the constraints $\alpha^T K_1 K_1^T \alpha = 1$ and $\beta^T K_2 K_2^T \beta = 1$. Therefore if we define $\Gamma_{11} = K_1 K_1^T$, $\Gamma_{22} = K_2 K_2^T$ and $\Gamma_{12} = K_1 K_2^T$ we solve the problem in the usual way: by forming matrix

$K = \Gamma_{11}^{-\frac{1}{2}}\Gamma_{12}\Gamma_{22}^{-\frac{1}{2}}$ and performing a singular value decomposition on it as before to get

$$K = (\gamma_1, \gamma_2, ..., \gamma_k)D(\theta_1, \theta_2, ..., \theta_k)^T \qquad (11.25)$$

where γ_i and θ_i are again the standardised eigenvectors of KK^T and K^TK, respectively, and D is the diagonal matrix of eigenvalues[1].

Then the first canonical correlation vectors in feature space are given by

$$\alpha_1 = \Gamma_{11}^{-\frac{1}{2}}\gamma_1 \qquad (11.26)$$

$$\beta_1 = \Gamma_{22}^{-\frac{1}{2}}\theta_1 \qquad (11.27)$$

with subsequent canonical correlation vectors defined in terms of the subsequent eigenvectors, γ_i and θ_i.

Now for any new values $\mathbf{x}_1$, we may calculate

$$\mathbf{w}_1.\Phi(\mathbf{x}_1) = \sum_i \alpha_i \Phi(\mathbf{x}_i)\Phi(\mathbf{x}_1) = \sum_i \alpha_i K_1(\mathbf{x}_i, \mathbf{x}_1) \qquad (11.28)$$

which then requires to be centered as before. We see that we are again performing a dot product in feature space (it is actually calculated in the subspace formed from projections of $\mathbf{x}_i$).

There are three particular aspects of the above algorithm which should be pointed out:

1. The optimal weight vectors are vectors in a feature space which we may never determine. We simply calculate the appropriate matrices using the kernel trick, e.g. we may use Gaussian kernels so that

$$K_1(\mathbf{x}_{1i}, \mathbf{x}_{1j}) = \exp(-|\mathbf{x}_{1i} - \mathbf{x}_{1j}|^2) \qquad (11.29)$$

   which gives us a means of calculating K_{11} without ever having to calculate $\Phi(\mathbf{x}_{1i})$ or $\Phi(\mathbf{x}_{1j})$ explicitly.
2. The method requires a dot product between members of the data set $\mathbf{x}_1$ and $\mathbf{x}_2$, and therefore the vectors must be of the same length. Therefore, for example, for the exam data, we must discard one set of exam marks.
3. The method requires a matrix inversion and the data sets may be such that one data point may be repeated (or almost) leading to a singularity or badly conditioned matrices. One solution is to add noise to the data set; this is effective in the exam data set, and is a nice solution if we were to consider biological information processors but need not always work. An alternative is to add μI, where I is the identity matrix to Γ_{11} and Γ_{22} – a method which was also used in [131]. This gives robust and reliable solutions.

---

[1] This optimisation is applicable for all symmetric matrices (Theorem A.9.2, [125]).

11.3.3 Simulations

Artificial Data

We repeat the experiment on the data set from Section 11.1.1 (the noisy circular manifold and line), which allows us compare the linear CCA network, the nonlinear CCA network and the Kernel CCA method.

The results from a Kernel CCA with radial basis function (RBF) Gaussian kernels on the same data is shown in Fig. 11.5: the solid contours are contours of equal correlation (with points on the line) in the space defined by points on (or near) the circle; the dotted lines are contours of equal correlation (with points on the circle) in the space defined by points on or near the line. While the results are not so easily interpreted as before, we see that areas of high correlation tend to be local areas in each space.

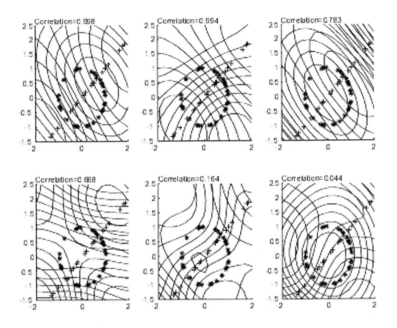

Fig. 11.5. The first six directions of principal correlation are shown here; all points on each contour have equal correlation with the points on the line. The solid lines show lines of equal correlation with the second data set; the dashed lines have points of equal correlation with the first data set.

Real Data

Our second experiment uses the data set comprising the 88 students' marks on five module exams discussed in Chapter 9. The correlations found by the kernel method are shown in Fig. 11.6; again we see that radial kernels give us larger maximal correlations within the data set than found with the linear CCA network. However, we have to add the caveat that it is entirely possible to achieve a correlation of 1 if we use radial kernels of infinite width. The nature of the correlations found here and the optimal kernel function are matters of ongoing research.

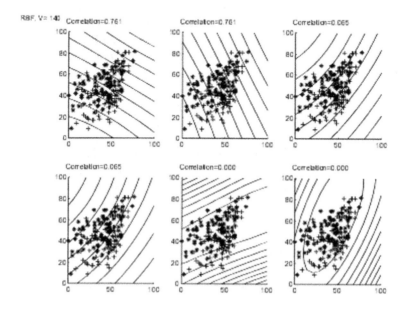

Fig. 11.6. The kernel canonical correlation directions found using radial kernels. The contour lines are lines of equal correlation. Each pair of diagrams shows the equal correlation contours from the perspective of one of the data sets.

11.3.4 ICA using KCCA

As we previously showed in Section 11.2, we may use nonlinear CCA to extract a single signal from a mixture of signals. We now use Kernel CCA for the same purpose.

Simulations

We repeat the ICA simulation in Section 11.2 in which a mixture of two sines is separated. The data set we use comprises three sinusoids in noise mixed linearly with a random mixing matrix; we keep the number of mixtures to two so that we do not stray too far from biological plausibility and so the second line of Fig. 11.7 has only two images. We use 80 samples and linear kernels. Each sample contains the mixture at time t and the mixture at times $t-1, ..., t-9$ so that a vector of 10 samples from the mixture is used at any one time. The underlying signals are shown in the top line of Fig. 11.7 and the correlation filtered data in the bottom line of that figure. The mixing matrix is shown at the top of the figure. We see that the three sinusoids have been separated with great accuracy. Similar results have been achieved with Gaussian and sigmoid kernels. There is a little beating in the higher frequency sinusoids which is not apparent when we separate two signals from two mixtures. This particular mixture was chosen since there is very little of the signal s_2 in the first mixture. This case is very readily treated with the technique of Minor Component Analysis (Chapter 3) but is much the most difficult case for KCCA.

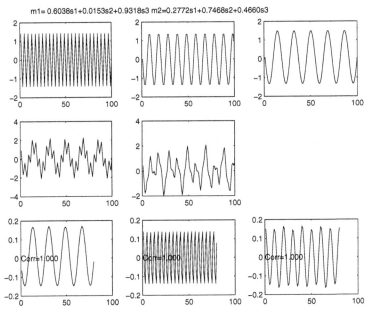

linear kernels, first, third and fifth

Fig. 11.7. The top line shows the three underlying sinusoids. The second line shows the two linear mixtures of these sinusoids presented to the algorithm. The bottom line shows three of the filters produced.

Nonlinear and Time-Varying Mixtures

We now consider one nonlinear mixture and one linear mixture of two sinusoids such as:

$$\text{Let } m_1 = s_1 + s_2$$
$$m_2 = (s_1 - s_2) * s_2$$

The results of linear KCCA are shown in Fig. 11.8. Again the signals are in the top line, the mixtures in the second line and the filtered data from the correlation process in the bottom line. We see that the two sinusoids have been found, but we have to sound a note of caution: the projections shown are on the second and seventh CCA directions. Each sinusoid appears in two filters (one π radians out of phase with the other), but even so, the second signal was not found until the seventh and eighth correlation directions. This finding of the signals in the lower-order filters is even more pronounced when we use Gaussian kernels; Fig. 11.9 shows the filtered data found by the nineth and tenth canonical correlation filters when we use Gaussian kernels.

The extraction of the individual sines becomes progressively harder the more nonlinear both mixtures become.

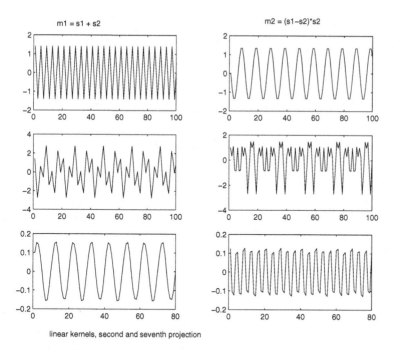

Fig. 11.8. The top line shows the underlying signals, the second line the data and the third line three kernel correlation projections when linear kernels are used.

Nonstationary mixtures are shown in Fig. 11.10. Now we create mixtures, m_1 and m_2, using

$$m_1 = 0.95 \sin(t/17)s_1 + 0.61 \sin(t/17)s_2$$

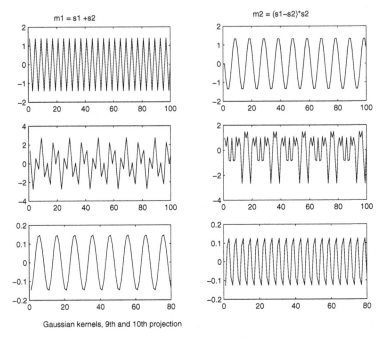

Gaussian kernels, 9th and 10th projection

Fig. 11.9. The top line shows the underlying signals, the second line the data and the third line three kernel correlation projections when rbf kernels are used.

$$m_2 = 0.23 \sin(t/17)s_1 + 0.49 \sin(t/17)s_2$$

where t is a parameter denoting time. The underlying frequencies are extracted by linear kernels (Fig. 11.10).

The rbf kernel, however, is also able to extract (Fig. 11.11) the mixing frequency in its 7th filter. This also happens when we use alternate sines and cosines (and with different frequencies) in our mixing matrix. The linear kernels also find the individual sines but do not find the mixing frequencies.

11.4 Relevance Vector Regression

The concept of Relevance Vectors was introduced by Tipping [177] in acknowledgement of McKay and Neal's contribution to the concept of Automatic Relevance Detection. The vectors found by the RVM method are prototypical vectors of the class types, which is a very different concept from the Support Vectors whose positions are always at the edges of clusters, thereby helping to delimit one cluster from another. In this chapter, we first review Relevance Vector Regression and then apply the concepts from this to unsupervised techniques which perform clustering. Relevance Vector Regression uses a dataset of input-target pairs $\{\mathbf{x}_i, t_i\}_{i=1}^N$. It assumes that the machine

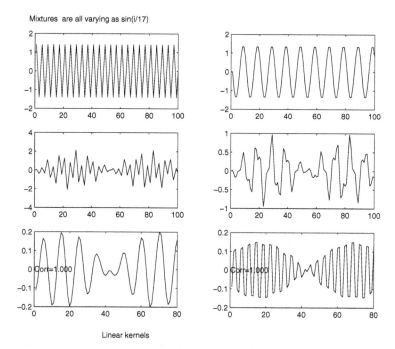

Fig. 11.10. The mixtures are time varying, (see text) but nevertheless the linear kernel method extracts the underlying frequencies. Each figure in the bottom line shows the amplitude of the signals varying as they do in the mixtures.

can form an output y from

$$y(\mathbf{x}) = \sum_{i=1}^{N} w_i K(\mathbf{x}, \mathbf{x}_i) + w_0 \qquad (11.30)$$

and $p(t|\mathbf{x})$ is Gaussian $N(y(\mathbf{x}), \sigma^2)$. The likelihood of the model is given by

$$p(\mathbf{t}|\mathbf{w}, \sigma) = \frac{1}{(2\pi\sigma^2)^{-\frac{N}{2}}} \exp\left\{-\frac{1}{2\sigma^2}\|\mathbf{t} - K\mathbf{w}\|^2\right\} \qquad (11.31)$$

where $\mathbf{t} = \{t_1, t_2, ..., t_N\}, \mathbf{w} = \{w_0, w_1, ..., w_N\}$, and K is the $N * (N + 1)$ design matrix. To prevent overfitting, an ARD prior is set over the weights

$$p(\mathbf{w}|\alpha) = \prod_{i=0}^{N} N(0, \alpha^{-1}) \qquad (11.32)$$

To find the maximum likelihood of the data set with respect to α and σ^2, we iterate between finding the mean and variance of the weight vector and then calculating new values for α and σ^2 using these statistics. We find that many

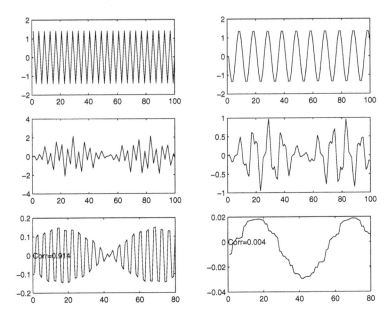

Fig. 11.11. The Gaussian kernel also finds the signals but also finds the underlying slowly changing mixing process in its seventh projection.

of the α_i tend to infinity, which means that the corresponding weights tend to 0. In detail, we have that the posterior of the weights is given by

$$p(\mathbf{w}|\mathbf{t},\alpha,\sigma^2) \propto |\Sigma|^{-\frac{1}{2}} \exp\left\{-\frac{1}{2}(\mathbf{w}-\mu)^T\Sigma^{-1}(\mathbf{w}-\mu)\right\} \qquad (11.33)$$

where

$$\Sigma = (K^TBK + A)^{-1}$$
$$\mu = \Sigma K^TB\mathbf{t} \qquad (11.34)$$

with $A = \mathrm{diag}(\alpha_0, \alpha_1, ..., \alpha_N)$ and $B = \sigma^{-2}I_N$.

If we integrate out the weights, we obtain the marginal likelihood

$$p(\mathbf{t}|\alpha,\sigma^2) \propto |B^{-1} + KA^{-1}K^T|^{-\frac{1}{2}} \exp\left\{-\frac{1}{2}\mathbf{t}^T(B^{-1} + KA^{-1}K^T)^{-1}\mathbf{t}\right\}$$
$$(11.35)$$

which can be differentiated to give at the optimum,

$$\alpha_i^{\mathrm{new}} = \frac{\gamma_i}{\mu_i^2} \qquad (11.36)$$

$$(\sigma^2)^{\mathrm{new}} = \frac{||\mathbf{t} - \mathbf{K}\mu||^2}{N - \sum_i \gamma_i} \qquad (11.37)$$

where $\gamma_i = 1 - \alpha_i\Sigma_{ii}$.

11.4.1 Application to CCA

We must first describe CCA in probabilistic terms. Consider two data sets $\Theta_1 = \{\mathbf{x}_1^i, i \in 1, \cdots, N\}$ and $\Theta_2 = \{\mathbf{x}_2^i, i \in 1, \cdots, N\}$ defined by probability density functions $p_1(\mathbf{x}_1)$ and $p_2(\mathbf{x}_2)$. Let there be some underlying relationship so that $y_1 = \mathbf{w}_1^T \mathbf{x}_1 + \epsilon_1$ is the canonical correlate corresponding to $y_2 = \mathbf{w}_2^T \mathbf{x}_2 + \epsilon_2$. Then y_1 can be used to predict the value of y_2, and vice versa. Let $y_1 = \rho y_2 + e_1$, where ρ is the correlation coefficient.

Then from the perspective of $\mathbf{x}_1$, the targets, $\mathbf{t}_1$ is given by the other input $\mathbf{x}_2$. i.e.

$$\mathbf{t}_1 = \mathbf{w}_2 K_2 \tag{11.38}$$

So we are using the current value of $\mathbf{w}_2$ to determine the target for updating the posterior probabilities for $\mathbf{w}_1$. Similarly, we create a target for updating the probabilities for $\mathbf{w}_2$ using the current value of $\mathbf{w}_1$. We then simply alternate between the two Relevance Vector Machines in a way which is reminiscent of the EM algorithm: from $\mathbf{x}_1$'s perspective, calculation of the new values of $\mathbf{w}_2$ corresponds to the E-step, while the calculation of the new values of $\mathbf{w}_1$ corresponds to the M-step and vice versa from $\mathbf{x}_2$'s perspective.

We have carried out experiment on the artificial data sets which have been used in Section 11.1.1. Thus, $\mathbf{x}_1 = \{x_{11}, x_{12}\}$ lies on or near the circular manifold while $\mathbf{x}_2 = \{x_{21}, x_{22}\}$ lies on or near the line. [115] have shown that the kernel method can find greater than linear correlations in such a data set. We now report that the current Relevance Vector Kernel method also finds greater than linear correlations but with a very sparse representation generally. Typically the resulting vectors will have zeros in all but one position with the single nonzero value being very strong. Occasionally, one vector, e.g. $\mathbf{w}_1$ will have a single nonzero value whose correlation with all the other points will be seen in its vector $\mathbf{w}_2$. However, in all cases we find a very strong correlation. Fig. 11.12 shows the outputs of a trained CCA-RVM network; the high correlation between y_1 and y_2 is clear. We may also use the CCA-RVM network to find stereo correspondences as before [114]; however, this task is relatively easy and does not use the full power of the CCA-RVM network.

11.5 Appearance-Based Object Recognition

For recognition of three-dimensional objects in two-dimensional gray-level images there exist two main approaches in computer vision: based on the result of a segmentation process or directly on the object's appearance. The segmentation approaches suffer from two disadvantages: segmentation errors and loss of information contained in the image caused by the segmentation.

Appearance-based approaches, in contrast, avoid these disadvantages. They use the image data, i.e. the pixel intensities, directly without a previous segmentation process. The simplest method is correlation of an image with an object template.

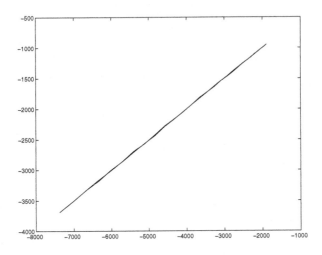

Fig. 11.12. The figure shows a graph of y_1 (horizontal axis) against y_2 (vertical axis) for a Relevance Vector CCA network trained on the negative distance kernel. The high correlation is obvious.

The aim of the present object-recognition system is pose estimation and classification of a rigid three-dimensional object from a two-dimensional gray-level image. Generally for this task, there are three degrees of freedom for the rotation $\phi = (\phi_x, \phi_y, \phi_z)^T$ and three for the translation $t = (t_x, t_y, t_z)^T$, and the transformation can be split into the internal transformation inside the image plane with $t_{int} = (t_x, t_y)^T$ and $\phi_{int} = \phi_z$ and the external transformations orthogonal to the image plane with $t_{ext} = t_z$ and $\phi_{ext} = (\phi_x, \phi_y)$. In the internal transformations, the object only changes its position in the image, whereas for the external transformations, the object varies its appearance. In this work, the objects were put on a turntable and the camera was fixed, so the distance of the object to the camera was fixed, i.e. $t_z = 0$, and we only have one external rotation ϕ_y, i.e. $\phi_x = 0$.

Columbia Object Image Library (COIL-20) is a database of gray-scale image of 20 objects (see Fig. 11.13). The objects were placed on a motorized turntable (see Fig. 11.14) against a black background. The turntable was rotated through 360 degrees to vary object pose with respect to a fixed camera. Images of the objects were taken at pose intervals of 5 degrees. This corresponds to 72 images per object. The images used have been size normalised. The object is clipped out from the black background using a rectangular bounding box. The bounding box is resized to 128 x 128 using interpolation decimation filters to minimise aliasing [136]. When resizing, the aspect ratio is preserved.

Fig. 11.13. Twenty objects in the Columbia Object Image Library(COIL-20). *Reprinted with permission from COIL.*

In this experiment, we used one set of 128 x 128 gray-scale images of toy objects. Each image set contained 72 images formed by rotating the object through 72 poses in 5 degrees steps; an example is shown in Fig. 11.15. Let $X = x_i, 1 \leq i \leq 72$ denote the set of images and $Y = y_i, y_i \in [0, 355]$ be the corresponding pose parameters (the camera position in degrees).

We performed kernel-CCA on the original image set and scalar output parameters using a RBF-kernel with $\sigma = 2$. Fig. 11.16 shows projections of all images onto the first canonical vector. The experiment indicates that kernel-CCA can be used to find optimal basis functions automatically.

Why did we not report results from the Relevance Vector Method on the real data? We found that the level of sparsity found by the relevance vector method on the real data was such that it precluded finding any useful canonical correlations in this data set.

Tipping [177] does state that one advantage of his method is that it gives a much sparser representation than even Support Vector Machines do. We have to report that this is a useful trait on our artificial data (in which we can always generate more data points) but one which is an encumbrance on the real data set: typically, in our Relevance vector implementation of CCA, we

Fig. 11.14. The objects were placed at the center of a motorized turntable. The turntable was rotated through 360 degrees. An image was acquired with a fixed camera at every 5 degrees of rotation. *Reprinted with permission from COIL.*

found that the CCA directions were typically a function of only a single data point, which may be useful in specific instances but is less useful in general on the wider data set.

Thus, our overall conclusion is that Kernel CCA is a useful method on general data sets and provides sufficient sparsity (but not too much) of itself to be useful for the task for which it was developed.

11.6 Mixtures of Linear Correlations

We now go on to demonstrate a locally linear version of the CCA network, using a lateral matrix of connections to order the linear correlations.

11.6.1 Many Locally Linear Correlations

In Section 9.3.6, we maximised

$$J = E\{(\mathbf{y}_1^T \mathbf{y}_2) + \Lambda_1(\mathbf{y}_1 \mathbf{y}_1^T - I) + \Lambda_2(\mathbf{y}_2 \mathbf{y}_2^T - I)\} \tag{11.39}$$

with $\Lambda_i, i = 1, 2$ a matrix of Lagrange multipliers, to get the learning rules

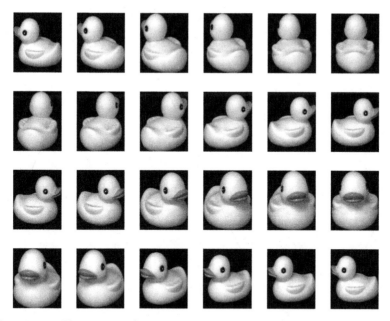

Fig. 11.15. The turning object image. *Reprinted with permission from COIL.*

$$\Delta W_1 = \eta \mathbf{x}_1 (\mathbf{y}_2 - \Lambda_1 \mathbf{y}_1)^T \qquad (11.40)$$

$$\Delta \Lambda_1 = \eta_0 (\mathbf{y}_1 \mathbf{y}_1^T - k_1 I) \qquad (11.41)$$

with W_1 the $m \times n$ matrix of weights connecting $\mathbf{x}_1$ to $\mathbf{y}_1$. Similarly with W_2, Λ_2. Now we will be interested to find a set of correlations which show gradual change over the set. We use ideas from Kohonen's [111] Self Organising Maps to create a topology-preserving mapping from the input data to the correlation finding output neurons. We do this by substituting

$$\Delta \Lambda_1 = \eta_0 (\mathbf{y}_1 \mathbf{y}_1^T - A) \qquad (11.42)$$

for (11.41) where A is a matrix whose largest values are round its main diagonal and which tends to zero away from that diagonal. Typically we use

$$A_{i,j} = \exp(-(i-j)^2/\sigma) \qquad (11.43)$$

Equation (11.42) may be derived from maximisation of

$$J = E\{(\mathbf{y}_1^T \mathbf{y}_2) + \Lambda_1 (\mathbf{y}_1 \mathbf{y}_1^T - A) + \Lambda_2 (\mathbf{y}_2 \mathbf{y}_2^T - A)\} \qquad (11.44)$$

11.6.2 Stone's Data

In the following experiments, we have used an artificial data set previously used by James Stone [174]; a description of Stone's method is found in Appendix B. This data simulates a moving surface with a slowly varying depth.

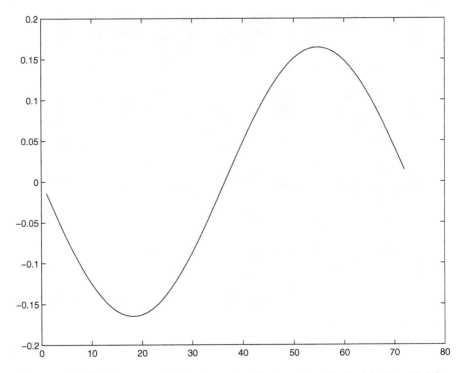

Fig. 11.16. The first canonical vector using the kernel method of Section 11.3.2.

The disparity between the left and right images of a stereo pair was generated by convolving a circular array of 1000 uniformly distributed random numbers with a Gaussian function, and then normalising these values to lie between ± 1. For example, consider an array, $\mathbf{r}$ of 1000 random numbers, and a 1000 $\times$ 1000 matrix $\mathbf{G}$ with values given by:

$$G_{a,b} = \frac{1}{\sigma}e^{\frac{-(\min(|a-b|,1000-|a-b|))^2}{2\sigma^2}} \tag{11.45}$$

The min( ) function in (11.45) ensures that the Gaussian wraps around the $\mathbf{r}$ array, which effectively makes it circular. Then, the un-normalised array of disparity values, $\mathbf{d}$, is given by the product in (11.46). Fig. 11.17 shows an example of an array of 1000 disparity values generated using this method.

$$d = Gr \tag{11.46}$$

One of the arrays of pixels (e.g. the left) is generated by randomising an array, $\mathbf{b}$, of 10 000 boolean values, each of which has a probability of $\frac{1}{6}$ of being 1. For each pixel, the mean of ten Boolean values in each group is then used as the magnitude of the pixel value. Subsequent pixels are given by the average

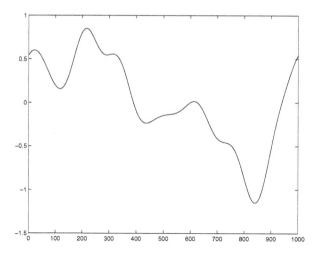

Fig. 11.17. Example of an array of 1000 disparity values generated using the method described above.

of nonoverlapping groups of Boolean values (11.47). To generate the other set of pixels (i.e. the right set), the same procedure is repeated, but this time, the positions of the mean values taken from the **b** array depend on the array of disparity values, **d**. The disparity values determine the shift along the **b** array that is different from the positions for the corresponding left pixels (11.48). For example, for a disparity value of 0.5, the right pixel would be taken over an average of ten random bits, five of which are also used to generate the left pixel.

$$x_i^{(1)} = \frac{1}{10} \sum_{j=10i}^{10i+9} \mathbf{b}_j \qquad (11.47)$$

$$x_i^{(2)} = \frac{1}{10} \sum_{j=10i+10d}^{10i+d_i+9} \mathbf{b}_j \qquad (11.48)$$

This procedure is illustrated in Fig. 11.18, which shows how the pairs of pixels are generated from the single array of random bits.

The locally linear CCA network was trained using this data set of 1000 pairs of pixels, with three left inputs and three right inputs. In each training iteration, a random position in the data set was chosen and both sets of inputs were taken from the same position in the left and right arrays. There is no reason to choose three-dimensional inputs over any other dimensionality since each input receives data with the same statistics. We therefore expect that all weights from a single output node should be similar (11.49), regardless of which input node it is connected to since all inputs receive the same data set, and therefore they have the same statistics:

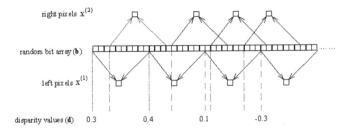

Fig. 11.18. Generation of Stone's data. The left pixels are taken from the average of sequential blocks of 10 random bits. The right pixels are generated in a similar way, but with disparity shifts implemented by shifts along the array of random bits.

$$w_{11}^i \approx w_{12}^i \approx w_{13}^i \approx w_{14}^i, \cdots \qquad (11.49)$$

Instead, the network should find relationships between the two data sets and therefore weights connecting a single output to two different modules will not necessarily be similar, for example, $\mathbf{w}_{11}^i$ and $\mathbf{w}_{21}^i$. This expectation was confirmed experimentally, as can be seen from the results in Fig. 11.19. This figure shows a plot of the weights of the network after 100 000 training iterations.

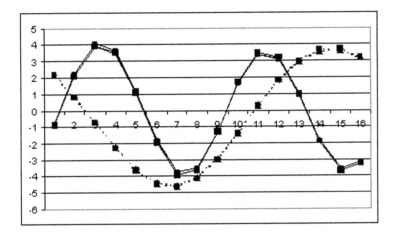

Fig. 11.19. The weights of the network after 100 000 training iterations. The horizontal axis shows the module number and the vertical axis shows the weight magnitude. The solid lines show the weights connected to the $\mathbf{y}_1^i$ outputs and the dotted lines indicate those connected to the $\mathbf{y}_2^i$ outputs.

The lateral matrix included wrap-around i.e. the modules of the network were considered to be in a ring, and node 1 was adjacent to nodes 2 and 16:

$$A_{a,b} = e^{\frac{\min(a-b,N-(a-b))^2}{-2\sigma^2}} \qquad (11.50)$$

Perhaps the most important finding to point out is that there is a smooth ordering of the modules i.e. the weights of neighbouring modules are similar, and we have successfully used the lateral weights to order the responses from the network. This smooth ordering has taken the form of two wave patterns: one formed by the y_1 vector and the other formed by y_2. Fig. 11.20 shows

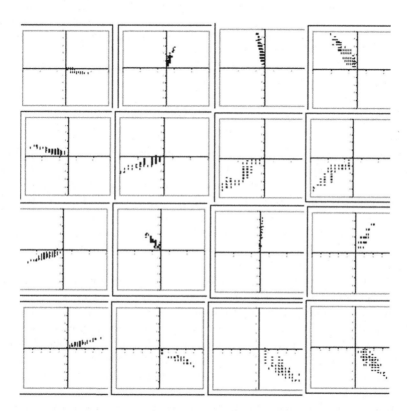

Fig. 11.20. Scatter-plots of each of the 16 pairs of outputs. After feeding forward to all modules, the responses of each module have similar orientation, which indicates the correlation found. The correlations found by the network are smoothly ordered as a result of using the lateral connections.

scatter-plots of the outputs of each of the modules, where the y_1-values are plotted on the horizontal axis and y_2 is plotted vertically. Each of the plots clearly shows an orientation, which indicates the correlation found by that particular module. The smooth ordering can clearly be seen as neighbouring modules show similar orientations to each other. Occasionally we get twists in the mapping so that, for example, the orientation may change gradually in an anticlockwise direction, but then it swings back into a clockwise rotation.

This may cause separate parts of a mapping to overlap, and this is a common (though undesirable) finding in topology-preserving maps such as the SOM [111] and the GTM [14].

The random dot stereogram data set has only positive values and therefore, only positive correlations can exist in the data. However, some modules of the network show anticorrelation between their outputs (see Fig. 11.20) which is clearly the opposite of what the objective function should achieve. This is an advantage of using wrap around in the lateral matrix, since the network has found a "ring" of local correlations and anti-correlations. If some anticorrelated data (i.e. positive and negative inputs simultaneously) were to be fed forward, the outputs of these modules would be correlated, satisfying our objective of causing module outputs to become correlated for local subsets of the entire data space.

11.6.3 Discussion

In this chapter, we have introduced a nonlinearity to our neural implementations of CCA. The nonlinear CCA rules were shown to find correlations greater in magnitude than could be found by linear CCA. We have also introduced kernel CCA and shown that again it can find correlations greater than linear correlations.

However, we should add the caveat that it is entirely possible to create correlations using these methods, and any use of such methods should be accompanied with a strong dose of common sense.

There is a case to be made that Kernel CCA and NLCCA are the same type of operation in that a kernel can be created out of many real-valued (nonlinear) functions; we have, for example, shown that both can be used for Independent Component Analysis. However the kernel approach seems to offer a new means of finding such nonlinear correlations and one which is very promising for future researchers. We also derived a new method for doing this based on relevance vectors. We have illustrated both methods on artificial data and the original kernel method on a real problem involving estimating pose from a set of images. The method based on relevance vectors seems to create too sparse a representation in many cases and we therefore find that the original kernel method is to be preferred on real data analysis tasks.

Finally, we note that local kernels may be developed which will find local correlations. This again suggests a strong tie-up between the mixture of linear CCA networks and the kernel CCA.

12

Exploratory Correlation Analysis

In this chapter, we present a neural method capable of extracting features from different data sources and combining these to form a jointly sparse coding. The difference between this and the statistically based dual stream neural architectures discussed in previous chapters is that the method that we propose is capable of searching for shared higher-order structure between data streams.

The group of methods which we derive in this chapter are all biologically plausible in that they are extensions of simple Hebbian learning and if used in the negative feedback framework, can all be implemented with locally learning connections. Nevertheless, they will go beyond the second-order methods of earlier chapters in that they will identify structure which is shared between two data streams and do so in a way which ignores the same (or, as we shall see, similar in type but even more pronounced) structure in a single data stream. This is a useful feature for biological information processing since

1. the most interesting structure in e.g. visual data tends to be identifiable with higher-order statistics, while
2. noise is typically "shot noise" which is very kurtotic, but due to this is unlikely to be present in more than one data stream at a time.

We will show that on real data our networks find sparsely firing filters of the data which are also found in biological networks.

12.1 Exploratory Correlation Analysis

The model we discuss here is an extension of the neural EPP algorithm (Chapter 6). We assume two data streams from which we wish to extract the common interesting features, i.e. both streams are assumed to have a set of common underlying factors, which are characterised by being non-Gaussian. Mathematically we can write this model as

$$\mathbf{y}_1 = W^T \mathbf{x}_1$$
$$\mathbf{y}_2 = V^T \mathbf{x}_2$$

The input streams are denoted by $\mathbf{x}_1$ and $\mathbf{x}_2$, the projected data by $\mathbf{y}_1$ and $\mathbf{y}_2$ and the basis vectors are the rows of the matrices W and V. Each input stream can be analysed separately by performing EPP and finding common statistical features that have maximum non-Gaussianity. However, if we know that the features we are looking for in both data sets have the same statistical structure, we can add another constraint which maximises the dependence between the outputs. This is depicted schematically in Fig. 12.1.

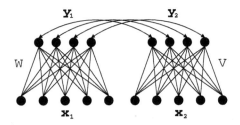

Fig. 12.1. The Exploratory Correlation Analysis network.

The simplest way to express this formally is by maximising $E(\mathbf{g}(\mathbf{y}_1)^T \mathbf{g}(\mathbf{y}_2))$. Additionally, we need to ensure the weights do not grow without bound, which we can achieve by adding weight constraints $W^T W = A$ and $V^T V = B$. Writing this as an energy function with Lagrange parameters $\lambda_{i,j}$ and $\mu_{i,j}$ we obtain

$$J(W, V) = E(\mathbf{g}(W^T \mathbf{x}_1)^T \mathbf{g}(V^T \mathbf{x}_2)) + \frac{1}{2} \sum_{i=1}^{N} \sum_{j=1}^{N} \lambda_{ij}(\mathbf{w}_i^T \mathbf{w}_j - a_{ij})$$

$$+ \frac{1}{2} \sum_{i=1}^{N} \sum_{j=1}^{N} \mu_{ij}(\mathbf{v}_i^T \mathbf{v}_j - b_{ij}) \tag{12.1}$$

where $\mathbf{v}_i$ and $\mathbf{w}_i$ are column vectors taken from the columns of matrices V and W, respectively.

The energy function (12.1) can be differentiated with respect to the weights v_{ij} and w_{ij} and Lagrange parameters, λ_{ij} and μ_{ij}. Setting the derivatives of (12.1) to 0, we obtain:

$$\frac{\partial(J(W, V))}{\partial W} = E(\mathbf{x}_1(\mathbf{g}(\mathbf{y}_2) \otimes \mathbf{g}'(\mathbf{y}_1))^T) + W\Lambda = 0 \tag{12.2}$$

$$\frac{\partial(J(W, V))}{\partial V} = E(\mathbf{x}_2(\mathbf{g}(\mathbf{y}_1) \otimes \mathbf{g}'(\mathbf{y}_2))^T) + VM = 0 \tag{12.3}$$

$$\frac{\partial(J(W,V))}{\partial \Lambda} = W^T W - A = 0 \qquad (12.4)$$

$$\frac{\partial(J(W,V))}{\partial M} = V^T V - B = 0 \qquad (12.5)$$

where Λ is a matrix consisting of the Lagrangian parameters λ_{ij}, M contains the Lagrangians μ_{ij} and A and B contain parameters, a_{ij} and b_{ij}, respectively. The $\otimes$ operator is defined as the elementwise multiplication of two vectors. The Lagrange multipliers can be calculated by premultiplying (12.2) and (12.3) by W^T and V^T, respectively. Inserting (12.4) and (12.5) results in:

$$\Lambda = -A^{-1} W^T E((\mathbf{g}(\mathbf{y}_2) \otimes \mathbf{g}'(\mathbf{y}_1)) \mathbf{x}_1^T)$$
$$M = -B^{-1} V^T E((\mathbf{g}(\mathbf{y}_1) \otimes \mathbf{g}'(\mathbf{y}_2)) \mathbf{x}_2^T)$$

Reinserting these optimal Lagrange parameters into (12.2) and (12.3) yields:

$$\frac{\partial(J(W,V))}{\partial W} = E(\mathbf{x}_1(\mathbf{g}(\mathbf{y}_2) \otimes \mathbf{g}'(\mathbf{y}_1))^T) - W^T A^{-1} W E(\mathbf{x}_1(\mathbf{g}(\mathbf{y}_2) \otimes \mathbf{g}'(\mathbf{y}_1))^T)$$

$$\frac{\partial(J(W,V))}{\partial V} = E(\mathbf{x}_2(\mathbf{g}(\mathbf{y}_1) \otimes \mathbf{g}'(\mathbf{y}_2))^T) - V^T B^{-1} V E(\mathbf{x}_2(\mathbf{g}(\mathbf{y}_1) \otimes \mathbf{g}'(\mathbf{y}_2))^T)$$

We typically set the A and B matrices to the identity matrix, which causes the weights W and V to converge to orthonormal weight matrices. By taking the instantaneous gradient of this and implementing a stochastic gradient descent we may derive the following weight update rules:

$$\Delta W = \eta[(\mathbf{x}_1 - W\mathbf{y}_1)(\mathbf{g}(\mathbf{y}_2) \otimes \mathbf{g}'(\mathbf{y}_1))^T] \qquad (12.6)$$
$$\Delta V = \eta[(\mathbf{x}_2 - V\mathbf{y}_2)(\mathbf{g}(\mathbf{y}_1) \otimes \mathbf{g}'(\mathbf{y}_2))^T] \qquad (12.7)$$

Equations (12.6) and (12.7) may be implemented in a negative feedback framework. In both equations, the input is fed forward, the outputs are fed back and subtracted from the input to form a residual. The weight update consists of multiplying the residual with functions of the values of both output streams.

Because of its origins in Exploratory Projection Pursuit and the simultaneous effect of searching for correlations in datastreams, we call this method Exploratory Correlation Analysis (ECA). The ECA network uses the third or fourth moments of the data sets to search for the shared higher order structure.

As with the neural EPP algorithm, it is convenient to replace the output functions with stable versions for the ECA algorithm. In contrast to the neural EPP algorithm, we not only require the derivative of the function to be maximised, but also the function itself. For kurtosis, we therefore need an additional stable function, whose truncated Taylor expansion is $g(y) = y^4$. The function we chose for the experiments in this chapter is $g(y) = \mu - \mu \exp(-\frac{1}{\mu} y^4)$ (see Fig. 12.2).

The parameter μ controls how accurately this expression approximates y^4. The greater μ is, the more accurately does the function represent the analytically derived y^4, but the greater the maximum of the function and thus the more unstable the ECA algorithm becomes. To examine the behaviour of the function in more detail we can expand $g(y)$ around 0:

$$g(y) = \mu - \mu \exp\left(-\frac{1}{\mu}y^4\right) = y^4 - \frac{y^8}{2!\mu} + \frac{y^{12}}{3!\mu^2} - \ldots \qquad (12.8)$$

Around $y = 0$, the dominant term after y^4 is $\frac{1}{2!\mu}y^8$ and therefore we can state that $g(y)$ approximates y^4 sufficiently, when $y^4 > \frac{1}{2!\mu}y^8$. This also makes it clear that as μ increases $\frac{1}{2!\mu}y^8$ decreases and that $g(y)$ approximates y^4 more accurately. It might appear at first sight that we are discussing the approximation wrongly: it is more usual to be approximating a more complex function with a simpler function (often by discarding the higher-order terms of a Taylor series). However in this case our theoretically derived function is a very simple function, y^4, which unfortunately is not robust to outlier terms. Thus, we are, in effect, adding in the higher-order terms to tame the simple but fragile function.

To choose a good value of μ, it is important that the approximation is valid for most of the data used. As our data is sphered, 95% of the data lies between -1.93 and 1.93 and therefore, $y^4 > \frac{1}{2!\mu}y^8$ must hold for these values, which is true for $\mu = 6.94$.

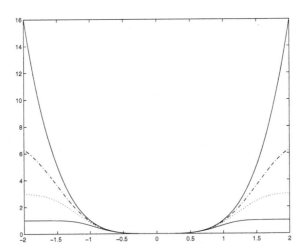

Fig. 12.2. Approximation of y^4 (solid line) with $\mu - \mu \exp(-\frac{1}{\mu}y^4)$ for $\mu = 1$ (dashed line), $\mu = 3$ (dotted line) and $\mu = 7$ (dot-dashed line)

12.2 Experiments

We begin with a simple experiment on artificial data to test the network. The artificial data set was generated from a mixture of kurtotic and normal data sources. We then go on to use the network for Indpendent Component Analysis.

12.2.1 Artificial Data

The inputs to the network are two three-dimensional input vectors as shown in Table 12.1. We used three types of data source, each with a different kurtosis value. Sources S_1 and S_2 were independently generated from a kurtotic source with kurtosis $= 20$ and source S_3 was generated from a kurtotic source with kurtosis $= 3$. Thus, the common data source S_3 is much less kurtotic than source S_1 or S_2. The last data source we used is S_4, which was taken from a normal distribution and is also shared by the two inputs. The variances of all sources were normalised to 1. In order to show the robustness of the network we added zero mean Gaussian noise with variance 0.25 to each of the inputs independently, which causes the correlations which exist between $\mathbf{x}_1$ and $\mathbf{x}_2$ to decrease.

Table 12.1. Artificial data set. S_1 and S_2 are more kurtotic than the common source S_3. S_4 is a common normal data source.

Input 1	Input 2	Kurtosis
$x_{11} = S_1 + N(0, 0.25)$	$x_{21} = S_2 + N(0, 0.25)$	20
$x_{12} = S_3 + N(0, 0.25)$	$x_{22} = S_3 + N(0, 0.25)$	3
$x_{13} = S_4 + N(0, 0.25)$	$x_{23} = S_4 + N(0, 0.25)$	0

The network was trained for 50 000 iterations with a learning rate of $\eta = 0.003$, which was annealed to 0. The weights converged to the values shown in Table 12.2. The network has clearly identified the common kurtotic data source and has ignored the common normal input and the independent input sources S_1 and S_2, although they are more kurtotic than S_3, thus identifying the shared higher-order structure in the data. A single stream EPP network is drawn to the more kurtotic sources S_1 and S_2.

Table 12.2. Weight vectors after training the ECA network on the artificial data of Table 12.1.

$\mathbf{w}$	0.0029	**1.0000**	0.0028
$\mathbf{v}$	0.0043	**1.0000**	-0.0182

12.2.2 Dual Stream Blind Source Separation

In this section, we describe an experiment which is an adaption of the blind source separation problem [97]. Using recordings from multiple microphones throughout a room, we try to isolate the voices of the individual speakers.

We assume that we can model the set of observed signals $\mathbf{x}$ as a mixture of unknown sources $\mathbf{s}$. These sources are mixed by an unknown mixing matrix A so that $\mathbf{x} = A\mathbf{s}$. The goal is to find the unmixing matrix W, so that we can recover the unknown sources, so that $\mathbf{s} = W\mathbf{x}$. Neural EPP has been used to find the unmixing matrix (Chapter 6).

Our adaptation of the original blind source separation problem uses two sets of inputs instead of one, which are both different linear mixtures of the same source signals. We used mixtures of three source signals, which were created artificially by randomly taking samples from a kurtotic source, with kurtosis $= 3$.

The mixing matrices, A and B, were randomly chosen; examples are:

$$A = \begin{pmatrix} 2\ 5\ 1 \\ 5\ 2\ 9 \\ 9\ 2\ 3 \end{pmatrix}, \qquad B = \begin{pmatrix} 3\ 6\ 1 \\ 9\ 4\ 7 \\ 1\ 5\ 3 \end{pmatrix}$$

The two sets of input signals were obtained by multiplying the source signals by A and B. The input data sets were subsequently sphered with sphering matrices $Q_1 = \Sigma_1^{-\frac{1}{2}}$ and $Q_2 = \Sigma_2^{-\frac{1}{2}}$, where Σ_1 and Σ_2 are the covariance matrices of the input streams. We trained the ECA network using a deflationary version of the neural ECA network, similar to Sanger's Generalised Hebbian Architecture (Chapter 2).

$$y_{1j} = \sum_i w_{ij} x_{1i}, \forall j \tag{12.9}$$

$$\Delta w_{ij} = \eta g'(y_{1j}) g(y_{2j}) \left(x_{1i} - \sum_{k=1}^{j} w_{ik} y_{1k} \right) \tag{12.10}$$

$$y_{2j} = \sum_i v_{ij} x_{2i}, \forall j \tag{12.11}$$

$$\Delta v_{ij} = \eta g'(y_{2j}) g(y_{1j}) \left(x_{2i} - \sum_{k=1}^{j} v_{ik} y_{2k} \right) \tag{12.12}$$

The network was trained for 150 000 iterations, using a learning rate of $\eta = 0.003$. To show the unmixing properties of the network, we examine the combined effect of the mixing, sphering and unmixing operations and therefore we display the product of the matrices WQ_1A and VQ_2B. In the ideal case, the combined effect of these matrices should be a permutation matrix:

$$WQ_1A = \begin{pmatrix} -0.0013 & -\mathbf{1.0006} & -0.0006 \\ \mathbf{1.0002} & 0.0311 & -0.0038 \\ -0.0021 & -0.0036 & \mathbf{1.0000} \end{pmatrix}$$

$$VQ_2B = \begin{pmatrix} -0.0021 & -\mathbf{1.0006} & -0.0017 \\ \mathbf{1.0002} & 0.0313 & -0.0031 \\ -0.0021 & -0.0036 & \mathbf{1.0000} \end{pmatrix}$$

As we can see, each row of these matrices contains one value close to one or negative one and the rest of the values are close to zero. We can also see that the positions of the ones (or negative ones) are the same for both matrices, indicating that the sources have been unmixed pairwise.

12.3 Connection to CCA

The one unit linear version of the ECA network is closely related to classical CCA, i.e. when the function $\mathbf{g}(\mathbf{y}) = \mathbf{y}$ is used in the weight update rules (12.6) and (12.7). This can be made more clear when we consider the situation where the network is fully converged so that the expected change in weights is zero [73]. Consider a single pair of output neurons, y_1 and y_2:

$$E(\Delta \mathbf{w}) = E(\eta(\mathbf{x}_1 - \mathbf{w}y_1)y_2)$$
$$= \eta E(\mathbf{x}_1 \mathbf{x}_2^T \mathbf{v} - \mathbf{w}y_1 y_2) = 0 \tag{12.13}$$
$$E(\Delta \mathbf{v}) = E(\eta(\mathbf{x}_2 - \mathbf{v}y_2)y_1)$$
$$= \eta E(\mathbf{x}_2 \mathbf{x}_1^T \mathbf{w} - \mathbf{v}y_1 y_2) = 0 \tag{12.14}$$

Writing $E(y_1 y_2)$ as λ, $E(\mathbf{x}_1 \mathbf{x}_2^T)$ as $\Sigma_{1,2}$ and $E(\mathbf{x}_2 \mathbf{x}_1^T)$ as $\Sigma_{2,1}$ we obtain

$$\Sigma_{1,2}\mathbf{v} = \lambda \mathbf{w} \tag{12.15}$$
$$\Sigma_{2,1}\mathbf{w} = \lambda \mathbf{v} \tag{12.16}$$

Substituting (12.15) in (12.16) and vice versa:

$$\Sigma_{1,2}\Sigma_{2,1}\mathbf{w} = \lambda^2 \mathbf{w} \tag{12.17}$$
$$\Sigma_{2,1}\Sigma_{1,2}\mathbf{v} = \lambda^2 \mathbf{v} \tag{12.18}$$

When the network is stable, the weight vectors will therefore be eigenvectors of $\Sigma_{1,2}\Sigma_{2,1}$ and $\Sigma_{2,1}\Sigma_{1,2}$. Classical CCA, however, requires the solutions to be eigenvectors of $\Sigma_{1,1}^{-1}\Sigma_{1,2}\Sigma_{2,2}^{-1}\Sigma_{2,1}$ and $\Sigma_{2,2}^{-1}\Sigma_{2,1}\Sigma_{1,1}^{-1}\Sigma_{1,2}$, which means that the ECA network is capable of performing CCA, if the data sources $\mathbf{x}_1$ and $\mathbf{x}_2$ are sphered prior to training the network. If we premultiply the data streams by $\Sigma_{1,1}^{-1/2}$ and $\Sigma_{2,2}^{-1/2}$, respectively, the matrices $\Sigma_{1,1}$ and $\Sigma_{2,2}$ will be identity matrices, and therefore their inverses will be identity matrices as well. Thus when the data is sphered, the eigenvectors of $\Sigma_{1,1}^{-1}\Sigma_{1,2}\Sigma_{2,2}^{-1}\Sigma_{2,1}$ will be equal to those of $\Sigma_{1,2}\Sigma_{2,1}$ and the eigenvectors of $\Sigma_{2,2}^{-1}\Sigma_{2,1}\Sigma_{1,1}^{-1}\Sigma_{1,2}$ will be equal to those of $\Sigma_{2,1}\Sigma_{1,2}$ and the resulting CCA weight vectors will be $\Sigma_{1,1}^{-1/2}\mathbf{w}$ and $\Sigma_{2,2}^{-1/2}\mathbf{v}$.

We use the students' marks dataset (with which the reader will, by now, be very familiar) to show that linear ECA is capable of performing CCA. We sphere the dataset by multiplying it with the sphering matrices $\Sigma_{1,1}^{-1/2}$ and $\Sigma_{2,2}^{-1/2}$. The ECA network was trained on this sphered dataset using 50,000 iterations with a learning-rate of 0.0005. We have displayed both the "true" results as well as the results obtained using the ECA-network in Table 12.3, where we display the vectors $\Sigma_{1,1}^{-1/2}\mathbf{w}_1$ and $\Sigma_{2,2}^{-1/2}\mathbf{w}_2$. We can see that the results are very close to the results obtained by statistical CCA, which demonstrates that linear ECA performed on sphered data performs CCA.

Table 12.3. Comparison between statistical CCA and linear ECA on the student exams dataset. Linear ECA needs to be sphered before it performs CCA and therefore we show the sphering and weight vectors combined. The values obtained by performing statistical CCA and those obtained by performing linear ECA are very similar.

Standard statistical CCA results		
$\mathbf{w}_1$ 0.0260 0.0518		
$\mathbf{w}_2$ 0.0824 0.0081 0.0035		
Linear ECA network		
$\Sigma_{1,1}^{-1/2}\mathbf{w}_1$ 0.0258 0.0515		
$\Sigma_{2,2}^{-1/2}\mathbf{w}_2$ 0.0826 0.0076 0.0032		

12.4 FastECA

In the previous sections, we maximised the objective function $E(\mathbf{g}(\mathbf{y}_1)^T\mathbf{g}(\mathbf{y}_2))$ by gradient ascent. In this section, we explore a different maximisation technique, based on Newton's method of root finding. Our derivation is similar to the derivation used for the FastICA method [85], in which Newton's method is used for ICA. We shall initially describe the FastECA method in terms of a single unit. Later we show how this algorithm may be extended to many units, using either a symmetrical or a deflationary method. Newton's method is a fixed-point method for finding roots of nonlinear functions numerically. If the function to be minimised is given by $\mathbf{f}(\mathbf{x})$, then Newton's method to find the root of $\mathbf{f}(\mathbf{x}) = \mathbf{0}$ is given by iterating:

$$\mathbf{x}_{t+1} = \mathbf{x}_t - F(\mathbf{x}_t)^{-1}\mathbf{f}(\mathbf{x}_t) \qquad t \geq 0 \qquad (12.19)$$

where $F(\mathbf{x})$ denotes the Jacobian of $\mathbf{f}(\mathbf{x})$.

For one-unit ECA, we have to maximise

$$E(g(\mathbf{w}^T\mathbf{x}_1)g(\mathbf{v}^T\mathbf{x}_2)) + \lambda(\mathbf{w}^T\mathbf{w} - 1) + \mu(\mathbf{v}^T\mathbf{v} - 1) \qquad (12.20)$$

where λ and μ are Lagrange parameters that constrain the magnitudes of the weights. In the following, we only consider maximisation with respect to $\mathbf{w}$; maximisation with respect to $\mathbf{v}$ can be done in the same way. (12.20) is at a local extremum when

$$E(\mathbf{x}_1 g'(\mathbf{w}^T \mathbf{x}_1) g(\mathbf{v}^T \mathbf{x}_2)) + 2\lambda \mathbf{w} = 0 \tag{12.21}$$

The Jacobian of the left-hand side of (12.21), which we denote by $F(\mathbf{w})$ is

$$F(\mathbf{w}) = E(\mathbf{x}_1 \mathbf{x}_1^T g''(\mathbf{w}^T \mathbf{x}_1) g(\mathbf{v}^T \mathbf{x}_2)) + 2\lambda \tag{12.22}$$

As the data is sphered, following the argument used by Hyvärinen [85], we approximate the first term in (12.22):

$$E(\mathbf{x}_1 \mathbf{x}_1^T g''(\mathbf{w}^T \mathbf{x}_1) g(\mathbf{v}^T \mathbf{x}_2)) \approx E(\mathbf{x}_1 \mathbf{x}_1^T) E(g''(\mathbf{w}^T \mathbf{x}_1) g(\mathbf{v}^T \mathbf{x}_2))$$
$$= E(g''(\mathbf{w}^T \mathbf{x}_1) g(\mathbf{v}^T \mathbf{x}_2))I$$

Thus the Jacobian matrix becomes diagonal and can be easily inverted. For the Newton-iteration, we now get

$$\mathbf{w}^+ = \mathbf{w} - \frac{E(\mathbf{x}_1 g'(\mathbf{w}^T \mathbf{x}_1) g(\mathbf{v}^T \mathbf{x}_2)) + 2\lambda \mathbf{w}}{E(g''(\mathbf{w}^T \mathbf{x}_1) g(\mathbf{v}^T \mathbf{x}_2)) + 2\lambda} \tag{12.23}$$

where $\mathbf{w}^+$ is the new estimate. We may simplify this algorithm by multiplying both sides of (12.23) by $(2\lambda + E(g''(\mathbf{w}^T \mathbf{x}_1) g(\mathbf{v}^T \mathbf{x}_2)))$, which is a scalar. In a second step we then normalise the size of the vector $\mathbf{w}^+$ to be unit-size. The FastECA algorithm then becomes:

1. Choose initial random weight vectors $\mathbf{w}$ and $\mathbf{v}$.
2. Calculate $\mathbf{w}^+ = \mathbf{x}_1 E(g'(\mathbf{w}^T \mathbf{x}_1) g(\mathbf{v}^T \mathbf{x}_2)) - \mathbf{w} E(g''(\mathbf{w}^T \mathbf{x}_1) g(\mathbf{v}^T \mathbf{x}_2))$
   and $\mathbf{v}^+ = \mathbf{x}_2 E(g(\mathbf{w}^T \mathbf{x}_1) g'(\mathbf{v}^T \mathbf{x}_2)) - \mathbf{v} E(g(\mathbf{w}^T \mathbf{x}_1) g''(\mathbf{v}^T \mathbf{x}_2))$.
3. $\mathbf{w} = \mathbf{w}^+ / \|\mathbf{w}^+\|$ and $\mathbf{v} = \mathbf{v}^+ / \|\mathbf{v}^+\|$.
4. If not converged go to step 2.

It should be noted that a common assessment of convergence is when the old and new values of $\mathbf{w}$ and $\mathbf{v}$ point in the same direction.

12.4.1 FastECA for Several Units

The single unit FastECA method proposed in the previous section can be extended to multiple outputs. We can do this by running a single unit FastECA method using several units with weight vectors $\mathbf{w}_1, \ldots, \mathbf{w}_n$ and $\mathbf{v}_1, \ldots, \mathbf{v}_n$. To prevent each unit from converging to the same maxima we must decorrelate the outputs $\mathbf{w}_1^T \mathbf{x}_1, \ldots, \mathbf{w}_n^T \mathbf{x}_1$ and $\mathbf{v}_1^T \mathbf{x}_2, \ldots, \mathbf{v}_n^T \mathbf{x}_2$ after each iteration.

One well-known technique to achieve this decorrelation is to use deflationary learning. To do this, we estimate the common signals individually, subtracting the projections of the estimates of each of the previous vectors in turn from the estimated projection before calculating the next. We then explicitly renormalise.

1. $\mathbf{w}_{1,n+1} = \mathbf{w}_{1,n+1} - \sum_{j=1}^{n} \mathbf{w}_{1,n+1}^T \mathbf{w}_{1,j} \mathbf{w}_{1,j}$.
2. $\mathbf{w}_{1,n+1} = \frac{\mathbf{w}_{1,n+1}}{\mathbf{w}_{1,n+1}^T \mathbf{w}_{1,n+1}}$.

We could also estimate each of the weights in parallel and perform a symmetric decorrelation after each iteration, using $W_{n+1} = (W_n^T W_n)^{-\frac{1}{2}} W_n$ [85].

12.4.2 Comparison of ECA and FastECA

We have compared the speed of convergence of ECA and FastECA using a simple artificial 10-dimensional dataset. The first nine dimensions were taken from a Gaussian distribution, and the tenth dimension was taken from a kurtotic source, with a kurtosis value of 8. This dataset was duplicated, and to each copy independent Gaussian noise was added. To compare the errors, we calculate the angular error of the weight vectors, given by

$$\text{Error} = \arccos\left(\frac{\mathbf{w}^T \mathbf{r}}{\| \mathbf{w} \| \| \mathbf{r} \|}\right) \qquad (12.24)$$

where $\mathbf{w} = \begin{pmatrix} \mathbf{w}_1 \\ \mathbf{w}_2 \end{pmatrix}$ is calculated with the relevant method and $\mathbf{r} = \begin{pmatrix} \mathbf{r}_1 \\ \mathbf{r}_2 \end{pmatrix}$ is the "right" answer.

To make a fair comparison, both networks were trained for on 200 samples at a time. The neural ECA method did 200 gradient ascent updates of the weights during each of these epochs, while the FastECA method only did one update, after calculating the averages of equation (12.23) over 200 samples. In order to improve the stability of the FastECA algorithm, we used a step-size in the update-rule, similar to the stabilised form of FastICA [85], where $\mathbf{w}_{n+1} = \mathbf{w}_n + 0.25 \mathbf{w}_{\text{calc}}$, where $\mathbf{w}_{\text{calc}}$ is the estimate of the weights.

Fig. 12.3 and Fig. 12.4 show the rates of convergence of the neural ECA network and the FastECA method, trained on the artificial data. The errors displayed are the angles between the theoretical solution and the weight obtained using ECA or FastECA. It can be clearly seen that the FastECA algorithm converges faster than the neural ECA method.

12.5 Local Filter Formation From Natural Stereo Images

This section will have a more biological focus. In particular, the statistical structures of images and ways the brain is thought to encode visual information is discussed. This area of research has recently received much interest from both the vision science community and the neural computation community.

12.5.1 Biological Vision

The brain is extremely effective in solving complex vision problems. In general, people easily outperform computers in tasks such as face recognition or

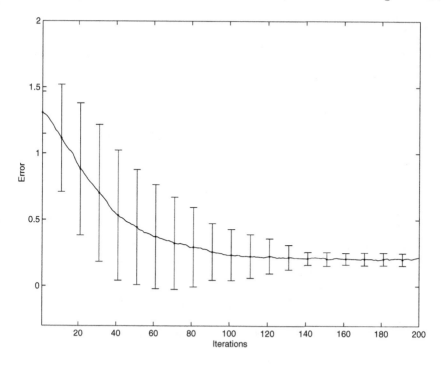

Fig. 12.3. Convergence of the neural ECA method.

reading hand-written text. We can therefore ask ourselves how the brain performs these complex tasks. In this section we will limit ourselves to the early processing stages of biological vision and in particular, how the brain encodes raw visual information.

When light hits the eye it is converted into electrical signals by a layer of photosensitive cells that cover the retina of the eye. This raw information undergoes some initial preprocessing by a small layer of cells in the eye, including the horizontal, bipolar, amacrine and ganglion cells and is then sent to a region of the brain called the striate cortex, which is considered vital in early visual processing [148]. The striate cortex is an approximately 2 mm thick sheet of neurons with a surface area of a few square inches containing about 200 million cells.

Insight into how this region works was gained in the late 1950s when Hubel and Wiesel [83] were able to measure the responses of the individual cells in the striate cortex. It was found that each individual cell fires most strongly when stimuli are presented in a small part of the field of vision, the so-called *receptive field* of that particular cell. Furthermore, it was found that each cell was tuned specifically to respond to a certain type of stimuli. In this regard, three different types of cells were distinguished. The *simple cells* respond most strongly to edges or lines of light at a certain position and

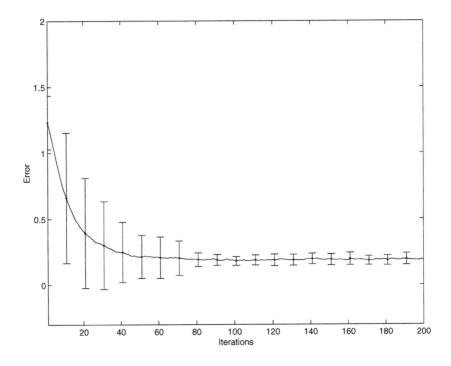

Fig. 12.4. Convergence of the FastECA method.

orientation. In Fig. 12.5, some examples of these types of receptive fields are shown. The receptive fields of the other two types of cells, the *complex* and *hypercomplex* cells, were much more difficult to determine as these cells have a highly nonlinear response and can be sensitive to motion.

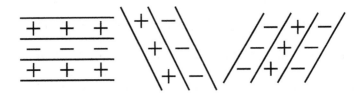

Fig. 12.5. Examples of the characteristics of simple cells in the striate cortex. The leftmost example is a light-dark-light line detector; the middle example responds most strongly to a light-dark edge and the right example is a dark-light-dark line detector. Each cell is tuned to respond to stimuli of a specific orientation.

After determining the responses of the cells in the visual cortex, a wide variety of theories were devised to explain why the striate cortex encodes visual information the way it does. One way to gain more insight into this problem is to investigate the local statistical structure of natural images. Olshausen and Field [145] wrote a seminal paper in this area in which experiments with natural images were used to show a possible link between the coding strategies of the striate cortex and sparse codes. In the following section we describe these experiments in more detail.

12.5.2 Sparse Coding of Natural Images

Natural images are images depicting landscapes, trees, people, etc., which constitutes the type of visual information the brain typically has to deal with. It is therefore assumed that the brain has evolved a coding scheme that facilitates processing this type of visual input. As the second order correlation between neighbouring pixels of natural images is high, one of the obvious first approaches is to investigate PCA as a possible coding scheme for natural images.

In order to reduce complexity, the images are typically divided into patches taken from random positions, and PCA is performed on these patches. Fig. 12.6 shows the results of this experiment, using 12×12 pixels image patches taken from nine grayscale images. Examining the results we see that some of the vectors display structure as found in the striate cortex, but only in the first few PCA vectors and none of them have local responses.

Olshausen and Field [145] therefore considered another coding scheme. They supposed that the simple cells in the striate cortex form a sparse coding of the visual input, which has been shown to have many advantages for an organism in coding its environment. They developed a method that can form sparse codes (i.e. search for transformations yielding outputs that have probability density functions with high kurtosis) and used that on patches taken from natural images [45]. Since then, more efficient and faster algorithms have been devised, and techniques such as ICA [85], have been able to obtain results similar to those presented by Olshausen and Field.

We have repeated this experiment using ICA to obtain a sparse coding. In these type of experiments, the visual data is first mean centred and then preprocessed, using a filter with frequency response $R(f) = f \exp(-(f/f_0)^4)$. This is a widely used whitening/low-pass filter that ensures the Fourier amplitude spectrum of the images is flattened and which also decreases the effect of noise by eliminating the highest frequencies. This preprocessing step is biologically plausible, as a similar preprocessing is thought to be performed by the ganglion cells in the retina.

We randomly sampled the preprocessed images by taking 12×12 pixel patches, on which FastICA with 100 outputs was used. Fig. 12.7 shows the weight vectors. Most of the features are very localised and have an oriented and wavelike structure, which is strikingly similar to the responses of the simple cells found in the striate cortex.

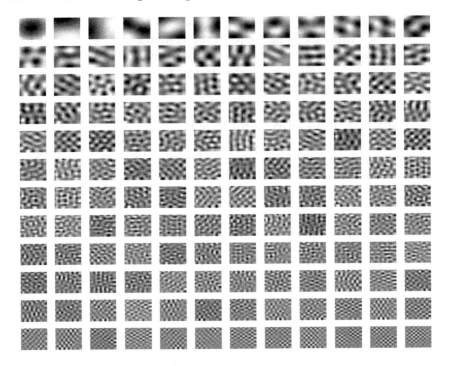

Fig. 12.6. PCA performed on image patches taken from natural images. The first 100 vectors are displayed in the figure in order. The weight vectors do not show a local structure, and only the first few weight vectors show a wavelike structure.

12.5.3 Stereo Experiments

We have extended the experiment in Section 12.5.2 to stereo images. Stereo images consist of two images: one part as seen through the left eye and another as seen through the right. Because both images are different views of the same scene, they share a number of features, which can be extracted with the ECA network.

Artificial Stereo Shift Experiments

In the first set of experiments, we introduce artificial stereo disparity in the images, by slightly shifting the position of the image patches. We used a set of 14 nonstereo natural images that were preprocessed as described in Section 12.5.2. Two sets of 12 × 12 image patches were extracted, where each patch from one set is a horizontally shifted version of the corresponding patch from the other set.

We used 14 x 5 000 pairs of patches, where the patches in one stream are sampled from the same vertical position as the other stream, but horizontally

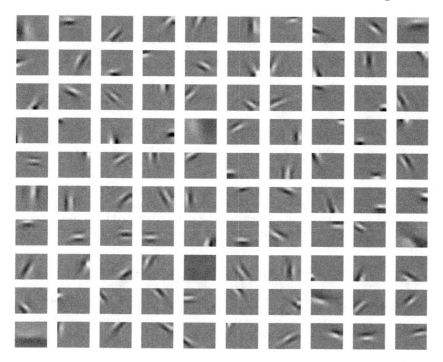

Fig. 12.7. Sparse coding of images patches taken from natural images. Each of the 100 patches represents a weight vector.

we sample the patch four pixels shifted to the right. The patches were sphered by projecting them onto the first 64 eigenvectors and each component was divided by the variance, thereby removing the second-order correlations from the data and reducing the dimensionality. The nonstabilised FastECA method was used to find the common higher-order structure, using the nonlinearities $\mathbf{g}(\mathbf{x}) = -\exp(-\mathbf{x}^4)$, $\mathbf{g}'(\mathbf{x}) = -4\tanh(\mathbf{x})$ and $\mathbf{g}''(\mathbf{x}) = -12\exp(-\mathbf{x}^2)$. After 350 iterations the network converged to the results displayed in Fig. 12.8.

In this figure we see that the features resemble the localised wavelike structures that are typically obtained from sparse coding experiments using natural images. In order to make the comparison between the features from the left and right view clearer, six features are enlarged and shown pairwise in Fig. 12.9. We can see that the patches from the left view are mostly shifted versions of those found in the right view. However, there are exceptions in which features are formed on the opposite sides of the patches, which can be attributed to the orthogonality constraint we imposed on the network. The amount of interesting-looking codes[1] that can be obtained using this data

---

[1] Using more outputs causes the additional features to become blurred and without visible structure.

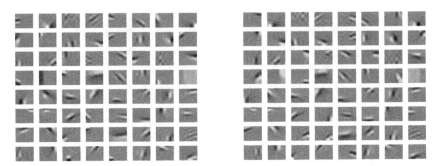

Fig. 12.8. Results of the sparse coding experiment using pairs of patches, shifted by four pixels. Most features are shifted by four pixels, although sometimes the features appear on the opposite side of a patch, which can be attributed to the orthogonality constraint. Some of the features are inverted, which is due to the positive objective function.

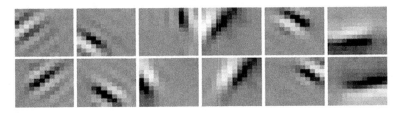

Fig. 12.9. Closer view of six pairs of weight vectors from Fig. 12.8. The features have a localised wavelike response, similar to that found in the single stream sparse-coding experiment. The features are generally shifted, but due to the orthogonality constraint are sometimes found on the opposite sides of the patch. They are sometimes inverted and it is only possible to obtain a smaller number of features; in this case 64 as opposed to 100 for the single stream experiment.

set is less than for single stream sparse coding experiments, which is due to the fact that the amount of shared information, i.e. the overlapping area, is smaller than one patch. The features are also sometimes inverted, which can be attributed to the positive-only objective function.

To examine whether this method allows the network to identify shift, we use the trained network on differently shifted image patches and we calculate the activation of the network at each shift. For these experiments, three trained networks were used, one where the network was trained on patches that were shifted two pixels, one where the patches were shifted four pixels and the last one where the network was trained on patches that were shifted six pixels. A training set was generated consisting of patches shifted from zero to eight pixels. For each shift 1000 patches were used and the average activation of the network, computed by $E(\mathbf{g}(W^T\mathbf{x}_1)^T\mathbf{g}(V^T\mathbf{x}_2))$, was plotted in three graphs (Fig. 12.10).

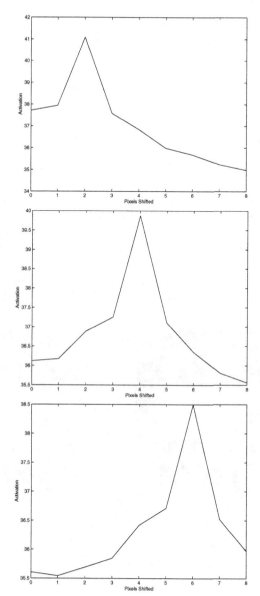

Fig. 12.10. Activation of the FastECA network trained on differently shifted image patches. Each maximum activation coincides with the shift the network was trained on, thereby allowing the network to identify the magnitude of the shift.

In each graph, we see a spike in the activation that coincides with the shift in the patches the network was trained on. This experiment therefore shows that a network trained in this manner, becomes specifically tuned to detect a specific amount of shift in image patches. This makes it theoretically possible to use sets of networks, each trained on a specific amount of shift to detect local stereo disparity in images.

Real Stereo Images

In this experiment, we use real stereo images for a two stream sparse coding experiment. We selected 14 natural stereo images, which were preprocessed as described in Section 12.5.2. An example of an unpreprocessed stereo image pair used is displayed in Fig. 12.11. The images were sampled by randomly taking 5000 12 × 12 patches from each image, which were used pairwise as input to the FastECA network. We again used the nonstabilised version of the FastECA method and the nonlinearities $\mathbf{g}(\mathbf{x}) = -\exp(-x^4)$, $\mathbf{g}'(\mathbf{x}) = -4\tanh(\mathbf{x})$ and $\mathbf{g}''(\mathbf{x}) = -12\exp(-\mathbf{x}^2)$. The resulting weight vectors of the trained network after 350 iterations are displayed in Fig. 12.12.

Fig. 12.11. Example of an unpreprocessed natural stereo image pair. *Reprinted with permission from Mark Blum; see http://www.undersea3d.com/.*

Examining the results, we find a number of interesting differences between the filters obtained from standard images and those obtained from stereo images. For better comparison, a closeup of eight pairs of weight vectors is displayed in Fig. 12.13. The first difference is that there are significantly fewer shared codes for stereo images. This can be explained by the fact that stereo

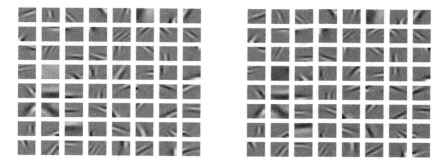

Fig. 12.12. Converged weight vectors when training the ECA network on natural stereo images. The left figure represents the 64 weight vectors of the ECA network that were trained on the left stereo image and the right figure respresents the corresponding weight vectors that were trained on the right stereo image.

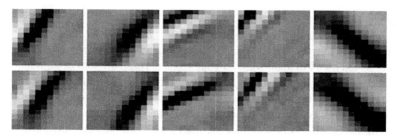

Fig. 12.13. Closer view of five pairs of weight vectors from Fig. 12.12. The features have a localised wavelike response, similar to that found in the single stream sparse coding experiment. The features are generally larger, are sometimes inverted and it is only possible to obtain a smaller number of features, in this case 64 as opposed to 100 for the single stream experiment.

images are not only views of a scene at slightly different angles, but also slightly shifted. This makes the "overlap" between two patches smaller and the amount of shared information less. For two input streams of both 12×12 pixels, we extracted 64 components, which was experimentally determined to give the most interesting results. Another difference is that the features found by the ECA network tend to have a wider variety of frequencies. Additionally, the features themselves tend to be larger.

When comparing the codes from both data streams, we can see many similarities in the codes, but there are also a number of interesting differences. A number of features are inverted versions of each other. This is a result of the positive-only kurtotic objective function $\mathbf{g}(\mathbf{y}) = \mathbf{y}^4$. Also, a number of features are shifted, which can be attributed to the stereo disparity between the left and the right images.

Stereo images have been analysed before using sparse coding methods [81]. Usually, two patches from each image are taken and both patches are used simultaneously as input to the sparse coder. The results in this chapter show that the network used here creates different types of filters when real stereo images as opposed to artificially generated stereo images are used. We take this also as a criticism of some of our own earlier work on stereo images, particularly that on the abstraction of random dot stereograms (Chapter 9).

12.6 Twinned Maximum Likelihood Learning

In this section, we find structure shared between two data streams using adaptions of the Maximum Likelihood network (Chapter 8). We will compare the results obtained with the twinned Maximum Likelihood method to results obtained with the ECA network above.

Method I

In this section we will extend the Maximum Likelihood learning algorithm to allow it to extract kurtotic sources that exist across two sets of signals. Since EPP and ECA use a function of the outputs to search for higher-order structure and the Maximum Likelihood method uses a function of the residuals, a first obvious idea is simply to join these two together. Although EPP as well as Maximum Likelihood both have strong theoretical backgrounds, combining these methods in a single rule can be considered to be somewhat ad hoc as it cannot be derived from a single objective function. We will combine Maximum Likelihood learning and ECA, to form a single learning rule and investigate its properties.

$$\Delta W = \eta[p|\mathbf{e}_1|^{p-1} \otimes sign(\mathbf{e}_1)(\mathbf{g}(\mathbf{y}_2) \otimes \mathbf{g}'(\mathbf{y}_1))^T] \qquad (12.25)$$

$$\Delta V = \eta[p|\mathbf{e}_2|^{p-1} \otimes sign(\mathbf{e}_2)(\mathbf{g}(\mathbf{y}_1) \otimes \mathbf{g}'(\mathbf{y}_2))^T] \qquad (12.26)$$

where $\mathbf{e}_1 = \mathbf{x}_1 - W\mathbf{y}_1$ and $\mathbf{e}_2 = \mathbf{x}_2 - V\mathbf{y}_2$.

To test this method, we consider a situation in which we have an underlying kurtotic signal which has been corrupted by sensor noise. The signal can be monitored with two sensors simultaneously. Some sensor noise will simply be white noise independent from any other noise source or signal, while some of the noise may be correlated across the two sensors. We wish to identify the common kurtotic signal as cleanly as possible. We have modeled this situation with two 10-dimensional data sets.

- The first dimension of both streams contains the common kurtotic signal (with a normalised kurtosis of 1) that we wish to extract.
- The second dimension contains two separate kurtotic signals, both with a normalised kurtosis of 5, that are uncorrelated between the two streams. Both of these signals are more kurtotic than the shared signal in the first dimension.

- The third dimension contains correlated Gaussian noise.
- The fourth to tenth dimensions contain independent white Gaussian noise.

We wish to extract the common kurtotic source, which is contained in the first dimension of the dataset, ignoring both the independent kurtotic source of dimension two and the normally distributed source of dimension three. Using a learning rate of 0.0005 and training for 50 000 iterations the results displayed in Table 12.4 were obtained. Examining these results, we can see that

Table 12.4. Converged weight vectors when using the combined ECA-Maximum Likelihood to extract a common kurtotic source.

w	v
−0.9772	0.9812
−0.1979	0.1543
0.0480	−0.0505
0.0178	0.0297
0.0418	0.0154
0.0076	0.0395
0.0201	−0.0611
0.0138	0.0035
−0.0260	0.0208
−0.0100	−0.0616

this method indeed extracts the common kurtotic source, but not as accurately as ECA on its own. Adding the extra Maximum Likelihood function to the original ECA network causes the network to place more emphasis on finding kurtosis and so the independent kurtotic sources are also identified, although the values of these weights are much smaller. We can therefore conclude that combining ECA and Maximum Likelihood does find common kurtotic sources, but is not very accurate as too much emphasis is placed on the kurtosis finding element of the algorithm.

Method II

As another approach to Twinned Maximum Likelihood, we can postulate that the outputs should be the same in each case and thus we may use each output as a target for the other. Considering the one output case, we postulate that y_1 is a noisy version of y_2 corrupted by Gaussian noise, which means that $p(y_1|y_2) \propto \exp(-(y_1 - y_2)^2)$. Now we wish to maximise this conditional likelihood and so, as with the Maximum Likelihood rules (Chapter 8), we define an objective function

$$K_1 = -\log(p(y_1|y_2)) \propto (y_1 - y_2)^2 \tag{12.27}$$
$$K_2 = -\log(p(y_2|y_1)) \propto (y_1 - y_2)^2 \tag{12.28}$$

on which we perform gradient descent. Thus we have the joint rules

$$\Delta \mathbf{w} \propto -\left(\frac{\partial J}{\partial \mathbf{w}} + \frac{\partial K_1}{\partial \mathbf{w}}\right) =$$
$$\eta y_1 (\text{sign}(\mathbf{e}_1) \otimes |\mathbf{e}_1|^{p-1}) + \gamma(y_1 - y_2) * \mathbf{x}_1 \qquad (12.29)$$

$$\Delta \mathbf{v} \propto -\left(\frac{\partial J}{\partial \mathbf{v}} + \frac{\partial K_2}{\partial \mathbf{v}}\right) =$$
$$\eta y_2 (\text{sign}(\mathbf{e}_2) \otimes |\mathbf{e}_2|^{p-1}) + \gamma(y_2 - y_1) * \mathbf{x}_2 \qquad (12.30)$$

where η and γ are constant learning parameters. Unfortunately the second term in each rule causes the network to become unstable as the weights are allowed to grow without bound. We can solve this problem by introducing Lagrangian parameters in equation (12.27) similar to the derivation of EPP and ECA:

$$K_1 \propto (y_2 - y_1)^2 + \lambda(1 - \mathbf{w}^T \mathbf{w}) \qquad (12.31)$$
$$K_2 \propto (y_2 - y_1)^2 + \mu(1 - \mathbf{v}^T \mathbf{v}) \qquad (12.32)$$

At the minima of K_1 and K_2 we obtain:

$$\frac{\partial K_1}{\partial \mathbf{w}} \propto -2(y_2 - y_1)\mathbf{x}_1 - 2\lambda \mathbf{w} = 0 \qquad (12.33)$$

$$\frac{\partial K_2}{\partial \mathbf{v}} \propto 2(y_2 - y_1)\mathbf{x}_2 - 2\mu \mathbf{v} = 0 \qquad (12.34)$$

Premultiplying (12.33) and (12.34) by $\mathbf{w}^T$ and $\mathbf{v}^T$ respectively results in

$$\lambda = -(y_2 - y_1)y_1 \qquad (12.35)$$
$$\mu = (y_2 - y_1)y_2 \qquad (12.36)$$

Inserting (12.35) and (12.36) into (12.33) and (12.34), we obtain the final gradient descent rules, which fit the negative feedback model:

$$\Delta \mathbf{w} \propto -\left(\frac{\partial J}{\partial \mathbf{w}} + \frac{\partial K_1}{\partial \mathbf{w}}\right) = \eta y_1(\text{sign}(\mathbf{e}_1) \otimes |\mathbf{e}_1|^{p-1}) + \gamma(y_1 - y_2) * \mathbf{e}_1 \quad (12.37)$$

$$\Delta \mathbf{v} \propto -\left(\frac{\partial J}{\partial \mathbf{v}} + \frac{\partial K_2}{\partial \mathbf{v}}\right) = \eta y_2(\text{sign}(\mathbf{e}_2) \otimes |\mathbf{e}_2|^{p-1}) + \gamma(y_2 - y_1) * \mathbf{e}_2 \quad (12.38)$$

These rules consist of two terms, the first term searching for non-Gaussianity and the second term ensuring that there is a relationship between the two sources. The learning parameters η and γ control the relative strengths of these objectives.

We test this method using the ten-dimensional data set from the previous section. We have extracted the common kurtotic source using three different methods: ECA, Twinned Maximum Likelihood and Twinned Minimum Likelihood. The results for ECA are displayed in Table 12.5 and the results for

Twinned Minimum and Maximum are displayed in Table 12.6. We can see that the first dimension, which was the common kurtotic source, has been identified by ECA and the Twinned Maximum and Minimum Likelihood method. All these methods also ignore the correlated Gaussian source and the uncorrelated kurtotic sources and the methods are almost equally accurate in this set of experiments. In the following sections, when we discuss Twinned Maximum Likelihood Methods, we are referring to the method of this section, i.e. method II.

Table 12.5. Converged weight vectors when using ECA to extract a common kurtotic source.

w	v
0.9974	**0.9983**
0.0094	−0.0070
0.0490	0.0480
0.0048	0.0114
−0.0176	−0.0100
−0.0136	0.0200
0.0353	0.0086
0.0069	0.0177
0.0239	0.0087
0.0252	0.0124

Table 12.6. Converged weight vectors when using Maximum Likelihood (left) and Minimum Likelihood (right) to extract a common kurtotic source.

w	v		w	v
−0.9987	**0.9974**		**0.9991**	**0.9981**
−0.0230	−0.0967		−0.0037	0.0077
0.0045	−0.0069		0.0412	0.0579
−0.0066	0.0248		0.0029	−0.0065
−0.0189	−0.0180		0.0013	−0.0065
−0.0255	0.0222		0.0125	−0.0163
0.0465	−0.0418		−0.0052	−0.0054
0.0326	0.0376		0.0021	0.0018
−0.0014	0.0427		−0.0074	−0.0056
0.0565	0.0201		−0.0102	−0.0011

12.7 Unmixing of Sound Signals

We will now use both the twinned maximum likelihood method as well as ECA on an adaption of the blind source separation problem. In this experiment we will use the same three sound signals as we used in Section 8.4.3. Additionally, we generate another two separate, independent signals of equal length as the voice signals, consisting of extremely kurtotic noise (kurtosis = 2600). We now create two mixtures; one mixture is created by mixing the three voice signals together with one of the noise signals using a mixing matrix A_1, and for the second mixture, we mix the same three voice signals with the other noise signal using mixing matrix A_2:

$$
A_1 = \begin{pmatrix}
0.3 & -0.3 & 0.4 & -0.015 \\
0.7 & 0.3 & -0.5 & 0.01 \\
-0.3 & -0.7 & 0.2 & -0.02 \\
-0.2 & 0.4 & -0.3 & 0.01
\end{pmatrix}
$$

$$
A_2 = \begin{pmatrix}
0.8 & 0.5 & -0.3 & 0.006 \\
0.3 & 0.5 & 1.0 & -0.007 \\
-0.2 & -0.1 & 0.7 & 0.005 \\
0.9 & -0.4 & 0.4 & -0.005
\end{pmatrix}
$$

We have deliberately kept the values of the last row small, to simulate kurtotic noise with low variance. Before applying any of the methods described in this section, we first sphere the data, with sphering matrices Q_1 and Q_2.

First we will examine the performance of single stream deflationary EPP on this set of signals. In order to examine the unmixing results we will look at $W * Q_1 * A_1$ and $V * Q_2 * A_2$, i.e. the weights learnt by the method times the sphering matrix times the mixture matrix. The results for EPP are shown in (12.39) and (12.40):

$$
W * Q_1 * A_1 = \begin{pmatrix}
0.002 & \mathbf{0.999} & -0.003 & -0.004 \\
-0.018 & -0.013 & -0.001 & \mathbf{0.999} \\
-0.004 & -0.005 & -\mathbf{1.000} & -0.002 \\
-\mathbf{1.000} & 0.005 & 0.007 & -0.006
\end{pmatrix} \tag{12.39}
$$

$$
V * Q_2 * A_2 = \begin{pmatrix}
-0.002 & -\mathbf{1.000} & 0.011 & -0.004 \\
0.003 & 0.010 & \mathbf{1.000} & 0.001 \\
0.007 & 0.006 & 0.003 & -\mathbf{1.000} \\
-\mathbf{1.000} & -0.005 & -0.001 & -0.012
\end{pmatrix} \tag{12.40}
$$

We used a learning rate of 0.0001 and 500 000 iterations (randomly sampling with replacement from the 40 000 samples) to extract each signal in a deflationary manner. We can see that EPP first extracts the fourth dimension in both mixtures, which represents the two most kurtotic signals, i.e. the noise signals, even though the the variance of these signals is quite low. After this signal has been identified and subtracted from the mixture, the two simulations extract the remaining signals in random order.

We now repeat the same experiment using deflationary ECA to unmix the signals. We use the same parameters as for the previous experiment. The results are shown in (12.41) and (12.42):

$$W * Q_1 * A_1 = \begin{pmatrix} -0.004 & \mathbf{0.999} & 0.029 & 0.0826 \\ \mathbf{0.999} & 0.006 & -0.029 & -0.186 \\ -0.027 & 0.026 & -\mathbf{0.999} & 0.118 \\ -0.010 & 0.005 & -0.003 & -0.036 \end{pmatrix} \tag{12.41}$$

$$V * Q_2 * A_2 = \begin{pmatrix} -0.004 & \mathbf{0.997} & -0.029 & 0.063 \\ \mathbf{0.999} & 0.006 & 0.029 & -0.169 \\ -0.027 & 0.026 & \mathbf{0.999} & 0.010 \\ 0.003 & 0.006 & -0.013 & -0.135 \end{pmatrix} \tag{12.42}$$

In these results we see that the independent kurtotic noise signals have been completely ignored and the three correlated, much less kurtotic, voice signals have all been extracted successfully. The last column shows no structure, indicating that there are no common kurtotic sources left in the mixtures. The ECA method therefore successfully removed the independent noise signals from the mixtures and unmixed the sound signals correctly. In the last experiment we use Maximum Likelihood Learning to solve the same twinned ICA problem. We use a value of $p = 1$ (a Laplacian distribution). The results are displayed in (12.43) and (12.44)

$$W * Q_1 * A_1 = \begin{pmatrix} -0.012 & -\mathbf{1.000} & -0.022 & -0.001 \\ -0.060 & 0.004 & -\mathbf{0.987} & -0.063 \\ \mathbf{1.001} & -0.009 & -0.073 & -0.013 \\ -0.014 & 0.003 & 0.055 & -\mathbf{0.997} \end{pmatrix} \tag{12.43}$$

$$V * Q_2 * A_2 = \begin{pmatrix} 0.007 & \mathbf{1.000} & 0.008 & 0.001 \\ 0.050 & -0.007 & \mathbf{0.999} & 0.001 \\ -\mathbf{1.005} & 0.002 & 0.052 & -0.006 \\ 0.001 & -0.011 & 0.006 & \mathbf{1.001} \end{pmatrix} \tag{12.44}$$

Again, the method has extracted the voice signals in the first three columns. Unlike the ECA method, however, the independent noise signal has now been extracted in the last column. This method extracts common kurtotic signals first, but when no common kurtotic signals are present in the mixture, it will continue to extract independent kurtotic signals.

12.8 Conclusion

In this chapter, we have developed a novel approach to uncover underlying signals between data sets. We have derived a neural algorithm, based on neural

EPP capable of uncovering higher-order shared structure between data sets, based on either the third or fourth moment. The method is closely related to CCA and therefore we called the first method Exploratory Correlation Analysis. We showed that ECA using linear outputs performs CCA, when applied to sphered data.

We have shown the validity of this network in a set of experiments, in which it was able to extract a common kurtotic source, even though there were unrelated kurtotic sources and normally distributed related sources present in the data sets. The network was also able to perform a dual stream blind source separation.

The basic method was improved by using a Newton iteration to minimise our objective function and we showed experimentally that this method converges more quickly than the neural method.

One of the applications for ECA is in the coding of natural stereo images. Our first investigation was performed using artificially shifted natural images. We developed filters for this data set and showed how a trained ECA network can detect shift. We proceeded to use real natural stereo images for our experiments and we were able to extract codes using the ECA algorithm that are similar to those found in single stream experiments, but we also found some interesting differences. On real image data, the wavelet-like filters were found to be not so localised as those obtained by the artificially generated shifted image data, and they vary far more in frequency.

One conclusion which we could draw from this set of experiments is that we must be very careful to qualify conclusions made from abstractions of real data sets. Thus the results from e.g. [7, 113, 174] on pulling structure from abstraction of random dot stereograms must be qualified by the statement that the same networks must be trained on real random dot stereograms to validate their conclusions. The results which researchers report when using ICA on single stream visual data must be similarly qualified.

The Maximum Likelihood method was also extended to a dual stream method that has similar properties to the ECA method. We showed how this new method can search for higher order structure shared by two sets of signals. Using an adaption of the normal blind source separation problem, we showed how these methods can be used to filter out unwanted, highly kurtotic noise from related sets of mixtures. Normal single-stream EPP networks find these unwanted signals first, resulting in uninformative projections.

The properties of the Twinned Maximum Likelihood method were compared to the properties of ECA. The major difference is that Twinned Maximum Likelihood extracts common kurtotic sources first, but when there are none left in the mixture, it will extract kurtotic sources that are not necessarily shared. On the other hand, ECA will simply not converge when there are no shared signals left in the mixtures.

We now have a range of tools for a variety of different situations.

1. At one extreme we have EPP which will look for structure (e.g. kurtosis) in a single data stream.
2. Slightly further along the correlation road we find Twinned Maximum Likelihood, which finds shared higher-order structure, but will identify individual higher-order structure in a single data stream when shared structure is exhausted.
3. Then we have ECA, which will identify shared higher-order structure in two data streams, but ignore higher-order structure in a single data stream.
4. Finally, we have CCA (and linear ECA) which identifies shared correlations alone.

Which tool to use in any particular situation will depend on the type of structure for which an investigator is searching.

Multicollinearity and Partial Least Squares

Multicollinearity is the effect of having too high correlation between variables. At its extreme, there may be perfect correlation which results in a singular (determinant equal to zero) correlation matrix. However, even when we do not reach this extreme, multicollinearity causes problems: because of the redundancy in the variables, we may have an ill-conditioned correlation matrix which is very prone to large variances which may be due to noise in the data set.

In this chapter we discuss the problems caused by internal dependencies within a data set. The neural network methods described in this book do not use explicit matrix inversions, but they do often involve implicit matrix inversions. Thus, the effects of such dependencies will be felt when we use these methods.

In Chapter 9, we showed that the canonical correlation directions $\mathbf{w}_1$ and $\mathbf{w}_2$ may be found using

$$\frac{d\mathbf{w}_1}{dt} = \Sigma_{12}\mathbf{w}_2 - f(\mathbf{w})\Sigma_{11}\mathbf{w}_1$$

$$\frac{d\mathbf{w}_2}{dt} = \Sigma_{21}\mathbf{w}_1 - f(\mathbf{w})\Sigma_{22}\mathbf{w}_2$$

Using the fact that $\Sigma_{ij} = E(\mathbf{x}_i\mathbf{x}_j^T), i, j = 1, 2$, we derived the instantaneous versions

$$\Delta\mathbf{w}_1 = \eta(\mathbf{x}_1 y_2 - f(\mathbf{w})\mathbf{x}_1 y_1)$$

$$\Delta\mathbf{w}_2 = \eta(\mathbf{x}_2 y_1 - f(\mathbf{w})\mathbf{x}_2 y_2)$$

which was shown to provide a family of networks capable of performing CCA.

In this chapter, we extend the method using ideas suggested by Ridge Regression. The resulting network is shown to operate on data sets which exhibit multicollinearity. We develop a second model which not only performs as well on multicollinear data but also on general data sets. This model allows us to vary a single parameter so that the network is capable of performing Partial

Least Squares Regression (at one extreme) to Canonical Correlation Analysis (at the other) and every intermediate operation between the two. Finally, we develop a second penalty term which acts on such data as a smoother in that the resulting weight vectors are much smoother and more interpretable than the weights without the robustification term.

13.1 The Ridge Model

The problem of multicollinearity arises in a regression problem whenever there is a linear dependency among the independent variables. That is, let $X = (\mathbf{x}_0, \mathbf{x}_1, \mathbf{x}_2, ..., \mathbf{x}_{p-1})$ where $\mathbf{x}_i$ is the $n \times 1$ vector of responses for the ith variable. The independent variables are said to have linear dependence whenever

$$\sum_{j=0}^{p-1} t_j \mathbf{x}_j = 0 \tag{13.1}$$

for $t_j \neq 0$. To solve $X\beta = \mathbf{y}$, the standard method is to multiply both sides of this equation by X^T and then solve for β to give $\beta = (X^T X)^{-1} X^T \mathbf{y}$. If condition (13.1) holds then $(X^T X)^{-1}$ does not exist. Seldom does the above linear dependency actually hold; rather, one nearly has linear dependency which implies that $(X^T X)^{-1}$ is ill-conditioned, hence any estimates using $(X^T X)^{-1}$ are "poor".

Ridge regression is a popular method for dealing with multicollinearity within a regression model [125]. The idea is fairly simple. Since the matrix $X^T X$ is ill-conditioned or nearly singular one can add positive constants to the diagonal of the matrix and ensure that the resulting matrix is not ill-conditioned. That is, consider the biased normal equations given by

$$[X^T X + kI]\beta = X^T y \tag{13.2}$$

where I is the identity matrix. This results in a biased estimate for β given by

$$\tilde{\beta} = [X^T X + kI]^{-1} X^T y \tag{13.3}$$

where k is called the shrinkage parameter. This has been shown to make the regression robust.

13.2 Application to CCA

The Canonical correlation coefficient is given by:

$$\rho = \frac{\mathbf{w}_1^T \Sigma_{12} \mathbf{w}_2}{[\mathbf{w}_1^T \Sigma_{11} \mathbf{w}_1]^{1/2} [\mathbf{w}_2^T \Sigma_{22} \mathbf{w}_2]^{1/2}} \tag{13.4}$$

which will clearly be difficult to calculate if the within class covariance matrices are singular or nearly so. Similarly, since the generalised eigenvectors found by solving

$$
\begin{bmatrix} 0 & \Sigma_{12} \\ \Sigma_{21} & 0 \end{bmatrix} \begin{bmatrix} \mathbf{w}_1 \\ \mathbf{w}_2 \end{bmatrix} = \rho \begin{bmatrix} \Sigma_{11} & 0 \\ 0 & \Sigma_{22} \end{bmatrix} \begin{bmatrix} \mathbf{w}_1 \\ \mathbf{w}_2 \end{bmatrix}
\tag{13.5}
$$

may be equally well defined as the eigenvectors found by solving

$$
\begin{bmatrix} \Sigma_{11} & 0 \\ 0 & \Sigma_{22} \end{bmatrix}^{-1} \begin{bmatrix} 0 & \Sigma_{12} \\ \Sigma_{21} & 0 \end{bmatrix} \begin{bmatrix} \mathbf{w}_1 \\ \mathbf{w}_2 \end{bmatrix} = \rho \begin{bmatrix} \mathbf{w}_1 \\ \mathbf{w}_2 \end{bmatrix}
\tag{13.6}
$$

we again will be in some difficulty if either Σ_{11} or Σ_{22} is singular or nearly so. In [179], Vinod used the canonical ridge model to deal with data which is nearly collinear. Just as in the multiple regression model, Vinod used $(\Sigma_{11}+k_1 I)$ and $(\Sigma_{22}+k_2 I)$ instead of Σ_{11} and Σ_{22} in CCA. Vinod [179] shows that the coefficients estimated from the usual canonical correlation analysis can be very unstable when the data are nonorthogonal, but after adding small constants to the diagonal of the correlation matrix of all variables before the usual canonical correlation analysis, a considerable improvement in the stability and reliability of regression coefficients is achieved. We can also consider this as a kind of smoothing for the data; in [116], a similar approach has been used to deal with functional data (see later). Thus, from the above, we have good reasons to believe that the penalty term kI can make CCA more robust.

Now if we use $\Sigma_{11} + k_1 I$ and $\Sigma_{22} + k_2 I$ instead of Σ_{11} and Σ_{22} in our neural implementation, we get

$$
\begin{bmatrix} 0 & \Sigma_{12} \\ \Sigma_{21} & 0 \end{bmatrix} \begin{bmatrix} \mathbf{w}_1 \\ \mathbf{w}_2 \end{bmatrix} = \rho \begin{bmatrix} \Sigma_{11} + k_1 I & 0 \\ 0 & \Sigma_{22} + k_2 I \end{bmatrix} \begin{bmatrix} \mathbf{w}_1 \\ \mathbf{w}_2 \end{bmatrix}
\tag{13.7}
$$

Taking $\mathbf{w} = [\mathbf{w}_1^T \mathbf{w}_2^T]^T$, we find the canonical correlation directions $\mathbf{w}_1$ and $\mathbf{w}_2$ using

$$
\frac{d\mathbf{w}_1}{dt} = \Sigma_{12}\mathbf{w}_2 - f(\mathbf{w}_1)(\Sigma_{11} + k_1 I)\mathbf{w}_1
$$

$$
\frac{d\mathbf{w}_2}{dt} = \Sigma_{21}\mathbf{w}_1 - f(\mathbf{w}_2)(\Sigma_{22} + k_2 I)\mathbf{w}_2
$$

Using the facts that $\Sigma_{ij} = E(\mathbf{x}_i \mathbf{x}_j^T), i, j = 1, 2$, and that $y_i = \mathbf{w}_i^T \mathbf{x}_i$, we may propose the instantaneous rules

$$
\Delta \mathbf{w}_1 = \eta(\mathbf{x}_1 y_2 - f(\mathbf{w}_1)\mathbf{x}_1 y_1 - f(\mathbf{w}_1) k_1 \mathbf{w}_1)
$$

$$
\Delta \mathbf{w}_2 = \eta(\mathbf{x}_2 y_1 - f(\mathbf{w}_2)\mathbf{x}_2 y_2 - f(\mathbf{w}_2) k_2 \mathbf{w}_2)
$$

For example, if we choose $f(\mathbf{w}) = \ln(\mathbf{w}^T \mathbf{w})$, we have:

$$
\Delta w_{1j} = \eta(x_{1j} y_2 - \ln(\mathbf{w}_1^T \mathbf{w}_1) x_{1j} y_1 - \ln(\mathbf{w}_1^T \mathbf{w}_1) k_1 w_{1j})
$$

$$
\Delta w_{2j} = \eta(x_{2j} y_1 - \ln(\mathbf{w}_2^T \mathbf{w}_2) x_{2j} y_2 - \ln(\mathbf{w}_2^T \mathbf{w}_2) k_2 w_{2j})
\tag{13.8}
$$

This algorithm, in fact, does perform an approximation to CCA (see below) but we have found experimentally that it does slow learning. The k_i parameters are optimal when rather large, but note that this has a detrimental effect in that it affects both the second term (which it was designed to do) and also the first term which causes growth towards the canonical correlation vectors. To restrict this effect, we restructure (13.7) to get

$$\begin{bmatrix} 0 & \Sigma_{12} \\ \Sigma_{21} & 0 \end{bmatrix} \begin{bmatrix} \mathbf{w}_1 \\ \mathbf{w}_2 \end{bmatrix} = \rho \begin{bmatrix} (1-k_1)\Sigma_{11} + k_1 I & 0 \\ 0 & (1-k_2)\Sigma_{22} + k_2 I \end{bmatrix} \begin{bmatrix} \mathbf{w}_1 \\ \mathbf{w}_2 \end{bmatrix}$$
(13.9)

Thus the update rule for the weights is

$$\Delta \mathbf{w}_1 = \eta(\mathbf{x}_1 y_2 - f(\mathbf{w}_1)(1-k_1)\mathbf{x}_1 y_1 - f(\mathbf{w}_1)k_1 \mathbf{w}_1)$$
$$\Delta \mathbf{w}_2 = \eta(\mathbf{x}_2 y_1 - f(\mathbf{w}_2)(1-k_2)\mathbf{x}_2 y_2 - f(\mathbf{w}_2)k_2 \mathbf{w}_2) \qquad (13.10)$$

This method of weight change is the first innovation in this chapter. We may, however, consider generalising this method by using different bias-inducing terms. For example, we may wish to produce smoothly changing CCA parameters (an example of such a dataset will be given later) and so we may wish to introduce a term which penalises roughness in the CCA weights.

Functional data analysis (FDA) has been developed for analyzing functional (or curve) data. In FDA, we treat the data as consisting of functions not of vectors. We take samples at time points $t_1, t_2, \ldots$ and regard $\{\mathbf{x}(t_j), j = 1, 2, \ldots\}$ as multivariate observations. In this sense the original functional $\mathbf{x}(t)$ can be regarded as the limit of $\{\mathbf{x}(t_j)\}$ as the sampling interval tends to zero and the dimension of multivariate observations tends to infinity. A central idea in performing FDA is to use a roughness penalty to incorporate smoothing. The most popular measure of roughness is the second derivative of the function which is a measure of the rapidity of the variability of the function f,

$$R(f) = \int (f''(x))^2 d(x) \qquad (13.11)$$

Since we do not care about the sign of the roughness, only its magnitude, we may define a penalty for roughness by

$$PEN_2(f) = \| D^2 f \|^2 = \int f(t) D^4 f(t) dt \qquad (13.12)$$

where D^2 is the second derivative operator and D^4 is the fourth derivative operator. So our penalised CCA uses the same driving force but a weighted version of this penalty in the learning rule.

Thus, we may use a fourth derivative operator D^4 instead of I in (13.9) to get:

$$\begin{bmatrix} 0 & \Sigma_{12} \\ \Sigma_{21} & 0 \end{bmatrix} \begin{bmatrix} \mathbf{w}_1 \\ \mathbf{w}_2 \end{bmatrix} = \rho \begin{bmatrix} (1-k_1)\Sigma_{11} + k_1 D^4 & 0 \\ 0 & (1-k_2)\Sigma_{22} + k_2 D^4 \end{bmatrix} \begin{bmatrix} \mathbf{w}_1 \\ \mathbf{w}_2 \end{bmatrix}$$

(13.13)

The terms $k_1 D^4$ and $k_2 D^4$ are the roughness penalties. For more details about the roughness penalty and functional data analysis please see [155].

We can replace the operator D^4 by a finite difference approximation, for example, a multiple of $F^T F$ where F is a circulant second difference matrix. Then we get a learning rule which can smooth the operation for functional data; the learning rule updates the ith weight value using:

$$\Delta \mathbf{w}_{1[i]} = \eta(\mathbf{x}_{1[i]}y_2 - f(\mathbf{w}_1)(1-k_1)\mathbf{x}_{1[i]}y_1 - f(\mathbf{w}_1)k_1 D^4_{[i]}\mathbf{w}_1)$$
$$\Delta \mathbf{w}_{2[i]} = \eta(\mathbf{x}_{2[i]}y_1 - f(\mathbf{w}_2)(1-k2)\mathbf{x}_{2[i]}y_2 - f(\mathbf{w}_2)k_2 D^4_{[i]}\mathbf{w}_2) \quad (13.14)$$

where the $D^4_{[i]}$ denote the ith row in the matrix D^4.

13.2.1 Relation to Partial Least Squares

The phrasing of the parameters in this way (13.9) allows us to consider a family of solutions found by varying the magnitude of the k_i parameters. For example, if $k_1 = k_2 = 1$ in (13.9), we revert to

$$A\mathbf{w} = \lambda B \mathbf{w}$$

which, with

$$B = \begin{bmatrix} I & 0 \\ 0 & I \end{bmatrix}$$

(13.15)

gives us the solution for Partial Least Square(PLS):

$$\begin{bmatrix} 0 & \Sigma_{12} \\ \Sigma_{21} & 0 \end{bmatrix} \begin{bmatrix} \mathbf{w}_1 \\ \mathbf{w}_2 \end{bmatrix} = \rho \begin{bmatrix} I & 0 \\ 0 & I \end{bmatrix} \begin{bmatrix} \mathbf{w}_1 \\ \mathbf{w}_2 \end{bmatrix}$$

(13.16)

The Partial Least Squares (PLS) regression method has been used extensively, especially for calibration tasks in chemometrics [57].

It may be helpful to compare PLS with canonical correlation analysis (CCA), which is a maximisation of the correlation between a similar pair of scores. PLS maximises covariance, not correlation. Interpreting the coefficients of canonical variates requires the usual stringent assumptions underlying multiple regression of either canonical variate upon the variables of the other block. Such assumptions are unlikely to obtain when predictors or outcomes are intentionally redundant. In contrast, by maximizing covariance between the latent variable scores, PLS optimizes the usefulness of the analysis for subsequent studies of intervention. Unlike the coefficients of a canonical correlations analysis, the saliences that PLS computes have meaning individually even when (indeed, especially when) the predictor block or the outcome block is intentionally multicollinear.

The new algorithm has other advantages: first, when we choose the ridge parameter k, we are able to choose a small number which is much more effective for smoothing the data. Further, by having this single parameter for adjusting the playoff between the two factors we can choose k appropriately so that we may vary any potential solution between the extremes of PLS and CCA dependent on the needs of any particular data set.

Finally, when the ridge parameter is close to the PLS value, then the new algorithm has a strong ability to deal with functional data and multicollinear data, because we can always use a small number for the ridge parameter (or the PLS parameter) and do not need any prior knowledge about the data set.

13.3 Extracting Multiple Canonical Correlations

For functional data, an algorithm which can extract multiple correlations is also very important. In order to extract multiple canonical correlations, we will employ a deflationary transformation. Instead of using explicit deflation, we will use an implicit deflationary approach, using lateral inhibition. Therefore, we just need to simply apply lateral inhibition to outputs y, noting that lateral inhibition is a more natural in-network method of deflation and has computational advantages over, e.g. Gram–Schmidt orthogonalisation. For example, if we want to extract the pth canonical correlation assuming that the previous $p-1$ have already been extracted, then the learning rule for second canonical correlation becomes

$$\Delta \mathbf{w}_{1p[i]} = \eta(\mathbf{x}_{1[i]}y'_{2p} - f(\mathbf{w}_{1p})(1-k_1)\mathbf{x}_{1[i]}y'_{1p} - f(\mathbf{w}_{1p})k_1 D^4_{[i]}\mathbf{w}_{1p})$$
$$\Delta \mathbf{w}_{2p[i]} = \eta(\mathbf{x}_{2[i]}y'_{1p} - f(\mathbf{w}_{2p})(1-k_2)\mathbf{x}_{2[i]}y'_{2p} - f(\mathbf{w}_{2p})k_2 D^4_{[i]}\mathbf{w}_{2p})$$

$$(13.17)$$

where y'_{1p} and y'_{2p} are defined as:

$$y'_{1p} = y_{1p} - \sum_{n<p} c_{1[pn]}y_{1n}$$
$$y'_{2p} = y_{2p} - \sum_{n<p} c_{2[pn]}y_{2n} \qquad (13.18)$$

We use Local Orthogonalization rule [40] to learn the lateral weights:

$$\Delta c_{i[pn]} = \eta(y_{ip}y_{in} - c_{i[np]}y^2_{in}) \qquad (13.19)$$

to update the weight strengths c which are local to the lateral synaptic connection. We use this algorithm to extract the second canonical correlation direction for the children's gait data, which is discussed in Section 13.4.3.

13.4 Experiments on Multicollinear Data

13.4.1 Artificial Data

We generate two artifical data sets $\mathbf{x}_1$ and $\mathbf{x}_2$,

$$\mathbf{x}_1 = (x_{11}, x_{12}, x_{13}, ..., x_{1n}) \tag{13.20}$$
$$\mathbf{x}_2 = (x_{21}, x_{22}, x_{23}, ..., x_{2n}) \tag{13.21}$$

in which we take $n = 20$; each x_{1n} is a linear combination of $\mathbf{b}_1$ and each x_{2n} is a linear combination of $\mathbf{b}_2$ where

$$\mathbf{b}_1 = (b_{11}, b_{12}, ..., b_{1p}) \tag{13.22}$$
$$\mathbf{b}_2 = (b_{21}, b_{22}, ..., b_{2p}) \tag{13.23}$$

with $p = 4$. Now, we have two data sets, each of which has very high internal correlations: the rank of both Σ_{11} and Σ_{22} is 4. Now we create a strong correlation between the two data sets by defining

$$c = \frac{x_{11} + x_{21}}{2} \tag{13.24}$$

and then set $x_{11} = x_{21} = c$. Now, the first elements in both data sets are exactly same, and each of these new first elements has a high correlation with other internal elements and also has a correlation with elements of the other data set. These provide the major correlations between the two data sets. We use two algorithms on this data set: one is our new algorithm (13.10) with the smoothing parameter, the other is the first algorithm in Chapter 9. The experimental result is shown in Table 13.1: we see that the existing neural algorithm [114] has had a great deal of difficulty with this data set while the new algorithm (13.10) has identified the major correlations very effectively.

13.4.2 Examination Data

Our second experiment uses the data set which was introduced in Chapter 9, the students' marks on five module exams. In Table 13.2, we see the correlations found by the CCA network for different values of k. There is rather little difference in the results, but a small trend towards increasing correlation as the value of k decreases can be seen. Also, the result from a standard statistical method (reported in [125]) was 0.660. We see that the neural methods tend to find slightly larger correlations though not as large as 0.692 which was reported in [114] with a previous neural method.

13.4.3 Children's Gait Data

The children's gait data has been used in [116] and was collected by the Motion Analysis Laboratory at the Children's Hospital, San Diego, California, (full

Table 13.1. The weights found on the artificial multicollinear Data. The left two columns are the results using the first neural algorithm from Chapter 9; the right two are those from (13.10).

	Existing algorithm		New algorithm	
	$\mathbf{w}_1$	$\mathbf{w}_2$	$\mathbf{w}_1$	$\mathbf{w}_2$
1	−0.6918	−0.5765	**−1.6644**	**−1.6222**
2	−0.0453	0.5527	0.0022	−0.0235
3	−0.8561	**0.946**	−0.0045	0.0048
4	0.4852	−0.2000	0.0031	−0.0290
5	−0.5817	−0.1928	0.0016	−0.0112
6	0.5947	−0.0139	0.0001	0.0239
7	−0.0364	0.6404	−0.0039	0.0117
8	−0.0373	−0.1642	−0.0006	−0.0059
9	0.1009	−0.7506	−0.0022	0.0037
10	0.4595	−0.2232	−0.0037	0.0057
11	−0.0454	0.0366	0.0042	0.0169
12	−0.2113	−0.1909	0.0001	−0.0071
13	0.0907	0.1842	0.0022	0.0051
14	−0.0137	0.3383	−0.0015	0.0007
15	−0.2990	0.4471	0.0038	−0.0059
16	−0.0754	−0.1469	−0.0034	−0.0073
17	−0.2422	−0.6754	0.0039	−0.0086
18	0.4185	0.7203	0.0000	0.0110
19	**1.8872**	−0.1287	−0.0040	−0.0052
20	−0.8019	0.6190	0.0037	0.0178

Table 13.2. The correlations found by the algorithm with different values of k. Standard statistical CCA yields 0.663.

k	1	0.9	0.8	0.7	0.6	0.5	0.4	0.3	0.2	0.1	0
Corr	0.659	0.660	0.662	0.665	0.668	0.671	0.675	0.679	0.684	0.688	0.690

details in [146]). The data set consist of the angular rotations in the sagittal plane of the hip and knee of 39 normal five-year-old children. The observations are taken over a gait cycle consisting of one double step taken by each child, and time is measured in terms of the cycle which has been discretized to a regular grid of 20 points; the data are illustrated in Fig. 13.1.

Fig. 13.2 (right) shows the results of the simulation in terms of the weight parameters with $k = 0.9$, while Fig. 13.2 (left) shows the results using the previous neural algorithm from Chapter 9. It is clear that Fig. 13.2 (left) is rather difficult to interpret while Fig. 13.2 (right) is much more interpretable. Also the smoothness of Fig. 13.2 (right) gives us somewhat greater confidence in the predictive power of this result since Fig. 13.2 (left) appears to be a noisy solution. Because we are not interested in this specific data, we do not analyse the experiment's results, but from Fig. 13.2, we can see the hip curve in the

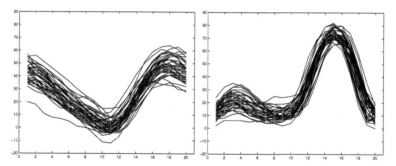

Fig. 13.1. Left: Angular rotations of the hip over one gait cycle in each of 39 subjects. Right: Angular rotations of the knee.

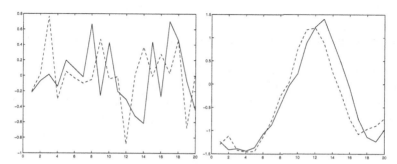

Fig. 13.2. Left: Canonical variate weights using the standard learning rule (Chapter 9), the solid line for hip, dashed line for knee. Right: Canonical variate weights using the new learning rule, the solid line for hip, dashed line for knee.

middle of the cycle occurs a little later than that in the knee curve, which concurs with the interpretation in [116]. Fig. 13.3 show the first and second weights value from the algorithm with the roughness penalty smoothing term (13.14) and (13.17). Both the first and second weights' values could be transformed roughly to being identical for the hip and the knee by speeding up the hip cycle relative to the knee cycle in the first half of the cycle and slowing it down in the second. Since the main interest is in comparing the curves, all of the weights value shown in Fig. 13.3 have been normalized so that the integral of their squares is equal to 1.

We also found experimentally that, if we use the simple ridge solution (13.8), we need to use a large k, (20 to 50), which has an adverse effect on the growth part of the algorithm and causes decay away from the optimal directions. The final result is that the estimate of canonical correlation is very poor. If we use the hybrid solution (13.10), we just need a number between 0 and 1.

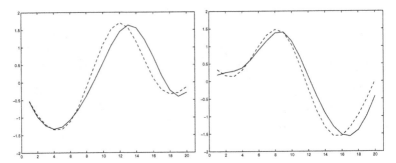

Fig. 13.3. Left: First pair of smoothed canonical weights, the solid line for hip, dashed line for knee. Right: Second pair of smoothed canonical weights, the solid line for hip, dashed line for knee.

13.5 A Neural Implementation of Partial Least Squares

The twinned network of the previous chapter contains the constraint (the orthonormalisation of the weights) which we require for PLS, and so we now assume two data streams from which we wish to extract the common interesting features, i.e. both streams are assumed to have a set of common underlying factors. We can write this model as

$$\mathbf{y}_1 = W_1^T \mathbf{x}_1$$
$$\mathbf{y}_2 = W_2^T \mathbf{x}_2$$

The input streams are denoted by $\mathbf{x}_1$ and $\mathbf{x}_2$, the projected data by $\mathbf{y}_1$ and $\mathbf{y}_2$ and the basis vectors are the rows of the matrices W_1 and W_2.

In the same way as the previous chapter, we may derive the following weight update rules:

$$\Delta W_1 = \eta[(\mathbf{x}_1 - W_1\mathbf{y}_1)\mathbf{y}_2^T] \qquad (13.25)$$
$$\Delta W_2 = \eta[(\mathbf{x}_2 - W_2\mathbf{y}_2)\mathbf{y}_1^T] \qquad (13.26)$$

Equations (13.25) and (13.26) may be implemented in a negative feedback framework. In both equations, the input is fed forward, and the outputs are fed back and subtracted from the input to form a residual. The weight update consists of multiplying the residual with functions of the values of both output streams. The sole difference between this network and that of the previous chapter is that we have removed the need for sphering.

13.5.1 Introducing Nonlinear Correlations

A neural implementation of PLS (or for that matter PCA, CCA or any other statistical technique) may be no more than a curiosity, engineeringwise [1] since

---

[1] Though, of course, it may be very interesting to neuroscientists.

efficient statistical techniques for these operations already exist. But when we introduce nonlinearities to such operations, no closed-form solutions may exist and so some iterative method such as an artificial neural network learning method becomes necessary.

As noted in [15], nonlinear projections are notoriously difficult to conceptualise and so, in practice, the nonlinearities which we use will have to be very circumscribed in order to maintain a tight control over the correlations found. In [67], we also wished to compare neural and kernel implementations of PLS and so we opted to maximise the function

$$J(W_1, W_2) = E((W_1^T \mathbf{g}(\mathbf{x}_1))^T W_2^T \mathbf{g}(\mathbf{x}_2)) + \frac{1}{2} \sum_{i=1}^{N} \sum_{j=1}^{N} \lambda_{ij}(\mathbf{w}_{1i}^T \mathbf{w}_{1j} - a_{ij})$$

$$+ \frac{1}{2} \sum_{i=1}^{N} \sum_{j=1}^{N} \mu_{ij}(\mathbf{w}_{2i}^T \mathbf{w}_{2j} - b_{ij}) \tag{13.27}$$

where typically the vector-valued function $\mathbf{g}()$ will be a radial basis function such as the Gaussian.

13.5.2 Simulations

Since the neural algorithms are novel, we illustrate the neural PLS methods on artificial data sets before discussing results on our financial data sets.

13.5.3 Linear Neural PLS

We draw 100 sample predictors from a uniform distribution in $[-2.5, 2.5] \times [-2.5, 2.5]$ and create target values from $X_2 = X_1 * A + \mu$ where $A = [3, 2]^T$ and μ is a vector of samples from zero mean Gaussian noise of variance 0.1. We initialise the learning rate to 0.001 and decrease this to zero during the course of the simulation. Each simulation was composed of 100 presentations of the training data in a random order. These parameters are somewhat ad hoc and the simulation results are robust to major changes in them. The learning rules are

$$\Delta \mathbf{w}_1 = \eta(\mathbf{x}_1 - \mathbf{w}_1 y_1)y_2 \tag{13.28}$$

$$\Delta \mathbf{w}_2 = \eta(\mathbf{x}_2 - \mathbf{w}_2 y_2)y_1 \tag{13.29}$$

Clearly the facts that the learning rule causes orthonormalisation of the weights and the second data set is one-dimensional means that $\mathbf{w}_2 = 1$ or -1 and this duly happens. Typical values of $\mathbf{w}_1^T$ were $[0.82\ 0.56]$ which is approximately in the ratio 3:2 whilst remaining of length 1.

We can also, from the training data, get an estimate of the correlation coefficient $\hat{\rho}$, which we use in creating our estimate on test data with the standard least squares estimate

$$\hat{y}_2 = (\mathbf{w}_2 * \mathbf{w}_2^T)^{-1} * \mathbf{w}_2 * (\mathbf{w}_1^T * \mathbf{x}_1 * \hat{\rho}) \qquad (13.30)$$

Fig. 13.4 graphs the actual output (horizontal axis) on noise-free test data against the estimated values (vertical axis). We see that the correlation is extremely high.

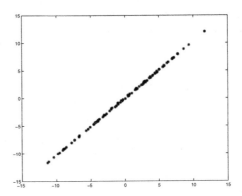

Fig. 13.4. Graph of the actual output (horizontal axis) against the estimated values on noise-free test data.

However standard statistics can perform this equally well and so we seek to go beyond this result. We first consider a scheme for creating mixtures of linear PLS regressors.

13.5.4 Mixtures of Linear Neural PLS

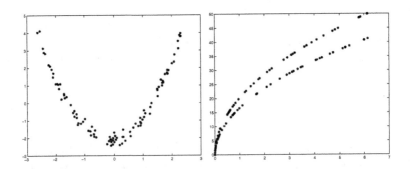

Fig. 13.5. The left diagram shows the training data, the horizontal axis being the predictors, the vertical axis being the targets. The right diagram shows the predicted outputs (vertical axis) against the actual (horizontal axis).

We begin with a very simple experiment: we draw predictors iid from a uniform distribution on $[-2.5, 2.5]$, and calculate targets from $X_2 = X_1^2 + \mu$ where μ is drawn from zero mean Gaussian noise of variance 0.2. Examples of one set of training data are shown in Fig. 13.5. We trained the network with exactly the same rules as in the previous section but with two enhancements: firstly, we had two pairs of outputs seeing the data and after the feedforward stage tested whether $y_1 y_2 > y_3 y_4$. If so, we updated $\mathbf{w}_1$ and $\mathbf{w}_2$ using (13.28) and (13.29). If not, we updated $\mathbf{w}_3$ and $\mathbf{w}_4$ using

$$\Delta \mathbf{w}_3 = \eta(\mathbf{x}_1 - \mathbf{w}_3 y_3) y_4 \tag{13.31}$$

$$\Delta \mathbf{w}_4 = \eta(\mathbf{x}_2 - \mathbf{w}_4 y_4) y_3 \tag{13.32}$$

where we have left the bold font to cater for the possibility that the input data might be more than unidimensional.

The second enhancement was necessitated by the requirement during testing that the new data must be allocated to either the $y_1 - y_2$ channel or the $y_3 - y_4$ channel. Thus we create two centres (one for each pair of channels) which determine which channel the current input should use. Thus, after each competition (as to whether $y_1 y_2 > y_3 y_4$) we also update the appropriate centre: if $y_1 y_2 > y_3 y_4$, $\mathbf{c}_1 = \mathbf{c}_1 + \eta(\mathbf{x}_1 - \mathbf{c}_1)$, else $\mathbf{c}_2 = \mathbf{c}_2 + \eta(\mathbf{x}_1 - \mathbf{c}_2)$. When a test point is presented to the network, we use the test: if $||\mathbf{x}_1 - \mathbf{c}_1|| < ||\mathbf{x}_1 - \mathbf{c}_2||$, estimate using (13.30), else use

$$\hat{y}_2 = (\mathbf{w}_4 * \mathbf{w}_4^T)^{-1} * \mathbf{w}_4 * (\mathbf{w}_3^T * \mathbf{x}_1 * \hat{\rho}_2) \tag{13.33}$$

with $\hat{\rho}_2$, the estimate of the correlation between y_3 and y_4. This method is clearly generalisable to any number of pairs of outputs in which while learning we simply choose the pair with greatest product and in testing select the channel whose centre is closest to the current input.

The right half of Fig. 13.5 shows the actual output (horizontal axis) against the estimated (vertical output). Clearly the major correlations are being captured but the quadratic nature of the mapping is clearly seen. We may state that the gross structure of the mapping is being captured as well as possible by a linear machine yet to go further we require a nonlinear machine.

13.5.5 Nonlinear Neural PLS Regression

We noted that radial kernels on the X_1 data set and linear kernels on the X_2 data set were the most successful kernels for forecasting. In the light of this, we use a specific radial nonlinearity in the experiments reported herein. We have also experimented with sigmoids but achieved no better results than those reported here.

We begin with the same type of data as in the last section - a noisy quadratic. The data was 100 samples from exactly the same distribution. However now the model is $X_2 = \phi(X_1).\mathbf{w} + \mu$. We take $\phi(\mathbf{t}) = \{\phi_1(\mathbf{t}), \phi_2(\mathbf{t}), ...,$

$\phi_M(\mathbf{t})\}$ where $\phi_i(\mathbf{t}) = \exp(-||\mathbf{t} - \mathbf{c}_i||^2/\sigma)$, a Gaussian with centre $\mathbf{c}_i$ and width parameter, σ. In the first experiment, we used $M = 5$ and positioned the centres evenly at -2.5, -1.25, 0, 1.25 and 2.5. $\sigma = 1$ though the results were robust to substantial changes in this. Results are shown in Fig. 13.6.

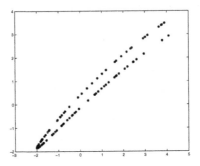

Fig. 13.6. The plot of actual value from the quadratic map (horizontal axis) against the estimated value (vertical axis).

We see that the two wings of the quadratic mapping are dealt with separately and that each has approximately found the optimal correlation. Further, while not all of the nonlinearity has been captured, Fig. 13.6 shows a much more linear relationship than does Fig. 13.5.

It must be pointed out that, in [67], a comparison between the linear and nonlinear PLS methods was performed on financial data sets and the nonlinear version was not able to perform better than the linear method on simple forecasting problems with the S&P 500 and on exchange rates. Also a method equivalent to FastECA was shown to exhibit very fast and accurate convergence.

13.6 Conclusion

In this chapter, we have used a basic idea from ridge regression to create an algorithm which is robust with respect to multicollinear data. We have shown the effectiveness of the algorithm on artificial data which was designed to be multicollinear and on a real data set which has a limited number of samples in relation to its dimensionality. With a second real data set – one which does not exhibit multicollinearity – we have shown that the addition of a robustification parameter does not materially affect the results for a wide range of parameter values.

We have introduced a penalty term which penalises roughness in the canonical correlation directions and shown that the resulting vectors are much more interpretable than the original. The resulting canonical correlation vectors are more suited to prediction than those achieved without the penalty term.

We have also used the network of the previous chapter but without sphering to perform PLS regression. The nonlinear version was shown to extract relationships exhibiting higher correlations than the linear version.

14

Twinned Principal Curves

In this chapter, we develop methods for combining information from two data sources which have some underlying correlation or dependency where this dependency may not be best viewed as a linear dependency. Our inspiration is biological information processing: sensory information from a single organism is sent to our brains via different sensory channels; however, at some stage we are able to integrate this information so that we can re-create much of the original sensory information from memory when we have a new input from a single sensory stream. We can view this as a form of forecasting – before we taste the orange we can visualise how it will taste. However, our main aim is to develop an engineering application in which extracting structure from two data sets simultaneously is, in some way, easier than extracting information from the individual data sets separately. Traditional statistics has thoroughly investigated linear correlations between two data sets (CCA); however, it is our aim to develop methods which do not rely on the underlying relationship between the data sets being linear.

In this chapter, we will develop an extension of Principal Curves which performs a type of nonparametric CCA and illustrate its use on artificial data and as a method for forecasting on a financial data set which we have previously [69] used to test other forecasting methods. We discuss which criterion is optimal for terminating the algorithm and investigate two alternative extensions of the basic algorithm which are shown to improve the ability of the algorithm to forecast.

14.1 Twinned Principal Curves

Principal Component Analysis (PCA) is a standard statistical technique for finding a lower-dimensional linear projection of high-dimensional data which gives minimum mean square error over all projections of this dimensionality. Principal Curves [36, 70, 106] is an extension of this method in which a nonlinear manifold can be used instead of the linear subspace determined by PCA.

However there is clearly a difficulty with this in that it is always possible to fit a finite training set with no error. There are several definitions of Principal Curves which constrain the curves in one way or another to overcome the problem of overfitting. In [70], every point, P, on the curve is the mean of the points that project onto P, which is known as self-consistency. The self-consistent curve with unit-speed – one whose derivative has norm 1 – is the Principal Curve. In [106], the Principal Curve is defined as the curve of a specific length which minimises the mean squared distance from the data.

We may extend the Principal Curve method so that we now find a non-linear manifold in each of two data sets simultaneously. A forward look at Fig. 14.1 may clarify our intentions: in this figure, we illustrate two data sets (left and right halves of the figure, respectively) which have an underlying non-linear relationship. Both data sets are inherently one-dimensional (though corrupted by noise) and as we move along the curve using the underlying one-dimensional parameter, we also move round the circle using the same parameter.

We use a nonparametric method to determine the two manifolds. Since we are drawing data iid from two data sets simultaneously, our method creates manifolds which exhibit a correlation between corresponding points on the manifolds which we can then use to forecast a sample from one data set given a sample from the other. The core of the algorithm performs a local smoothing operation on both data sets where the smoothing is over all neighbours of the point which have

- projections close to the projections of the chosen point, and
- projections of their corresponding points in the other data set satisfying the same constraint with respect to the second data set. Note that these projections will be to different curves.

Thus, the algorithm in outline is the following:

1. Initialise d_1^i and d_2^i, with the projections of $\mathbf{x}_1^i$ and $\mathbf{x}_2^i$ $\forall i$ onto the first principal component of each data set.
2. With the current projections d_1^i and d_2^i, $\forall i$, select $\mathbf{x}_1^i$ from the first data set and the corresponding point, $\mathbf{x}_2^i$ from the second data set.
3. If d_1^i is the projection of $\mathbf{x}_1^i$ and d_2^i is the projection of $\mathbf{x}_2^i$, then $S_i = \{k : |d_1^k - d_1^i| < \epsilon_1 \text{ and } |d_2^k - d_2^i| < \epsilon_2\}$.
4. Find the local average of points projecting close to $\mathbf{x}_1^i$ and $\mathbf{x}_2^i$. i.e. $d_1^i(\text{new}) = $ mean of $d_1^j, j \in S_i$ and $d_2^i(\text{new}) = $ mean of $d_2^j, j \in S_i$.
5. Return to step 2 until all points have been selected.
6. $d_1^i = d_1^i(\text{new})$ and $d_2^i = d_2^i(\text{new})$, $\forall i$.
7. If stopping criteria is not met, return to step 2.

The algorithm iterates until a stopping criteria is met: either the algorithm repeats for a set number of rounds or until the number of nodes to which the data is projected reaches a certain number (see below) or until the mean square error reaches a particular value. In the remainder of this chapter, we will describe these projected nodes as "knot points" and they will be used as the basis for a linear interpolation of the data set.

Clearly there are extensions which can be made to this algorithm. For example it is possible to change the value of the width parameters ϵ_1 and ϵ_2 during the course of the iterations, though this is not implemented in the simulations discussed in this paper for reasons which will become clear in the next section. We might also consider the use of a weighted average rather than a simple average to improve the accuracy of the new projections; however, simulation results have shown no improvement with this method. Finally, the algorithm tends to draw data from the extremes of the principal curve and so some additional local averaging may be useful at these points.

A reviewer has pointed out that the method described above has more in common with vector quantisation than with curve fitting. This is to some extent true and a similar criticism can be made of the original Principal Curves algorithm. However, note that the smoothing/quantising of the data set is followed by a joining up of the knot points to create an approximation to a curve by local linear projections, which, in this chapter, are of three main types. Thus, for example, when we use this algorithm to forecast, we project a data point onto curve 1 at a position which lies between two knot points. Since each knot point has a corresponding knot point on the second curve, we can identify any intermediate position on the first curve with an intermediate position on the second curve. Thus, the curve aspect of the algorithm plays an important part in the forecasting procedure.

14.2 Properties of Twinned Principal Curves

Canonical Correlation Analysis (CCA) is a purely linear method (Chapter 9): it will find the basis of each of two spaces which captures the greatest correlations between two data sets. In previous chapters, we have developed neural implementations of CCA and have been able to add nonlinearities to these so that they can "bend" through data sets. In Chapter 11, we developed kernel CCA (based very much on the pioneering work of [166] in developing kernel principal component analysis). The advantage that the current work has over these methods is one of clarity – we can directly visualise both where we are in the data space and the local direction of the principal curve.

14.2.1 Comparison with Single Principal Curves

The Twinned Principal Curve Algorithm is a somewhat different algorithm from that suggested by [70] or [106]. It has a rather nice property of sparsification of the projections: the local averaging provides a smoothing of the data set and since we keep the values of ϵ_1 and ϵ_2 constant during the course of the simulation this smoothing progressively works out from each data point resulting in fewer and fewer projections onto the principal curve (compare the central two rows in Fig. 14.1). We may use this property to allow the number of distinct nodes we seek to determine the value of ϵ_1 and ϵ_2 (or vice versa).

It is worth noting also that this algorithm is able to deal with data sets which standard Principal Curve algorithms find difficult: the very fact of having two data sets with which to work simultaneously alleviates several problems. For example, since we initialise with a PCA and one of our data sets is circular, any diameter of the circle may be a Principal Component direction. This unfortunately means that points on opposite sides of the circle project onto the same part of the eigenvector and so we often have an initial twisting of a single Principal Curve as it moves from the centre of mass on one side of the circle to the centre of mass on the other side, these centres of mass being caused by the finite numbers of samples. However, with Twinned Principal Curves, we only consider points to be local to the current point if they are local in both projections. This makes it much less likely that false neighbours will be chosen.

We have been asked whether we can simply use a single $(m + n)$-dimensional Principal Curve rather than twinning an n-dimensional curve with an m dimensional curve. That this is not necessarily an equivalent procedure can be seen from part 3 of our algorithm in Section 14.1: a single $(m + n)$-dimensional parameter would allow one criterion to be breached if the other was comfortably achieved. To illustrate this, let us compare the creation of a single line in a $2d$-dimensional space and two lines in two separate d-dimensional space. Let us consider the simple case where noise in the data set is drawn iid from a uniform distribution in each dimension from $[-r, r]$. Then the smoothing algorithm above, if using the Euclidean norm, demands a sphere of radius at least r. In the case of r = 0.5 (a value used in the experiments described in the next section) with d = 10, V_{20}, the volume of a 20-dimensional sphere is 2.4611×10^{-8} while V_{10}, the volume of a 10-dimensional sphere is 0.0025.

However, even more importantly, the proportion of V_{20} which lies within 0.1 of the surface of the sphere is 98.8% while the proportion of V_{10} which lies within 0.1 of the surface of the sphere is 89.3%. *Much more of the smoothing is being done for points which are far from the centre of the sphere in the higher-dimensional sphere.* Further, the above assumes that there is equal noise in all dimensions: we actually want to have as little smoothing as possible in the dimension in which the underlying curve is actually lying, and sufficient (i.e. up to width r) smoothing in the other dimensions. Below, we show that the twinned algorithm is useful when the structure determining the curve results in a curve which intersects itself. By smoothing in all dimensions equally, we are smoothing in space which may be occupied by a curve returning close to the point where we are currently smoothing: if the next approach to the current point is principally within one of the d-dimensional subspaces, the above algorithm will identify it as a separate part of the curve while an algorithm which uses the $2d$-dimensional space may smooth the new points to a single curve. Therefore, just as it is possible to view the two weight vectors in standard statistical CCA to be a single weight vector in a combined space, it is possible to view the resulting curve in Twinned Principal Curves as a one-dimensional curve in a combined space. However, this is not a useful view in either case and ignores the distinctive "twoness" in the method for creating the pair of vectors/curves.

Finally, CCA maximises the correlation between two data sets under the constraint that the variances of $y_1 = \mathbf{w}_1^T \mathbf{x}_1$ and $y_2 = \mathbf{w}_2^T \mathbf{x}_2$ are both 1. Twinned Principal Curves can still meet this criterion; having found our sum of linear approximators, we may project new samples onto these Twinned Principal Curves and calculate the variance of the resultant projections. In calculating new correlations, we may simply then divide each of y_1 and y_2 by their corresponding standard deviations.

14.2.2 Illustrative Examples

Artificial data

We first create two sets of two-dimensional artificial data which are known to have a correlation from $x_1(t) = \sin(t) + \mu_1, y_1(t) = \cos(t) + \mu_2, x_2(t) = t + \mu_3, y_2(t) = \frac{t}{3} + \sin(t) + \mu_4$, where t is drawn from a uniform distribution in $[0, 2\pi]$ and $\mu_i \sim N(0, 0.2)$ is Gaussian noise. Examples of this data are shown in the top row of Fig. 14.1.

Fig. 14.1 also shows the thinning which takes place in data set 2 after 1, 2 and 10 iterations and in data set 1 after 10 iterations. The sparsification discussed above is clearly evident.

Now we may use these projections to predict the position of a point, $\mathbf{x}_2$, in data set 2 given its corresponding point $\mathbf{x}_1$ in data set 1. Typically we will approximate the principal curves with the sum of linear projections given by joining the sparse points as shown in the last row of Fig. 14.1. To forecast, we project $\mathbf{x}_1$ onto the current principal curve of the first data set and use the corresponding point on the current principal curve of the second data as the predictor of $\mathbf{x}_2$. Typical results are shown in Fig. 14.2, the "*" on the curve being the predictor while the "+" shows the point's actual position.

Real Data

We now investigate if we can forecast on real data: given the last few days' exchange rates (U.S. dollar against British pound), is it possible to forecast the next day's exchange rate with some degree of accuracy? We have previously [69] used a variety of methods (Principal Component Analysis, Factor Analysis, Independent Component Analysis and Complexity Pursuit) to find the underlying factors in this data set. We then used a standard multilayered perceptron using backpropagation to predict future values of the time series; the inputs to the multilayered perceptron were the projections of previous values of the time series onto the factors found by each method. To test our

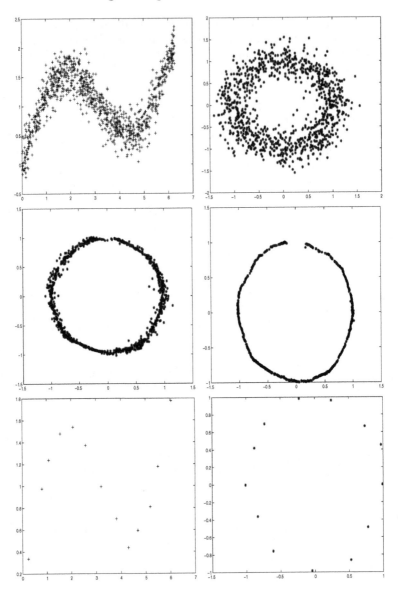

Fig. 14.1. The top two diagrams show samples from the data sets: the first column shows samples and results from $x_1(t) = \sin(t) + \mu_1, y_1(t) = \cos(t) + \mu_2$ and the second column shows samples and results from $x_2(t) = t + \mu_3, y_2(t) = \frac{t}{3} + \sin(t) + \mu_4$. The second row shows the first and second projections of the second data set. The third row shows the projections of both data sets after 10 iterations.

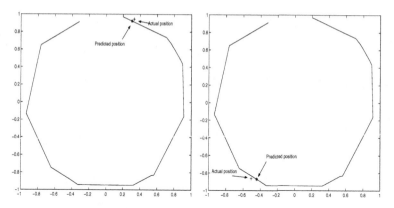

Fig. 14.2. The results of forecasting the positions of points in data set 2 given only the position of the corresponding point in data set 1.

multilayered perceptron, we have split the data set into two sets: 1706 samples were used as the training data and 1706 for the test data. Each training input comprised a particular day's exchange rate plus the previous n days' exchange rates where values of n ranged from 5 to 25. With the Twinned Principal Curves algorithm, we can simultaneously forecast as many days in advance as we wish, since our second principal curve can be as high dimensional as we wish. Typical results in terms of mean absolute percentage error on the test set are given in Table 14.1.

Table 14.1. The first column gives the number of knot points and the others give the mean absolute percentage error on a test data set predicting 1 to 5 days ahead.

Knot points	Day 1 Day 2 Day 3 Day 4 Day5
57	1.0006 1.1086 1.2103 1.3035 1.4022
408	0.7413 0.9158 1.0685 1.1863 1.2887
607	0.6711 0.7939 0.9018 1.0197 1.0880

14.2.3 Intersecting Curves

One of the limiting factors for Principal Curves is that the curve (and hence the data set) should not intersect with itself. If this happens, the direction of the maximum rate of change will not be unique at that point and so the Principal Curve cannot be found uniquely. However, when we have two data sets such intersections are permissible *provided* intersections in both data sets do not occur at the same time in both data sets.

Consider the data shown in the top line of Fig. 14.3: it comprises two sets of two-dimensional artificial data from $x_1(t) = \sin(t) + \mu_1, y_1(t) = \cos(t) +$

$\mu_2, x_2(t) = t + \mu_3$, if $t \in [0, 2\pi]$ else $x_2(t) = (4\pi - t) + \mu_3, y_2(t) = \frac{t}{3} + \sin(t) + \mu_4$ if $t \in [0, 2\pi]$ else $y_2(t) = \frac{4\pi - t}{3} + \sin(4\pi - t) + \mu_4$ where t is drawn from a uniform distribution in $[0, 4\pi]$ and $\mu_i \sim N(0, 0.2)$ is Gaussian noise. The knot points after 10 iterations of the twinned principal curves algorithm are shown in the second line of this figure. The third line of this figure shows two examples of prediction using this method. A point is generated from data set 1 and the equivalent projection onto data set 2 is calculated. The real data point 2 is also shown in these figures. We see that both points are close together.

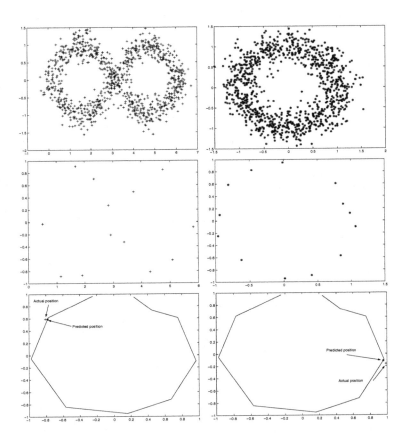

Fig. 14.3. The top two diagrams show samples from the data sets. The second row shows the projections of both data sets after 10 iterations. The last row shows the results of forecasting the positions of points in data set 2 given only the position of the corresponding point in data set 1.

14.2.4 Termination Criteria: MSE

Deciding when to stop the iterations of the Twinned Principal Curves method has been ignored in the above. In order to gain some insight into the performance of the curve at different iterations, we can calculate the MSE on a separate data set at each iteration. In the following experiments, we have recreated a new test set for each iteration so that the decision as to when to stop was taken on a totally independent set each time. Each parameter set was investigated with multiple training runs. We are interested in two functions of the training regime: the number of knot points and the sum of the squared errors for the given knot points. An example of three runs is given in Table 14.2.

We see that as the number of knot points decreases, the error initially decreases before beginning to increase again. The later increase is due to an increase in bias – the number of knot points is not sufficient to adequately represent the data. The initial decrease is due to a decrease in variance as the noise is removed due to the smoothing effect of the algorithm. Table

Table 14.2. The sum of the squared errors on a test set of artificial data. Each pair of columns is a separate run of the algorithm with a different width (ϵ) parameter.

Iter	Knots	SSE	Knots	SSE	Knots	SSE
1	973	58.8	950	73.7	976	80.5
2	690	41.1	638	51.5	888	58.4
3	483	48.8	428	59.3	609	52.0
4	316	52.4	268	76.3	318	53.9
5	173	60.9	165	88.7	113	80.7
6	72	71.9	73	97.9	51	88.5
7	37	75.6	33	105.3	31	117.6

14.3 shows the decreasing number of knot points in a simulation based on the dollar–pound exchange rate (3497 data points). We see that the number of knot points decreases to 18 and then remains stable for the last three iterations; also, unsurprisingly, the mean absolute percentage error increases as the number of knot points decreases and as we try to forecast further into the future.

14.2.5 Termination Criteria: Using Derivative Information

Another measure which we might use as a stopping criterion is the second derivatives of the Principal Curves found. We anticipate that the initial estimates of the Principal Curves are liable to be extremely noisy and so the curves will tend to wiggle about a lot: the second derivatives of the curves will be very high. Subsequently, the local averaging smoothes the curve and

Table 14.3. The knot points (from a data set of 3497 exchange rates) after 1,2,...,10 iterations of the Twinned Principal Curves algorithm and the corresponding mean absolute percentage errors. The simulation converges to a stable 18 knot points.

Knot points	Day 1	Day 2	Day 3	Day 4	Day 5
1459	0.907	1.008	1.128	1.216	1.328
915	0.920	1.025	1.145	1.237	1.346
571	0.935	1.037	1.162	1.248	1.360
297	1.044	1.143	1.252	1.335	1.430
134	1.194	1.278	1.373	1.441	1.539
42	1.255	1.328	1.421	1.491	1.579
25	1.330	1.397	1.489	1.548	1.632
18	1.349	1.416	1.504	1.566	1.644
18	1.349	1.416	1.504	1.566	1.644
18	1.349	1.416	1.504	1.566	1.644

the second derivatives will tend to decrease. We have investigated this in two ways:

1. We have calculated the gradients of the curves between successive knot points. To estimate the gradients, we have found the position of each knot point in terms of distance along the curve, and then estimated the rate of change in each dimension using

$$\frac{dx_j}{dt}\Big|_i \approx \frac{\Delta x_j}{\Delta t}\Big|_i \tag{14.1}$$

where Δx_j is the difference between the ith and $(i+1)^{th}$ point in the jth input dimension and Δt is the distance along the Principal Curve again between the ith and $(i+1)^{th}$ point. The scalar product, $\frac{dx_j}{dt}\Big|_i \cdot \frac{dx_j}{dt}\Big|_{i+1}$, tends to 1 during the course of the simulation as consecutive gradients become close to one another (see Fig. 14.4).

2. An alternative is to reuse the same gradient information to calculate an estimate of the second derivative:

$$\frac{d^2 x_j}{dt^2}\Big|_i \approx \frac{\Delta x_j|_i - \Delta x_j|_{i-1}}{\Delta t|_i} \tag{14.2}$$

where $\Delta t|_i$ is the difference along the line between the mean of the ith and $(i+1)^{th}$ segments. The average of the absolute values of the estimates of the second derivative begin rather high but tends towards 0 during the course of the simulation (see Fig. 14.5). The increase at iteration 2 is rather typical of our results and is caused by the shortening of the line segments $\Delta t|_i$ in (14.2) as the smoothing takes place.

Comparing the results of these two methods (Fig. 14.4 and Fig. 14.5), we find that the first rises quickly and levels off while the second tends to take longer to tend to 0. Given the MSE results, we consider that the knee on the graph of the first method is the best predictor of when to stop iterating the algorithm.

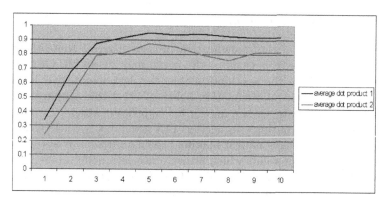

Fig. 14.4. The average dot product between consecutive gradient vectors. We see that the dot product tends to 1 during the course of the simulation. The actual highest value is, of course, dependent on the curvature of the data set.

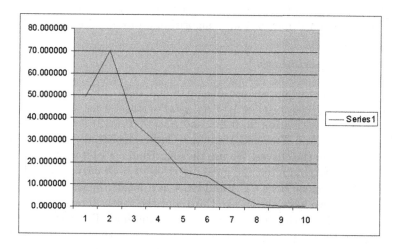

Fig. 14.5. The average second derivative of the curves in various iterations. We see that this tends towards 0 during the course of the simulation.

14.2.6 Alternative Twinned Principal Curves

The similarity between this algorithm and Canonical Correlation Analysis suggests other methods of creating the twinned curves rather than simply joining knot points. In this section, we propose an alternative method of approximating the principal curve by using a locally linear method similar to Local Principal Component Analysis (LPCA)[98].

Local Linear Approximation

The twinned nature of the sparsification process suggests that other means of finding linear approximations to the curve may be useful. Of course, all such methods rely on the assumption that the data manifold is locally flat enough that the linear approximation holds; whether this assumption is true of a particular real data set is a matter for empirical testing.

The algorithm as outlined in Section 14.1 generates two sets of matching nodes that describe the nonlinear manifold the data lies on. It also provides a set of the original data points which project onto each node. These points can be used to perform a Local Canonical Correlation Analysis (LCCA) which will give us a locally linear projection of each data set which will approximate a globally nonlinear manifold. Of course, we lose the simplicity of the previous method and we must take care at the junctions of consecutive LCCA vectors, but we will show that the LCCA forecasts provide lower mean square errors than the original method. We will also use a local version of Exploratory Correlation Analysis (Chapter 12) which will also be shown to outperform the above Twinned Principal Curve algorithm on forecasting problems.

Illustrative Experiments

In our first experiment, we use the artificial data set: $x_1(t) = \sin(t) + \mu_1, y_1(t) = \cos(t) + \mu_2, x_2(t) = \frac{1}{3}t + \mu_3, y_2(t) = \sin(t) + \mu_4$, where t is drawn from a uniform distribution in $[0, 2\pi]$ and $\mu_i = N(0, 0.12)$ is Gaussian noise.

We used the local Twinned Principal Curve algorithm to determine the nodes which lie on the curves. We use the algorithm for 10 iterations and after each iteration, we computed the local direction of both curves at each of the matching nodes, using both CCA and ECA as a locally linear approximation.

In Fig. 14.6 we have shown the third and the tenth iteration of the algorithm using CCA as the linear approximation method. The nodes are shown as circles and the estimated directions of the curve at the nodes are indicated by small line segments. During convergence, the number of nodes decreases, which simplifies the description of the principal curve. The same experiment has been repeated using ECA as the linear approximation method and those results are shown in Fig. 14.7. To measure performance, we predict the location of a set of samples that lie on the principal curve of the second data set, given the corresponding samples that lie on the first. The error is measured as the mean of the Euclidean distances between the locations of the predicted data points and the locations of the actual data points. The results of the error estimations are shown in Fig. 14.8.

The dotted line shows the error using the original twinned principal curve algorithm. The dashed line shows the error when the data point is projected onto the locally estimated CCA vector. The solid line shows the error when the ECA vectors are used in the prediction.

As we can see from these error-measurements, both the local ECA and CCA predictions have a significantly lower error than the prediction of the original method. As the number of nodes decreases, the error of the original method increases, whereas the errors of local ECA and CCA predictions remain more constant. Note that the original method approaches the more complex methods at iteration 2, giving us another reason to investigate termination criteria of the original algorithm.

The predictions using ECA are slightly more accurate than the corresponding CCA predictions. This can also be seen in Fig. 14.6 and Fig. 14.7, where the CCA vectors track the nonlinear manifold less accurately than the ECA vectors.

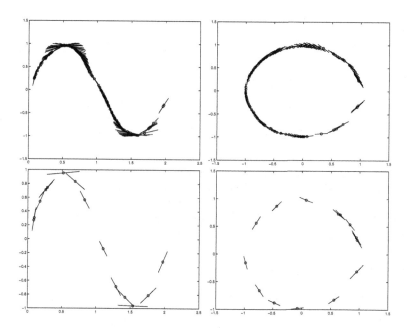

Fig. 14.6. Results on the artificial data sets using CCA as a local approximation: the left column shows the local linear approximation on the data set $x_1(t) = \sin(t) + \mu_1, y_1(t) = \cos(t) + \mu_2$ at three iterations (top) and at ten iterations (bottom). The right column shows the local linear approximation on the data set $x_2(t) = \frac{1}{3}t + \mu_3, y_2(t) = \sin(t) + \mu_4$ at three and ten iterations.

Financial Forecasting

In order to test the performance of the locally linear method on real data, we repeat the forecasting problem as described in Section 14.2.2. After each iteration of training, we calculate the mean absolute percentage error for each

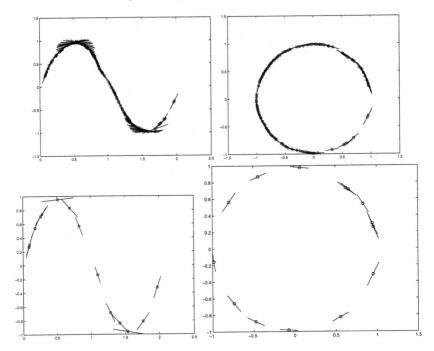

Fig. 14.7. Results on the artificial data sets using ECA as a local approximation: the left column shows the local linear approximation on the data set $x_1(t) = \sin(t) + \mu_1, y_1(t) = \cos(t) + \mu_2$ at three iterations (top) and at ten iterations (bottom). The right column shows the local linear approximation on the data set $x_2(t) = \frac{1}{3}t + \mu_3, y_2(t) = \sin(t) + \mu_4$ at three and ten iterations.

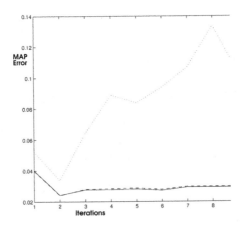

Fig. 14.8. The dotted line shows the errors when using the original method, the dashed line shows the errors using CCA as a local approximation and the solid line shows the errors using ECA.

forecasted day, using both the original method and the local CCA and local ECA methods.

The results are shown in Table 14.4, Table 14.5 and in Table 14.6. As we can see from the tables, both the local CCA and the local ECA methods outperform the original method. When we compare the performance of local CCA to local ECA on this real data set, we see that in this case local CCA outperforms local ECA; this is probably due to noise in the data set, which provides a variance term to which the ECA algorithm responds.

Table 14.4. Mean square error forecasting exchange rates between the dollar and the pound, using the original principal curve algorithm.

Iteration	Nodes	Day 1	Day 2	Day 3	Day 4	Day 5
1	777	0.230	0.321	0.424	0.505	0.615
5	96	0.446	0.554	0.686	0.790	0.932
10	71	0.462	0.583	0.736	0.860	1.007

Table 14.5. Mean square error forecasting exchange rates between the dollar and the pound, using the local ECA twinned principal curve algorithm.

Iteration	Nodes	Day 1	Day 2	Day 3	Day 4	Day 5
1	777	0.219	0.315	0.411	0.493	0.598
5	96	0.299	0.396	0.520	0.619	0.751
10	71	0.303	0.416	0.566	0.685	0.816

Table 14.6. Mean square error forecasting exchange rates between the dollar and the pound, using the local CCA twinned principal curve algorithm.

Iteration	Nodes	Day 1	Day 2	Day 3	Day 4	Day 5
1	777	0.159	0.261	0.361	0.447	0.560
5	96	0.164	0.268	0.399	0.502	0.637
10	71	0.169	0.288	0.451	0.580	0.716

14.3 Twinned Self-Organising Maps

The connection between Principal Curves and Self-organising Maps (SOM) has been discussed often in the literature e.g. [134]. This suggests that the SOM might be used in a similar manner to the Twinned Principal Curves algorithm. This may be conceptually thought of as two SOMs linked via the method of determining the winning neuron. Thus, if the centres in our first space (the last 10 days data, for example) are given by $\mathbf{w}_i$ and the centres in the second space (the 5 days ahead, which we wish to predict) are given by $\mathbf{v}_i$, then we can select our winner using

$$c = \arg\min\{||\mathbf{x}_1 - \mathbf{w}_i|| + ||\mathbf{x}_2 - \mathbf{v}_i||\} \tag{14.3}$$

and then updating our individual centres each with the standard learning rules for a SOM,

$$\Delta\mathbf{w}_i = \eta\Lambda(c, i)(\mathbf{x}_1 - \mathbf{w}_i)$$
$$\Delta\mathbf{v}_i = \eta\Lambda(c, i)(\mathbf{x}_2 - \mathbf{v}_i)$$

where η is a learning rate and $\Lambda(c, i)$ is the neighbourhood function which in our case was a simple Gaussian. To test how accurate the trained model is on new data, we determine the winner only on the $\mathbf{x}_1$ data set

$$c = \arg\min\{||\mathbf{x}_1 - \mathbf{w}_i||\} \tag{14.4}$$

and use

$$|\mathbf{x}_2 - \mathbf{v}_c| \tag{14.5}$$

as a measure of the error of the prediction. Results on the same financial data set as previously are shown in Table 14.7 which should be compared with Table 14.1. We see that the SOM easily outperforms the Twinned Principal Curves algorithm for this task for an equivalent number of centres or knot points.

Table 14.7. Using the same data set of 3497 exchange rates, the mean absolute percentage error when using twinned Self-Organising Maps.

Centres	Day 1	Day 2	Day 3	Day 4	Day 5
25 (SOM)	0.6552	0.6671	0.6768	0.6908	0.6933

14.3.1 Predicting Student's Exam Marks

Our second experiment uses the 88 students' marks on five module exams (Chapter 9). The exam results can be partitioned into two data sets: two exams were given as closed-book exams while the other three were open-book exams. We have used the two methods above to attempt to predict the students' open-book exams from their closed-book exams. The results (in terms of mean absolute percentage errors) are shown in Table 14.8. The Principal Curve method has a slight advantage over the Twinned SOM method, but this advantage is reversed for Exam 3. It is very difficult to analyse why these results take the form that they do, though clearly the results merit further study.

The Twinned Principal Curve method used only two iterations through the data set while the Twinned SOM was trained with 100 000 samples (with replacement, clearly) from the 88 data points.

Table 14.8. The mean absolute percentage error when using the two methods to predict 88 students' open-book exams from their closed-book exams.

SOM	0.1453	0.3055	0.3601
P Curve	0.2573	0.2494	0.1420

14.4 Discussion

The Twinned Principal Curves algorithm is a somewhat different algorithm from that suggested by [70] or [106] in that it iteratively uses a kernel smoother rather than attempting to approximate a principal curve by a mixture of straight lines. However, it has a rather nice property of sparsification of the projections: the local averaging provides a smoothing of the data set and since we keep the values of ϵ_1 and ϵ_2 constant during the course of the simulation this smoothing progressively works out from each data point resulting in fewer and fewer projections onto the principal curve (compare the central two rows in Fig. 14.1). We may use this property to allow the number of distinct nodes we seek to determine the value of ϵ_1 and ϵ_2 (or vice versa).

It is worth noting also that this algorithm is able to deal with data sets which standard Principal Curve algorithms find difficult: the very fact of having two data sets with which to work simultaneously alleviates several problems. For example, since we initialise with a PCA and one of our data sets is circular, any diameter of the circle may be a Principal Component direction. This unfortunately means that points on opposite sides of the circle project onto the same part of the eigenvector and so we often have an initial twisting of the Principal Curve as it moves from the centre of mass on one side of the circle to the centre of mass on the other side, these centres of mass being caused by the finite numbers of samples. However, we only consider points to be local to the current point if they are local in both projections. This makes it much less likely that false neighbours will be chosen.

CCA maximises the correlation between two data sets under the constraint that the variance of $y_1 = \mathbf{w}_1^T \mathbf{x}_1$ and $y_2 = \mathbf{w}_2^T \mathbf{x}_2$ are both 1. Twinned Principal Curves can still meet this criterion; having found our sum of linear approximators, we may project new samples onto these Twinned Principal Curves and calculate the variance of the resultant projections. In calculating new correlations, we may simply then divide each of y_1 and y_2 by their corresponding standard deviations.

We have been asked if the algorithm can be viewed as a single Principal Curve algorithm which has dimensionality equal to the sum of the dimensionality of $\mathbf{x}_1$ and $\mathbf{x}_2$. The answer is really "no" in that we use the criteria of closeness in each space independently and so simply having a single Principal Curve which joins together the two points $\mathbf{x}_1$ and $\mathbf{x}_2$ would give different results. It must be noted, however, that the SOM algorithm which does precisely this appears to work rather well.

The Future

We stated at the start of the book that it would be based on several PhD theses. We have done this, but a book which did justice to all these theses would be much longer than the current text. Thus, if any part of this book has proved especially interesting to a reader, he or she may consult the original thesis from the University of Paisley:

- Dr. Mark Girolami (1998): Chapter 6, Independent Component Analysis
- Dr. Darryl Charles (1999): Chapter 5, Factor Analysis
- Dr. Stephen McGlinchey (2000): Chapter 7, Topology-preserving networks
- Dr. Donald MacDonald (2001): Chapters 6 and 8, Exploratory data analysis
- Dr. Emilio Corchado (2002, Universidad de Salamanca): Chapter 8, Maximum Likelihood learning
- Dr. Pei Ling Lai (2000): Chapters 9, 10 and 11, Canonical Correlation Analysis
- Dr. Zhenkun Gou (2003): Chapters 9 and 13, Canonical Correlation Analysis
- Dr. Jos Koetsier (2003): Chapters 12 and 14, Higher-order correlations
- Dr. Ying Han (2004): Chapters 12, 13 and 14, Time series analysis

In this final chapter, we review the work which has been covered in this book, give pointers to work which is not included and discuss current and future work in this area.

15.1 Review

The aim of our investigations has been to develop artificial neural networks which can self-organise in order to extract structure from their environment.

Our major goal in developing unsupervised learning procedures is to build self-organizing network modules which capture important regularities of the environment in a simple form suitable for further perceptual processing. We would like an unsupervised learning procedure to extract and explicitly represent progressively higher-order features. If, at one stage, the algorithm could learn to explicitly represent continuous real-valued parameters such as relative depth, position and orientation of features in an image, subsequent learning could then discover higher-order relations between these features, representing, for example, the location of object boundaries.

We began with a negative feedback network which self-organises on the basis of simple Hebbian learning. We showed that this network was capable of performing a Principal Component Analysis i.e. it could find the filters of the data set which contained most of the variance in the data set or, equivalently, compressed the data to achieve the least mean squared error when compressed to a linear basis of a particular dimensionality.

This network was shown to be equal to other networks which only feed forward the activation but use weight decay terms in their learning rules [138, 160]. However, we believe that putting the negative feedback of activation as a separate phase in the self-organising process yields advantages in terms of providing us with a mental image which enables us to visualise processes which would not otherwise occur to us. Thus, for example, the competition which we introduced in Chapter 7 is difficult to contemplate in a solely feedforward architecture.

We have introduced a number of data exploration tools which all self-organise to find structure of one type or another in data sets. Two of the networks, those of Chapter 6 and 8, we explicitly linked to the statistical technique of Exploratory Projection Pursuit, and indeed, we combined these two techniques so as to develop an artificial neural network which exhibited fast and accurate convergence to filters which reveal some interesting structure in a data set where the term "interesting" is used as a highly specialised and mathematically precisely defined concept.

We have also introduced networks which cluster the data: we have been especially interested in those clusterings which maintain in the coding, some topological relations within the data. We discussed three separate networks which performed this way, each having somewhat different properties from existing topological networks and from each other.

The second half of this book has discussed the development of artificial neural network algorithms which can be used for processing information from more than one data stream at a time. This is especially helpful when each data stream contains information which is related to the contents of the other data stream. We related the two sets of basic techniques in Chapter 9 to Canonical Correlation Analysis and showed that the resulting networks gave results commensurate with statistical CCA results.

We went on to show that similar networks could be derived from different starting points and that the resulting networks all shared a similar structure: they self-organise using Hebbian learning between each output and the opposite input and anti-Hebbian learning between each output and the input which gave rise to the output. Our conjecture is that this very general finding may give clues to natural information processing, but this is a conjecture which we do not have the technical expertise to validate.

Methods to find greater than linear correlations were introduced in Chapter 11. We did this in two ways:

1. By introducing a nonlinearity into our neural CCA methods.
2. By transforming the data into a feature space, and performing linear CCA in this feature space.

The second of these methods is not truly a neural method, but the use of kernels is widespread in neural network journals and conferences and we are influenced by our environment as are all researchers. These methods were shown to provide correlations which exceed those which statistical CCA (or linear neural implementations of CCA) can find. Yet we have stopped short of stating that we are approaching mutual information. Our mental model for the first of these methods is rather like that of Principal Curves which takes the first Principal Component direction from a data set and bends it through the major (nonlinear) axis of a data set; thus, by analogy, we imagine that our nonlinear CCA is rather like performing a linear CCA of a twinned data set and bending both canonical correlation filters through the major (nonlinear) axes of the data set to maximise correlations. This is, as yet, a less well defined model than we would like, and future research will undoubtedly tackle this problem.

A further change allowed us to tackle the difficult problem of multicollinear data, which may provide less-than-robust solutions to CCA. The methods proved to be extremely informative with functional data with the resulting filters providing far more interest than the standard methods. Chapter 12 took two single stream methods for extracting structure from data sets and adapted them so that they sought joint, higher-order structure in two data streams at once. A comparison at the end of Chapter 12 discusses the various aspects of the rules developed. The FastECA method of Chapter 12, again, is not a truly neural method, but again, the FastICA method on which it is based, is well known in neural network circles and is, in fact, the tool of choice for researchers interested in performing an Independent Component Analysis of a data set. Exploratory Correlation Analysis was performed on visual data: results on artificially created twinned data were commensurate with those found by researchers with single stream data; however, when used with real binocular data, the results were less convincing than previously, suggesting that it is possible that artificial neural network researchers have been over-optimistic in equating their results to those found in biology.

A final chapter investigated the twinning of the Principal Curve algorithm. We used this family of methods initially to forecast financial data, but concluded that the method was more appropriate for exploratory data investigation.

15.2 Omissions

In order to make the current exposition readable, we have tended to keep each topic separate, however several students have used mixtures of these methods and have interesting results from these networks. Thus for example, we have not discussed lateral weights other than those in Chapter 4; however, in the theses of Charles [24] and Corchado [29], we derive lateral connections from the Rectified Gaussian Distribution and use these for imposing specific structures on the outputs. In [29], for example, we derive several methods for identifying separate sets of horizontal and vertical bars from mixtures of bars using these lateral connections and the maximum likelihood learning method.

We have also, in this book, tended to give credit to one individual for a specific topic. However, inevitably, when students are working together in a group, they influence each other and ideas migrate throughout the group. For example, we have credited Dr. McGlinchey with the work on topology preservation but Dr. MacDonald also has performed substantial research on this topic [121].

Other specific topics missing from this book include

- In [66], we have shown that the non-negativity constraints may be applied to both the bi-gradient algorithm [180] and a class of algorithms derived from the generalised eigenproblem [188] and that the resulting network has properties similar to those discussed in Chapter 5.
- We have not discussed unsupervised single stream kernel methods [56] at all, but these feature very strongly in several theses, particularly that of Dr. MacDonald [121].
- Inevitably, there is much more in the way of demonstrations of the methods described herein in the various theses; however, our focus in this volume has been on the development of methods and so we have had to omit extended experimental simulations.

However, we have given pointers to the various theses and so they may be accessed subsequently.

15.3 Current and Future Work

Perhaps the most interesting aspect of this book is that using a very simple negative feedback architecture, we can create self-organising artificial neural networks which will find different types of structure in a data set dependent on the actual rules which are used to change the weights. One of the areas that merits further attention is the search for nonlinear manifolds in data spaces. An open question is to what extent the negative feedback architecture can contribute to this. There are two current streams to our current work on negative feedback networks:

1. Combining the negative feedback networks with other exploratory data analysis techniques. Thus, for example, we are investigating [147] combinations of Andrews' Curves and the Exploratory Projection Pursuit network of Chapter 6. We are also investigating [152] combining networks using statistical techniques such as bagging [19] and boosting [49].
2. The methods discussed in this book give linear projections except for those in Chapter 7. However Fig. 6.1 (left) shows a one-dimensional distribution whose structure cannot be best represented by a projection onto a straight line. Thus, one stream of research is finding whether it is possible to identify a generic nonlinearity which will best describe a data set. Some work has been done with Principal Curves and Self-organising Maps [68] but much more is both possible and necessary.

We are also aware of the great deal of interest in deriving unsupervised artificial neural networks from probabilistic models (see Chapter 8 and Appendix B) and anticipate that this will also form one strand of our future work. How much more we can develop the negative feedback network is an open question, but, with many talented young researchers coming through, this is a challenge which we relish.

A

Negative Feedback Artificial Neural Networks

Because this book is based on a negative feedback network which is derived from the Interneuron Model [153], we give pride of place to this model before going on to investigate other negative feedback models. We review separately those models where the dynamics of the network settling to an attractor state have been important to the value of the state reached and those models which have considered only the transfer of activation as a single event. Finally, we consider the relationship between the negative feedback model of this book and biological models.

A.1 The Interneuron Model

Plumbley [153] has developed a model of Hebbian learning which is based on the minimisation of information loss throughout the system.

He develops these interneurons in two ways, suggesting that he is giving two different views of the same network. However, we will see that these interneurons have different capabilities depending on which network is used.

In both networks the interneurons are developed as anti-Hebbian neurons with the additional property of trying to optimise information transfer within limited power constraints. Plumbley notes that the best information transfer rate will be found when the outputs are decorrelated; however, he also attempts to equalise the variance of the outputs to ensure that they are then carrying equal information.

The dynamics of the network are described by

$$\mathbf{z} = V^T \mathbf{y}$$

where z_j is the activation of the interneuron, y_i is the output from the network, and V_{ij} is the weight joining the ith ouput neuron to the jth interneuron.

This makes the output response

$$\mathbf{y} = \mathbf{x} - V\mathbf{z}$$

where $\mathbf{x}$ is the original input value. Plumbley concentrates on the information preserving properties of the forward transformation between inputs and outputs and shows

$$\mathbf{y} = (I + VV^T)^{-1}\mathbf{x}$$

A weight decay mechanism is used in the learning:

$$\Delta v_{ij} = \eta(y_i z_j - \lambda v_{ij})$$

This is equivalent to a learning rule in the limit of

$$\frac{d}{dt}\,\mathbf{v}(t) = (C - \gamma I)\mathbf{v}$$

A solution to this equation is

$$\mathbf{v}(t) = A\exp(C - \gamma I)t$$

Therefore, the weights will increase without limit in directions where the eigenvalue of the correlation matrix exceeds γ. Thus the weights will never tend to a multiple of the principal eigenvector and no selectivity in information transfer will be achieved. Note that there are fixed points on the eigenvectors but these are not stable.

The crucial difference between this model and Oja's model (Chapter 2) is that in Oja's model the decay term is a function of the weights times the weights. In this model, the decay term is not strong enough to force the required convergence.

Equally, the anti-Hebbian learning rule does not force convergence to a set of decorrelated outputs.

$$\Delta v_{ij} = \eta(y_i z_j - \lambda v_{ij})$$

does not mean that

$$(\Delta v_{ij} = 0) \implies (E(y_i z_j) = 0).$$

However, in taking "another view of the skew-symmetric network", Plumbley uses the interneurons as the outputs to the network.

In this model, we have forward excitations U and backward excitations V where

$$\mathbf{z} = U^T \mathbf{y}$$

$$\mathbf{y} = \mathbf{x} - V\mathbf{z}$$

i.e.

$$\mathbf{z} = U^T(I + VU^T)^{-1}\mathbf{x}$$

where the weight update is done using the same update rule

$$\Delta v_{ij} = \eta(y_i z_j - \lambda v_{ij})$$

Since the output is from the interneurons, we are interested in the forward transform from the x values to the z values.

$$y_i = x_i - \sum_k u_{ki} z_k$$

Now,

$$\Delta u_{ij} = \eta(y_i z_j - \lambda u_{ij}) = \eta \left(\left(x_i - \sum_k u_{ki} z_k \right) z_j - \lambda u_{ij} \right)$$

Plumbley states that the last term is the weight decay term. In fact, as can be seen from the above equations, the second term is the important weight decay term, being a form of Multiplicative Constraint (Chapter 2). There is an implicit weight decay built into the recurrent architecture.

However, if we consider the network as a transformation from the x values to the y values we do not find the same implicit weight decay term:

$$z_i = \sum_j u_{ij} y_j \tag{A.1}$$

$$= \sum_j u_{ij} \left(x_j - \sum_k u_{kj} z_k \right) \tag{A.2}$$

$$= \sum_j u_{ij} x_j - \sum_k z_k \left(\sum_j u_{ij} u_{kj} \right) \tag{A.3}$$

And so,

$$\Delta u_{ij} = \eta(y_i z_j - \lambda u_{ij}) \tag{A.4}$$

$$= \eta \left(y_i \left(\sum_j u_{ij} x_j - \sum_k z_k \left(\sum_j u_{ij} u_{kj} \right) \right) - \lambda u_{ij} \right) \tag{A.5}$$

Using this form, it is hard to recognise the learning rule as a Hebb rule, let alone a decaying Hebb rule of a particular type. However, as we have seen elsewhere in this book, the negative feedback in Plumbley's first network is an extremely valuable tool.

A.2 Other Models

As with many classication systems, it is possible to classify the artificial neural network models which are similar to the negative feedback model of this book in a number of ways. We have chosen to split the group into static models (next section) and dynamic models (Section A.2.2).

A.2.1 Static Models

The role of negative feedback in static models has most often been as the mechanism for competition (see e.g.[22, 110] for summaries) often based on biological models of activation transfer e.g.[124] and sometimes based on psychological models e.g. [28, 62, 64]

An interesting early model was proposed by Kohonen [110] who uses negative feedback in a number of models, the most famous of which (at least of the simple models) is the so-called "novelty filter".

In the novelty filter, we have an input vector $\mathbf{x}$ which generates feedback gain by the vector of weights, M. Each element of M is adapted using anti-Hebbian learning:

$$\frac{dm_{ij}}{dt} = -\alpha x_i' x_j' \tag{A.6}$$

$$\text{where } \mathbf{x}' = \mathbf{x} + M\mathbf{x}' \tag{A.7}$$

$$= (I - M)^{-1}\mathbf{x} = F\mathbf{x} \tag{A.8}$$

"It is tentatively assumed $(I - M)^{-1}$ always exists." Kohonen shows that, under fairly general conditions on the sequence of $\mathbf{x}$ and the initial conditions of the matrix M, the values of F always converge to a projection matrix under which the output $\mathbf{x}'$ approaches zero although F does not converge to the zero matrix i.e. F converges to a mapping whose kernel ([120], page 125) is the subspace spanned by the vectors $\mathbf{x}$. Thus, any new input vector $\mathbf{x}_1$ will cause an output which is solely a function of the novel features in $\mathbf{x}_1$.

Other negative feedback-type networks include William's Symmetric Error Correction (SEC) Network [183] where the residuals at y were used in a symmetric manner to change the network weights. The SEC network may be easily shown to be equivalent to the network described in Chapter 3.

A second reference to a negative feedback-type network was given in [117]. Levin introduced a network very similar to Plumbley's network and investigated its noise resistant properties. He developed a rule for finding the optimal converged properties and, in passing, showed that it can be implemented using simple Hebbian learning.

A third strand has been the adaption of simple Elman nets([6, 43, 44, 78, 107]) which have a feedforward architecture but with a feedback from the central hidden layer to a "context layer". Typically, the Elman nets use an error-descent method to learn; however, Dennis and Wiles [37, 38] have modified the network so that the feedback connection uses Hebbian learning. However, the Hebbian part of the network uses weight decay to stop uncontrolled weight growth and the other parts of the network continue to use back propagation of errors to learn.

More recently, Xu [185] has analysed a negative feedback network and has provided a very strong analysis of its properties. While he begins by considering the dynamic properties of a multilayer network (all postinput

layers use negative feedback of activation), it is clear from his discussion that
the single-layer model which he investigates in detail is similar to the network
in this book.

An interesting feature is Xu's empirical investigation into using a sigmoid
activation function at the negative feedback networks; he reveals results which
show that the network is performing a PCA and suggests that this feature
enabled the network to be more robust i.e. resistant to outliers, a finding in
agreement with other researchers (e.g. [103, 140, 144]).

A.2.2 Dynamic Models

The negative feedback of activation has most often been used in those models
of artificial neural networks which are based on a dynamic settling of activation. These are generally called Hopfield nets [73] after John Hopfield [79] who
performed an early analysis of their properties though earlier work on their
properties was performed by other researchers e.g. following Grossberg [63],
we note that there are two types of on-center off-surround networks possible
using inhibition. It is possible to generate the following:

- Feed forward inhibition: the activation transfer rule is

$$\frac{dy_i}{dt} = -Ay_i + (B - y_i)x_i - y_i \sum_{k \neq i} x_k \qquad (A.9)$$

  where A, B are constants and x_i is the input to the ith neuron. This is
  clearly not a biological model as it requires each cell to have information about all inputs to all other neurons $x_k, k \neq i$. Grossberg points out
  though, that, if the activation is allowed to settle, this model has a stationary point ($\frac{dy_i}{dt} = 0$) when

$$y_i = \frac{x_i}{\sum_k x_k} * \frac{B \sum_k x_k}{A + \sum_k x_k} \qquad (A.10)$$

  Possibly of most interest is its self-normalisation property, in that the total
  activity

$$\sum_k y_k = \frac{B \sum_k x_k}{A + \sum_k x_k} \qquad (A.11)$$

  is a constant.
- Feedback inhibition: we use here Grossberg's term though we tend to make
  a distinction between feedback inhibition between layers (as in Plumbley's
  network) and lateral inhibition between neurons in the same layer. Here
  Grossberg discusses the activation passing equation

$$\frac{dy_i}{dt} = -Ay_i + (B - y_i)[x_i + f(y_i)] - y_i \left[J_i + \sum_{k \neq i} f(y_k) \right] \qquad (A.12)$$

where $J_i = \sum_{k \neq i} x_k$. The most interesting properties from this model develop when the activation function, $f()$, is a sigmoid which has the property that it forms a winner-take-all network which suppresses noise, and quantises the total activity. Again these properties arise from an analysis of the dynamic properties of the negative feedback acting on the network activations.

A.3 Related Biological Models

In keeping with our overall aim, we would like to link our networks with those of biology. The overall aim for an early processing network has been described as the minimisation of redundancy so that the further network can be developed as a "suspicious coincidence" [5] detector. The decorrelation of inputs formed by projection onto the Principal Components clearly achieves this. The network most like that described above was devised by Ambrose-Ingerson et al. [1] in which a network which uses negative feedback between layers attempts to simulate the transfer of olfactory information in the paleocortex. The sole difference between that network and the negative feedback network is that the network uses a competitive activation transfer arrangement; the authors conjecture that a form of PCA is taking place.

Murphy and Sillito [135] have shown that LGN neurons seem to be inhibited by the V1 cells (in the visual cortex) which they excite. Pece [151] has developed a model based on negative feedback which simulates the reduction in redundancy in an information-transferring network.

As an interesting aside, we note that Robinson [156] has shown that negative feedback cannot be used to control the visuomotor system in a continuously operating closed-loop system with a finite delay term. He shows that the negative feedback in the system can be made stable if the system is refractory: each eye saccade is followed by a short period when it will not respond to another change in target position (caused by the sampling rate having a finite frequency). Because of this, we can think of such a system as running on open-loop dynamics for much of the time, which is equivalent to having discrete time intervals in which activation is passed forward and back. It is results like this which underlie our conviction that the negative feedback network is based on cybernetic principles.

A network very much like the network which we have investigated, has been developed in [95] in which inhibition is specifically used in an artificial neural network to model the cerebellum. The network appears identical to that of Chapter 3 but is considered as a dynamic model where the activation is allowed to pass round the network until settling takes place. However, since Jonker makes "the biologically plausible assumption that the characteristic time-scales in the evolution of interactions are much larger than the time-scales involved in the neuronal dynamics", ([95], page 87) it is not surprising that the emergent properties of the network are very similar to those which are developed in Chapter 3 from a static network.

We have discussed only algorithms which have been defined as artificial neural networks. However, adapting parameters incrementally is not confined to this field and one may find similar models in the fields of statistics, control and mathematics.

B

Previous Factor Analysis Models

In this appendix, we discuss several models which also find individual factors underlying a data set and, where possible, relate them to the algorithm and experiments of Chapter 5. Not all the models in this section were explicitly related to factor analysis in the original papers, but all were used to find factors underlying a data set.

B.1 Földiák's Sixth Model

The major difference between this model which is shown in Fig. B.1, and Földiák's [46] previous models is that the neurons are nonlinear units, each with an adjustable threshold over which its activation must be, before the neuron can fire. In order to get a sparse coding, each neuron tries to keep down its probability of firing by adjusting its own threshold. If a neuron has been firing frequently, its threshold will increase to make it more difficult for the neuron to fire in the future; if it has been firing only occasionally, its threshold will decrease, making it easier for the neuron to fire in future. This mechanism does have some biogical plausibility in that neurons do become habituated to inputs and stop responding so strongly to repeated sets of inputs.

Let there be n inputs and m output neurons (the representational layer). Then

$$y_i = f\left(\sum_{j=1}^{n} q_{ij}x_j + \sum_{j=1}^{m} w_{ij}y_j - t_i\right) \tag{B.1}$$

where q_{ij} is the weight of the feedforward connection from the jth input x_j, w_{ij} is the weight of the lateral connection from the jth output neuron to the ith in that layer and t_i is the adjustable threshold for the ith output neuron. Both sets of weights and the threshold are adjustable by competitive-type learning:

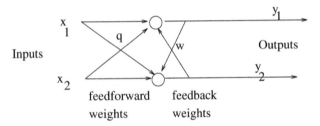

Fig. B.1. Földiák's sixth model: we have feedforward weights from inputs to outputs and then feedback (or lateral) connections between the outputs.

$$\Delta w_{ij} = \begin{cases} -\alpha(y_i y_j - p^2), & \text{if } i \neq j, \\ 0, & \text{if } i = j \text{ or } w_{ij} < 0. \end{cases}$$

$$\Delta q_{ij} = \beta y_i (x_j - q_{ij})$$

$$\Delta t_i = \gamma(y_i - p)$$

where α, β, γ are adjustable learning rates. The feedforward weights, q_{ij}, use simple competitive learning. The lateral learning rule for the w weights will stabilise when $E(y_i y_j) = p^2$, i.e. each pair of units will tend to fire together a fixed proportion of the time. This rule will interact with the rule which changes the threshold: the long-term effect of this rule should be to set the threshold t_i to a set value to ensure $E(y_i) = p$. By choosing the value of p appropriately we can determine the level of sparseness of the firing pattern. For example, suppose $p = \frac{1}{10}$. Each neuron will fire about $\frac{1}{10}$ of the time and will fire at the same time as each of its neighbours about $\frac{1}{100}$ of the time. So if we have 500 output neurons, then each input pattern would elicit a response from about 50 output neurons, which is a distributed representation and not a grandmother cell response, but one in which the number of neurons firing at any one time is much less than the number of output neurons.

B.1.1 Implementation Details

Földiák actually describes his feed forward rule with the differential equation

$$\frac{dy_i^*}{dt} = f\left(\sum_{j=1}^{n} q_{ij} x_j + \sum_{j=1}^{m} w_{ij} y_j^* - t_i\right) - y_i^* \tag{B.2}$$

which is viewing the neuron as a dynamically changing neuron responding to a mixture of feedforward inputs and lateral inputs which are themselves changing in time. It can be shown that, provided the feedback is symmetric, the network is guaranteed to settle after a brief transient. Földiák simulates this transient by numerically solving the differential equations. He uses initial conditions

$$y_i^*(0) = f\left(\sum_{j=1}^{n} q_{ij}x_j - t_i\right) \tag{B.3}$$

with the function $f(u)$ equal to the logistic function, $\frac{1}{1+\exp(-\lambda u)}$. The values at the stable state are now rounded to 1 or 0 depending on whether $y_i^* > 0.5$ or not. Feedforward weights are initially random but normalised so that $\sum_j q_{ij}^2 = 1$ and the feedback weights are zero.

B.1.2 Results

On the bars data (Chapter 5), the network learns to find the bars so that the feed forward weights match a single possible bar. The code generated in this case is optimal in that there is no redundancy between neurons — every neuron represents a different bar — and all information in the input is preserved — the set of output neurons will identify exactly which of the input squares is on by identifying which bars are represented in the pattern.

An extension to this experiment was the presentation of input patterns consisting of images of letters presented in a fixed position on an 8×15 raster display. During training, the letters were presented in a random order with the same probabilities that they appeared in a piece of English text. The results were as you might have hoped (Table B.1) had you hand-designed a code (e.g. á la Huffmann): frequent letters had fewer active bits than infrequent ones since otherwise the correlations introduced by simultaneous frequent firing would force the decorrelating lateral weights to increase inhibition between the neurons. Another feature was that no two frequent letters were assigned to the same output though some rare letters did share an output. Finally, the code has the nice property that similar inputs e.g. O and Q are mapped to similar output neurons.

Table B.1. Some of the codes found by Földiák's network after being presented with 8 000 letters. The letter t appears frequently in the text and so t uses a small number of firing neurons, while J appears infrequently and so can use a larger number of firing neurons. Note also that i and ! share a code.

Network output	Input pattern
1000000000000000	t
0000100000000000	i or !
0010011000001000	J

B.2 Competitive Hebbian Learning

White [181] has attempted to play off the extreme focusing of competitive learning with the broad smearing of Hebbian learning with a model which attempts to combine both effects.

Consider a one layer network with outputs

$$y_i = f\left(\sum_j w_{ij} x_j\right) - \frac{1}{2} \tag{B.4}$$

where $f(u) = \frac{1}{1+\exp(-u)}$ and x_j is the jth input. Then each output satisfies $-\frac{1}{2} \le y_i \le \frac{1}{2}$. The information given by the outputs is proportional to

$$F = \sum_{i=1}^{N} y_i^2 \tag{B.5}$$

If we simply wish to maximise F we can use gradient ascent so that

$$\Delta w_{ij} = \frac{\partial F}{\partial w_{ij}} = 2f_i' y_i x_j \tag{B.6}$$

Now since $2f_i'$ is constant for the whole weight vector into output neuron i, we can ignore this factor (it alters the magnitude of the change but not the direction) to get simple Hebbian learning:

$$\Delta w_{ij} = \alpha y_i x_j \tag{B.7}$$

where α, the learning rate, contains the $2f_i'$ factor. But we know that this causes every output neuron to move to the principal component of the input data, i.e. all are covering the same information. We introduce a penalty term to discourage this and the simplest is that the mean square correlation between output neurons should be as low as possible: if we get totally decorrelated outputs, then

$$E(g_{ik}) = E((y_i y_k)^2) = 0 \tag{B.8}$$

We can incorporate the constraint into the optimisation by using the method of Lagrange multipliers: now we try to maximise the function G given by

$$G = F + \frac{1}{2} \sum_{i=1}^{N} \sum_{k=1, k\neq i}^{N} \lambda_{ik} g_{ik} \tag{B.9}$$

which gives us

$$\Delta w_{ij} = \frac{\partial G}{\partial w_{ij}} \propto \alpha y_i x_j \left\{ 1 + \sum_{k\neq i} \lambda_{ik} y_k^2 \right\} \tag{B.10}$$

White shows that one solution to this (for the λ values) occurs when all λ_{ik} are equal to -4. Thus the weight change equation becomes

$$\Delta w_{ij} = \alpha y_i x_j \left\{ 1 - 4 \sum_{k \neq i} y_k^2 \right\} \qquad (B.11)$$

This rule is known as the Competitive Hebbian Rule: it is used with a hard limit to the overall length of the vector of weights into a particular output neuron and in addition we do not allow the term in brackets to become negative. There are therefore three phases to the learning in this network:

- Neuron outputs are weak. In this case there is essentially no interaction between nodes (the term in the bracket $<< 1$) and all nodes will use simple Hebbian learning to learn the first Principal Component direction. But note that when they have done so, (or are doing so) they enter a region where the interaction is not insignificant and so some interaction between output neurons begins to take place — the neurons have learned their way out of the uniform region of weak outputs.
- Neuron outputs are in the intermediate range. This is the effective range of learning for the Competitive Hebbian Algorithm. A winning node — one which happens to be firing strongly — will tend to turn off the learning for the other nodes. This will result in different neurons learning different regions of the input space.
- Neuron outputs are too strong. When the competitive learning factor (the bracketed term) becomes negative, no learning takes place since we simply make the last term 0. If a system gets to this stage, there is no way it can learn its way out of the region and we must simply stop the algorithm and restart.

The Competitive Hebbian Algorithm has some success on our horizontal and vertical bars problem but has proved difficult to stabilise and indeed the actual parameters used in the algorithm have to be hand-tuned for a particular problem to get the best results.

B.3 Multiple Cause Models

The algorithms of this type are sometimes known as "Multiple Cause Models" since the aim is to discover a vocabulary of independent causes or generators such that each input can be completely accounted for by the action of a few of these possible generators e.g. that any set of the 2^{16} possible patterns from the 8×8 bars problem (Chapter 5) can be accounted for by some combination of the 16 output neurons, each of which represents one of the 16 possible independent inputs.

B.3.1 Saund's Model

Saund [161] has focussed his attention on the activation function. His network is shown in Fig. B.2. We have a binary input vector $\{x_1, x_2, ..., x_j, ..., x_n\}$ which is joined by weights w_{ij} to an output/coding layer $\{y_1, ..., y_i, ..., y_m\}$. The weights are thought of as associating an input vector with activity at a cluster centre determined by the weights into that centre. This is similar to competitive learning which is the simplest way to view this part of the process.

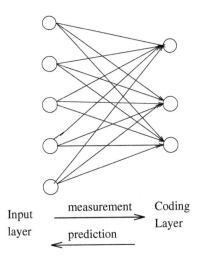

$$
\begin{array}{ccc}
\text{Input} & \xrightarrow{\text{measurement}} & \text{Coding} \\
\text{layer} & \xleftarrow{\text{prediction}} & \text{Layer}
\end{array}
$$

Fig. B.2. Saund's model has activation propagating forwards and prediction propagating backwards.

The cluster centre then corresponds to an output neuron which is in a position to predict the input that caused it to react. This prediction is then returned to the input layer. Saund views the outputs as performing a voting rule associating each prediction with the firing of the various output neurons: if the ith neuron is firing strongly, then it must be because the inputs to it approximate the data vector which is currently being presented to the network. We can view such firing as the probability that it is caused by an item of data which is close to the weights into that neuron. Thus, he defines the prediction of the network for the jth input value as

$$r_j = 1 - \prod_k (1 - w_{kj} . y_k) \tag{B.12}$$

Clearly if all weighted sums of the outputs are close to zero then the prediction is close to zero. On the other hand, as soon as a single weighted feedback is close to 1, the prediction of the network goes to one — rather like the neuron saying "I've found it". The prediction function for two-dimensional inputs is shown in Fig. B.3. It is sometimes known as a noisy-OR function since it is a smoothing of the Boolean OR function.

The prediction function

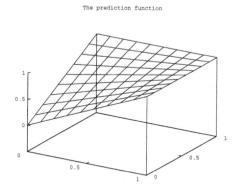

Fig. B.3. The prediction function for a two-dimensional data set.

Saund identifies an objective function equal to the log likelihood

$$g_i = \sum_j \log \left(x_{i,j} r_{i,j} + (1 - x_{i,j})(1 - r_{i,j}) \right) \qquad (B.13)$$

where the i subscript identifies the ith pattern and the j the jth input neuron. If $x_{i,j}$ and $r_{i,j}$ simultaneously approach 1 or 0, then the objective function tends to zero. At all other times the objective function will be negative. A global objective function is merely the sum of these objective functions over all input patterns and the optimal weights are found by batch gradient ascent.

Saund shows that the network will identify the independent horizontal or vertical lines, and points out the interdependencies in the training regime: the optimal weights cannot be calculated independently for each hidden unit but are dependent on one another's predictions.

On the left-hand side of Figure B.4, we show a single neuron with cluster centre (1,1,1) which is attempting to predict the pattern (1,1,0). The best it can do is to predict (0.666,0.666,0.666) which minimises the error function. However, a second neuron with centre (1,1,0) can minimise this error (in fact send it to zero) when it comes online, leaving the first neuron free to continue responding maximally to (1,1,1). Note that each neuron has responded to a different pattern and each is minimising the error for a different part of the input space.

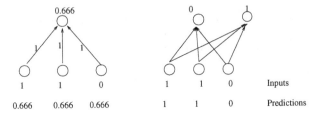

Fig. B.4. If there is only a single cluster centre at (1,1,1) the output neuron cannot respond accurately to the input (1,1,0) since the incorrect prediction of a 1 on the third input would give an error of magnitude 1. The best that can be done is to predict $\frac{2}{3}$ for each input. But if we have two centres, the second centre can concentrate its resources on this input, leaving the other centre to react to the pattern (1,1,1).

B.3.2 Dayan and Zemel

Dayan and Zemel [35] view the problem as one of finding a set of prior probability and conditional probabilities that (approximately) minimize the description length of a set of examples drawn from the distribution: the minimum description length refers to the number of output neurons required to describe the input in an invertible code. Their first model uses a backpropagation network with a single hidden layer for autoassociation. They use, however, a cross-entropy error to judge the accuracy of the reconstructions at the output layer. Using this on the bars data, they had some success but found that the network tended to get stuck in a local minima about 73% of the time in which a particular output would have responsibility for more than one bar.

B.4 Predictability Minimisation

Schmidhuber and Prelinger [162] have developed a network (Fig. B.5) for extraction of independent sources based on the principle of predictability minimisation. The core idea is that each prediction layer (the last layer in the figure) attempts to predict each code neuron's output based only on the output of the other code neurons while simultaneously each code unit is attempting to become as unpredictable as possible by representing input properties which are independent from the features which are being represented by other code neurons.

Notice that each prediction neuron is connected in a feedforward manner to all code neurons *except* the code neuron whose output it is attempting to predict. Let the output of the ith code neuron be y_i. Then

$$y_i = f\left(\sum_j w_{ij} x_j\right) \tag{B.14}$$

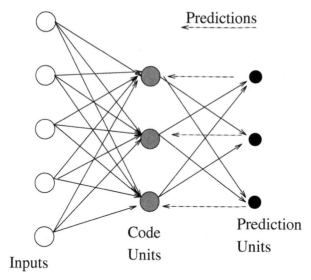

Fig. B.5. Schmidhuber's Predictability Minimisation network: input patterns are coded across the (large) code neurons. Each code neuron feeds forward to all the other prediction units *but not its own*. Each prediction unit attempts to predict the corresponding code unit's output. Each code neuron is meanwhile attempting to avoid being predictable.

where $f(u) = \frac{1}{1+\exp(-u)}$ and x_j is the jth input. So $0 \le y_i \le 1$. The output of the prediction units is given by

$$P_k = \sum_{j \neq k} v_{kj} y_j \qquad (B.15)$$

So P_k sees only the response of the other code neurons and not the code neuron which it is trying to predict. Now we have two-pass training (both using conventional backprop).

Pass 1. Each set of weights is trained to minimise the mean squared prediction error over all sets of inputs, i.e. to minimise $\sum_p \sum_i (P_i^p - y_i^p)^2$, where P_i^p is the output of the ith prediction neuron to the pth input pattern etc. If it does minimise this function, the P_i will then be the conditional expectation $E(y_i|\{y_k, k \neq j\})$ of the prediction unit given the output of all the other code neurons. We can see that this conditional expectation will not be the same as actually predicting the value of the code unit, e.g. if the code unit fires 1 one-third of the time and 0 the other two-thirds, the conditional expectation would be 0.333 in this context.

Pass 2. Now in the second pass the weights are again adjusted so that the code units are attempting to maximise essentially the same objective function which was previously being minimised, i.e. to make them as unpredictable as possible. But now we only need change the w weights into the code units.

The two criteria co-evolve by fighting each other. Note that each w weight is maximising only the local prediction error while each v weight is being updated to minimise the global prediction error.

Schmidhuber states that the code evolved by his network is quasi-binary: each coding neuron is either 0/1 for each pattern or responds with a constant value to each input pattern in which case it is simply giving a "don't know" response to each pattern.

B.5 Mixtures of Experts

A famous method is the mixtures of experts [92, 96]: the desire is to have each expert (which is typically itself a backpropagation neural network) learn to take responsibility for a particular subset of the input data. The experts are then assigned a weight which can be thought of as the confidence that a gating network has in each expert's responsibility for responding to that input data.

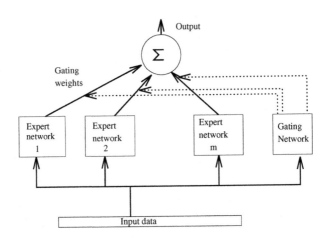

Fig. B.6. Each expert and the gating network sees all input data. Each expert learns to take responsibility for a particular set of input data and the gating expert learns how much confidence to have in each network's output for that input.

Consider the network shown in Figure B.6. Each expert sees all the input data and is itself a network which can be trained by supervised learning (error descent) methods. The gating network also sees the input data and is trained by supervised learning. Its responsibility is to put a probability on the chance that each network is the one to take responsibility for that input.

A First Attempt

A first attempt might be to use error descent on

$$E^c = \| \, \mathbf{t}^c - \sum_i p_i^c \mathbf{y}_i^c \, \|^2 \tag{B.16}$$

where $\mathbf{y}_i^c$ is the output vector of expert i on case c, p_i^c is the proportional contribution of expert i to the combined output vector and $\mathbf{t}^c$ is the target output on case c. Notice that it is the gating expert which determines p_i^c while the experts independently calculate their outputs, $\mathbf{y}_i$.

The problem with this error function is that this introduces a strong coupling between the experts, causing them to act together to try to reduce the error. So each expert attempts to minimise the residual errors when all the other experts have had their say, which tends to lead to situations where several experts are combining to take responsibility for each case.

An Improvement

What we really require is for our gating expert to make the decision about which single expert should be used in each case. This suggests an error term of the form

$$E^c = E(\| \, \mathbf{t}^c - \mathbf{y}_i^c \, \|^2) = \sum_i p_i^c \, \| \, \mathbf{t}^c - \mathbf{y}_i^c \, \|^2 \tag{B.17}$$

Here the error is the expected value of the residual error where each expert is expected to provide the whole of the approximation to the target and the gating network evaluates how close this whole approximation is. There will be still some indirect coupling since when another expert changes its weights the gating network may alter its assessment of responsibility apportionment, but this is much less direct than before. Simulations have shown that this network can evolve to find independent sources.

When an expert network gives less error than the weighted average of the errors over all experts, its responsibility for that case is increased (the gating network increases the value of p_i^c) and if it does worse, its responsibility is decreased.

There is still a difficulty with this error measure, however, which we see when we consider the rate of change of the error with the output,

$$\frac{\partial E^c}{\partial \mathbf{y}_i^c} = -2p_i^c(\mathbf{t}^c - \mathbf{y}_i^c) \tag{B.18}$$

from which we can see that the rate of change for terms which are far away from the current target is greater than those closer. While this is gated by the probability vector, it still has an adverse effect on convergence.

The Final Error Measure

Consider the error measure

$$E^c = -\log \sum_i p_i^c \exp\left(-\frac{1}{2} \parallel \mathbf{t}^c - \mathbf{y}_i^c \parallel^2\right) \tag{B.19}$$

which can be derived by assuming that all output vectors can be described by a Gaussian distribution and we are maximising the negative log probability of their independent joint distributions. When we now calculate the derivative of the error function with respect to the experts' outputs we get

$$\frac{\partial E^c}{\partial \mathbf{y}_i^c} = -\frac{p_i^c \exp\left(-\frac{1}{2} \parallel \mathbf{t}^c - \mathbf{y}_i^c \parallel^2\right)}{\sum_j p_j^c \exp\left(-\frac{1}{2} \parallel \mathbf{t}^c - \mathbf{y}_j^c \parallel^2\right)}.(\mathbf{t}^c - \mathbf{y}_i^c) \tag{B.20}$$

This term has been shown to have much better performance due to the fact that the first fractional term takes into account how well the ith expert is performing compared to all other experts.

B.5.1 An Example

Jacobs et al. [92] performed an experiment in which four vowel sounds from a 75 speakers were to be discriminated by an ANN. They compared the results using the Mixture of Experts network with a backpropagation network and showed that the latter typically took about twice as long to train.

The actual convergence of the weights in the network is interesting: initially each expert in the network attempts to minimise all the errors in the network over every example. There is, at this stage, no cooperation in the network and each network is attempting to deal with the whole problem itself. However a stage is reached when one network begins to receive more cases from a particular class than the others and quickly assumes responsibility for this class. This leaves the other experts free to concentrate on error minimisation for the other classes.

Finally, we see an exaggerated version of converged weights in Figure B.7 where we have experts and the gating network working together to model the problem. It can be seen that each of the networks has a region of the input space where it is effective and a region where it is ineffective. The gating network has learned these regions and assigns responsibility for the network's output to the expert for problems in the area where the expert is effective.

B.6 Probabilistic Models

Hinton and colleagues (e.g. [75, 76]) have developed models which attempt to use a generative model which consists of top-down connections from underlying reasons to the input image i.e. the top-down connections create the image

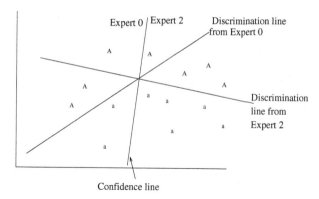

Fig. B.7. An exaggerated view of how the gating network and the mixtures of experts work together. The almost vertical line is the gating networks' view of the experts: on examples to the left of the line it assigns responsibility to expert 0 while to the right it assigns responsibility to expert 2. Each of these experts have a linear discriminant function which is accurate for its region of responsibility but inaccurate outside. In this way the combined network acts together to give a non-linear boundary between classes.

from an abstraction of the image. This is based on the view that "Visual perception consists of inferring the underlying state of the stochastic graphics model using the false but useful assumption that the observed sensory input was generated by the model."

Learning is done by maximising the likelihood that the observed data came from the generative model. The simplest generative model is the Mixtures of Gaussians model.

B.6.1 Mixtures of Gaussians

In essence, the process is the following:

- Each data point has an associated probability of being generated by a mixture of Gaussian distributions.
- Given the current parameters of the model, we calculate the posterior probability that any data point came from the distributions.
- The learning process adjusts the parameters of the model — the means, variances and mixing proportions (weights) of the Gaussians — to max-imise the likelihood that the model produced the points.

So when we generate a data point, **d**, we

- Pick a hidden neuron (the underlying cause of the data). Give it a state of 1, set all other hidden neurons' states to 0.
- Each hidden neuron will have a prior probability of being picked of π_j.

- Feedback to input neurons through weight vector, $\mathbf{g}_j$. The $\mathbf{g}_j$ is the center of the Gaussian $= \{g_{j1}, g_{j2}, ..., g_{jn}\}$.
- Add local independent zero mean Gaussian noise to each input.
- This means that each data point is a Gaussian cloud with mean $\mathbf{g}_j$ and variance σ_i^2.

This gives us

$$p(\mathbf{d}) = \sum_j \pi_j \prod_i \frac{1}{\sqrt{2\pi}\sigma_i} e^{-(d_i - g_{ji})^2 / 2\sigma_i^2} \tag{B.21}$$

Interpreting Data: Expectation Step

The process is:

1. Compute the probability density for each data point (assuming the model is correct):

$$p(\mathbf{d}|s_j = 1) = \prod_i \frac{1}{\sqrt{2\pi}\sigma_i} e^{-(d_i - g_{ji})^2 / 2\sigma_i^2} \tag{B.22}$$

2. Weight these with the prior probabilities π_j.
3. Use Bayes theorem to calculate the probability that the cause, s_j generated the data,

$$p(s_j = 1|\mathbf{d}) = \frac{\pi_j p(\mathbf{d}|s_j = 1)}{\sum_k \pi_k p(\mathbf{d}|s_k = 1)} \tag{B.23}$$

We have now calculated the posterior probabilities of the hidden states given the data ("perceptual inference") — the E-step of the EM Algorithm and we now perform the M step, which involves changing the parameters to maximise the Expectation.

Learning as Expectation Maximisation

We use the existing parameters to calculate the optimal new parameters:

$$\mathbf{g}_j = \frac{E\{p(s_j = 1|\mathbf{d})\mathbf{d}\}}{E\{p(s_j = 1|\mathbf{d})\}}$$

$$\sigma_i^2 = \frac{E\{p(s_j = 1|\mathbf{d})(d_i - g_{ji})^2\}}{E\{p(s_j = 1|\mathbf{d})\}}$$

$$\pi_j = E\{p(s_j = 1|\mathbf{d})\}$$

We have now a means of maximising the expectation — the M-step — but we can also use incremental learning, which involves gradient ascent such as

$$\Delta g_{ji} = \epsilon p(s_j = 1|\mathbf{d})(d_i - g_{ji}) \tag{B.24}$$

B.6.2 A Logistic Belief Network

A logistic belief network is composed of multiple layers of binary stochastic neurons whose state s_j is based on top-down expectations $\hat{s}_j$ from the layers above:

$$p(s_j = 1) = \hat{s}_j = \sigma\left(g_{0j} + \sum_k s_k g_{kj}\right) \tag{B.25}$$

If the configuration of a network is α,

$$P^\alpha = \prod_i p(s_i^\alpha | pa(i, \alpha)) \tag{B.26}$$

where $pa(i, \alpha)$ is the states of i's parents in configuration α and s_i^α is the state of the ith neuron in configuration α.

Using negative log probabilities as an energy measure, we have

$$E^\alpha = -\ln P^\alpha = -\sum_u (s_u^\alpha \ln \hat{s}_u^\alpha + (1 - s_u^\alpha) \ln(1 - \hat{s}_u^\alpha)) \tag{B.27}$$

where $\hat{s}_u^\alpha$ is the top-down expectation for unit u.

We use the ratio of

$$\Delta E_u^\alpha = E^{\alpha | s_u = 0} - E^{\alpha | s_u = 1} \tag{B.28}$$

to chose the new state of the uth neuron:

$$p(s_u = 1 | \alpha) = \sigma(\Delta E_u^\alpha) \tag{B.29}$$

Hinton shows that this can be factorised into top-down effects and knock-on effects from below.

Given Gibbs sampling,

$$\Delta g_{ji} = \epsilon s_j (s_i - \hat{s}_i) \tag{B.30}$$

B.6.3 The Helmholtz Machine and the EM Algorithm

As prelude to the discussion in the next section, a discussion of the Helmholtz machine [33, 34]will be useful. The stochastic Helmholtz machine consists of a pair of Bayesian networks that are fitted to training data using an algorithm that approximates generalised Expectation Maximisation (EM).

The EM algorithm is a general approach to iterative computation of maximum-likelihood estimates when the observed data can be viewed as incomplete data. The term "incomplete data" has two implications:

1. The existence of two sample spaces X and Y represented by the observed data vector x and the complete data vector y, respectively.
2. The one-to-many mapping $y \rightarrow x(y)$ from space Y to X.

The complete vector y is not observed directly, but only through the vector x. The EM algorithm is executed in two steps: first an expectation step (E), followed by a maximisation step (M). During the E-step the "complete-data" log-likelihood, $P(Y|\mathbf{x}, M)$, given the observed data vector $\mathbf{x}$ and the current model, M, is calculated.

The Helmholtz machine then, based on a generalised EM algorithm, has a generative network $P(\mathbf{x}, \mathbf{y}|V)$ and a recognition network $Q(\mathbf{y}|\mathbf{x}, W)$, where V and W may be thought of as generative and recognition weight parameters, respectively. The recognition model is used to infer a probability distribution over the underlying causes from the sensory input. The separate generative model is used to train the recognition model.

B.6.4 The Wake-Sleep Algorithm

The Wake-Sleep algorithm [74] was designed as an improvement on the Helmholtz machine. The main disadvantage of the Helmholtz machine is that a recognition network that is compatible with the generative network must be estimated and this is often a difficult task. In the Wake-Sleep algorithm, rather than using Gibbs sampling, a separate set of bottom-up recognition connections are used to pick binary states for units in the layer below. The learning for the top-down generative weights is the same as for a Logistic Belief Net. This learning rule follows the gradient of the penalised log-likelihood where the penalty term is the Kullback–Liebler divergence between the true posterior distribution and the distribution produced by the recognition process. The penalised log-likelihood acts as a lower bound on the log-likelihood of the data and the effect of learning is to improve this lower bound. In attempting to raise the bound, the learning tries to adjust the generative model so that the true posterior distribution is as close as possible to the distribution actually computed. The recognition weights are learned by introducing a sleep phase in which the generative model is run top-down to produce fantasy data. The network knows the true causes of this fantasy data and attempts to maximise the log-likelihood of recovering these causes by adjusting the recognition weights. Frey [48] provides a clear description of the mathematical process for the Wake-Sleep algorithm which may be summarised by the following analysis.

The Rectified Gaussian Belief Net

Hinton and Ghahramani's [76] network is based on the Rectified Gaussian Belief Network (RGBN) which is an improvement in some ways on the Wake-Sleep algorithm. The RGBN uses units with states that are either positive real values or zero, so it can represent real-valued latent variables directly. The main disadvantage of the network is that the recognition process requires Gibbs sampling.

The generative model for the RGBN consists of multiple layers of units, each of which has a real-valued unrectified state, y_j, and a rectified state, $[y_j]^+ = \max(y_j, 0)$. The value of y_j is Gaussian distributed with a standard deviation σ_j and a mean $\hat{y}_j$ that is determined by the generative bias g_{0j}, and the combined effects of the rectified states of units, k, in the layer above:

$$\hat{y}_j = g_{oj} + \sum_k [y_k]^+ g_{kj} \tag{B.31}$$

Given the states of its parents, then the rectified state $[y_j]^+$ has a Gaussian distribution above zero, but all of the mass that falls below zero is concentrated in an infinitely dense spike at zero. This form of density is a problem for sampling and so Gibbs sampling is performed on the unrectified states. Now, consider a unit in an intermediate layer of a multilayer RGBN. With the unrectified states of all the other units in the network, then Gibbs sampling is performed to select a value for y_j according to its posterior distribution given the unrectified states of all the other units. In terms of energies, which are defined as the negative log probabilities, then the rectified states of the units in the layer above contribute a quadratic energy term by determining $\hat{y}_j$. The unrectified states of units, i, in the layer below contribute nothing if $[y_j]^+$ is 0, and if $[y_j]^+$ is positive then they each contribute because of the effect of $[y_j]^+$ on $\hat{y}_j$. The energy function may then be written as

$$E(y_j) = \frac{(y_j - \hat{y}_j)^2}{2\sigma_j^2} + \sum_i \frac{y_i - \sum_k [y_k]^+ g_{ki}}{2\sigma_i^2} \tag{B.32}$$

where h is an index over all of the units in the same layer as j including j itself; so y_j influences the right-hand side of this energy function by $[y_j]^+ = \max(y_j, 0)$. Hinton and Ghahramani show that learning rules for generative, recognition and lateral connections may be formed that not only identify the underlying causes in a data set but also that a topographical mapping may also be formed on data sets such as in the stereo disparity problem. Because sampling is used, this method is considerably slower than the methods of Chapter 5. Additionally, because of the top-down mechanism of learning it must be assumed that either horizontal mixes or vertical mixes of bars are present (in the case of the bars data) at the inputs; that is, there must not be a mix of both types in the data.

Attias [3] provides a review of current probabilistic approaches to Factor Analysis and related areas and Frey [48] provides a comprehensive overview of the probabilistic theory and techniques related to this area of research.

B.6.5 Olshausen and Field's Sparse Coding Network

The network of Olshausen and Field [145] is essentially a linear network that attempts to minimise the least mean square error in the network between actual images to the network and reconstructions of the image by the network. A prior belief that the output neurons should have sparse responses is incorporated into the network, which introduces a nonlinearity so that local features may be identified in the visual data.

A generative model is assumed in which the output values are fixed then fed back through a matrix of weights and used to determine the mean squared error between the image generated by the model and the actual image which is then used in a weight update rule. Error descent is performed on the following energy function with respect to both the output values y and for the weights W:

$$E = -\frac{1}{2}[(\mathbf{x} - W^T\mathbf{y})^2 + S(\mathbf{y}) \tag{B.33}$$

where $S(y_i)$ is a function that ensures a sparse prior distribution on $p(y_i) = \frac{1}{Z}\exp(-S(y_i))$ The weight update rule is essentially the same as in the negative feedback network of Chapter 5 but, unlike that network, gradient descent is performed to find values for the outputs given the current values on the weights.

C

Related Models for ICA

In this appendix, we review a few ICA models, particularly concentrating on the negative feedback model of Jutten and Herault [97], partly because it is a negative feedback model and partly because it was so important in generating interest in this problem in the artificial neural networks community. Bell and Sejnowski's [10] algorithm similarly stimulated the community and is also included. We also include a section on nonlinear PCA [101] and penalized error reconstruction [185] because of their obvious importance to the topic of the book before concluding with Hyvarinen's FastICA [85] because this is generally accepted currently to give the best performance.

C.1 Jutten and Herault

Jutten and Herault [97] proposed a neural network architecture (Figure C.1) similar to Földiák's first model. The feedforward of activation and the lateral inhibition are defined by:

$$y_i = x_i - \sum_{j=1}^{n} w_{ij} y_j$$

As before, in matrix terms, we can write

$$\mathbf{y} = \mathbf{x} - W\mathbf{y}$$
$$\text{And so, } \mathbf{y} = (I + W)^{-1}\mathbf{x}$$

This looks very similar to Földiák's decorrelating model, which we discussed in Chapter 8, but we shall see that critically Jutten and Herault introduce a nonlinearity into the learning rule.

Before we look at whether we can find a learning rule which will adaptively solve the problem, let us first consider if it is possible that such a network can separate two mixed sources.

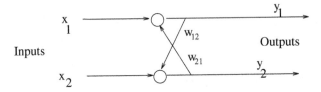

Fig. C.1. Jutten and Herault's model.

C.1.1 An Example Separation

Consider the 2×2 problem. Let us have two signals $s_1(t), s_2(t)$ which are functions of time, and let them be mixed using a matrix A to get the mixed signals $x_1(t), x_2(t)$. Then

$$x_1(t) = a_{11}s_1(t) + a_{12}s_2(t)$$
$$x_2(t) = a_{21}s_1(t) + a_{22}s_2(t)$$

Now we have outputs from the network satisfying

$$y_1 = x_1 - w_{21}y_2$$
$$y_2 = x_2 - w_{12}y_1$$

Substituting we get

$$y_1 = x_1 - w_{21}(x_2 - w_{12}y_1)$$
$$\text{and rearranging gives } y_1 = \frac{x_1 - w_{21}x_2}{1 - w_{21}w_{12}}$$
$$\text{Similarly, } y_2 = \frac{x_2 - w_{12}x_1}{1 - w_{21}w_{12}}$$

Substituting for the x_i values gives

$$y_1(t) = \frac{(a_{11} - w_{21}a_{21})s_1 + (a_{12} - w_{21}a_{22})s_2}{1 - w_{21}w_{12}}$$
$$y_2(t) = \frac{(a_{21} - w_{12}a_{11})s_1 + (a_{22} - w_{12}a_{12})s_2}{1 - w_{21}w_{12}}$$

From this we can see two pairs of solutions at which the y values are a function only of a single signal s value:

- If $w_{21} = \frac{a_{11}}{a_{21}}$ and $w_{12} = \frac{a_{22}}{a_{12}}$ then we get the solution:

$$y_1(t) = \frac{(a_{12} - w_{21}a_{22})s_2}{1 - w_{21}w_{12}}$$
$$= \frac{(a_{12} - \frac{a_{11}}{a_{12}}a_{22})s_2}{1 - \frac{a_{11}}{a_{12}}\frac{a_{22}}{a_{12}}}$$

$$= \frac{a_{12}(a_{12}a_{12} - a_{11}a_{22})}{a_{12}a_{12} - a_{11}a_{22}} s_2$$

$$= a_{12}s_2(t)$$

Similarly, $y_2(t) = \dfrac{(a_{21} - w_{12}a_{11})s_1}{1 - w_{21}w_{12}}$

$$= a_{21}s_1(t)$$

- Alternatively, if $w_{21} = \frac{a_{12}}{a_{22}}$ and $w_{12} = \frac{a_{21}}{a_{11}}$ then we get the solution:

$$y_1(t) = \frac{(a_{11} - w_{21}a_{21})s_1}{1 - w_{21}w_{12}} = a_{11}s_1(t)$$

$$y_2(t) = \frac{(a_{22} - w_{12}a_{12})s_2}{1 - w_{21}w_{12}} = a_{22}s_2(t)$$

In either case the network is extracting the single signals from the mixture - each output is a function of only one $s_i(t)$. However, this only shows that the solution is possible; it does not give us an algorithm for finding these optimal weights. We must find a learning algorithm which will adapt the weights during learning to find these optimal weights.

C.1.2 Learning the Weights

Having created a network which is theoretically capable of separating the two sources, we need to find a (neural network) algorithm which will adjust the parameters until the separation actually happens.

The learning rule Jutten and Herault use is

$$\Delta w_{ij} = -\alpha f(y_i)g(y_j), \text{ for } i \neq j \tag{C.1}$$

which is clearly anti-Hebbian learning.

Notice that if we use identity functions for $f()$ and $g()$ we are using exactly simple anti-Hebbian learning which we know will decorrelate the outputs. Recall that when we decorrelate the outputs, $E(y_i y_j) = 0$. If we have two independent sources the expected value of all joint higher-order moments will be equal to the product of individual higher-order moments. i.e. $E(y_i^n y_j^m) = E(y_i^n)E(y_j^m)$ for all values of m and n.

Jutten and Herault suggest using two different odd functions in the learning rule C.1. Since the functions $f()$ and $g()$ are odd functions, their Taylor series expansion will consist solely of the odd terms e.g.

$$f(x) = \sum_{j=0}^{\infty} a_{2j+1}x^{2j+1}, \text{ and } g(x) = \sum_{j=0}^{\infty} b_{2j+1}x^{2j+1} \tag{C.2}$$

Therefore the change due to the learning rule for a two-output network is of the form

$$\Delta w_{ij} = -\alpha f(y_1)g(y_2)$$
$$= -\alpha \sum_j \sum_k a_j b_k y_1^{2j+1} y_2^{2k+1}$$

Convergence is reached when all the moments $E(y_1^{2j+1} y_2^{2k+1}) = 0, \forall j, k$. Now statistical independence occurs when

$$E(y_1^{2j+1} y_2^{2k+1}) = E(y_1^{2j+1})E(y_2^{2k+1}) \qquad (C.3)$$

Jutten and Herault state that since most audio signals have an even distribution, their odd moments are zero and hence at the above state of convergence we have the independence criterion (C.3) satisfied.

In practice the signal separation properties seem to work when separating two or perhaps three voices from a mixture but no more than that; also, the process is not robust and requires careful parameter setting.

Another example of using the network was given in the image processing field: the input data were words written on card but with a sloping style. The network successfully converted the image to one in which the writing was parallel to the edges of the paper. The result is due to the fact that a sloped line introduces dependencies between the x and y coordinates. This dependency is minimised when the lines are either horizontal or vertical. In fact, if the original writing is closer to the vertical than horizontal orientation, the output will be a vertical line of text.

C.2 Nonlinear PCA

Oja's Subspace Algorithm was shown earlier to find the Principal Components of the input data which we know means decorrelation rather than independence of the outputs. The learning rule is repeated here for convenience:

$$\Delta w_{ij} = \alpha \left(x_i y_j - y_j \sum_k w_{ik} y_k \right) \qquad (C.4)$$

Since it finds an approximation to true PCA and PCA gives us the least error in reconstruction of the data from a linear operation, the Oja network can be thought of as finding the best linearly compressed form of the data.

We discussed in Chapter 5 how Karhunen and Joutsensalo [101] have derived from (C.4) a nonlinear equivalent:

$$\Delta w_{ij} = \alpha \left(x_i f(y_j) - f(y_j) \sum_k w_{ik} f(y_k) \right) \qquad (C.5)$$

as an approximation to the best non-linear compression of the data. While there is no 100% secure derivation of this algorithm as a solution to the ICA of a data set, it has been found experimentally that the algorithm does indeed find independent sources of some mixtures of signals (see below). Also, the addition of a nonlinearity breaks the symmetry which we found using the original subspace algorithm: with the original algorithm, the individual principal components were not found (indeed it is experimentally found that the algorithm tends to divide the variance evenly between the output neurons). Therefore, the original linear algorithm finds only a basis of the subspace, *not* the actual principal components themselves. However, this nonlinear algorithm (C.5) finds the independent sources exactly, not just a linear combination of the sources.

C.2.1 Simulations and Discussion

Karhunen and Joutsensalo have shown that the algorithm derived above is capable of separating signals into their significant subsignals. As an example, we repeat their experiment to separate samples of a sum of sinusoids into its component parts: the experimental data consists of N samples of a signal composed of the sum of two sinusoids in noise:

$$x(t) = \sum_{j=1}^{2} A_j \cos(2\pi f_j t - \theta_j) + \omega_t \tag{C.6}$$

The amplitudes, A_j, frequencies, f_j, and phases, θ_j, are unknown and must be estimated by the algorithm. We use initially white noise, $\omega_t \sim N(0, 0.05)$ where t denotes time.

Our input vector is a vector comprising a randomly drawn instance of $x(t)$ and that of the 14 subsequent times, $t + 1, \cdots, t + 14$. We can show that a network whose output neurons use a nonlinear function are better able to separate the input signal into its component parts while those using linear functions are less able to differentiate the individual subsignals. The original signal is shown in Fig. C.2 while the output of the non-linear output neurons are shown in Fig. C.3. This capability is not affected by coloured noise. Clearly the output neuron has identified one of the independent sinusoids. But in general, this network's performance on e.g. voice data has not yielded very good results.

C.3 Information Maximisation

Bell and Sejnowski [10] have developed a network based on the desire to maximise mutual information between inputs X and outputs Y:

$$I(X;Y) = H(Y) - H(Y|X) \tag{C.7}$$

They reason, however, that $H(Y|X)$ is independent of the weights W and so

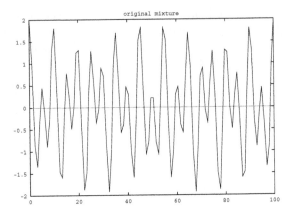

Fig. C.2. The original signal comprising a mixture of sinusoids.

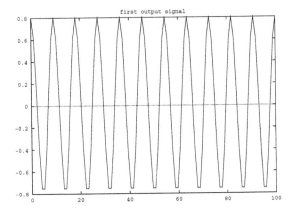

Fig. C.3. The output from the first interneuron after training when the output neuron's output is a non-linear function of its inputs.

$$\frac{\partial I(X;Y)}{\partial w} = \frac{\partial H(Y)}{\partial w} \tag{C.8}$$

Now comes the interesting part: the entropy of a distribution is maximised when all outcomes are equally likely. Therefore we wish to choose an activation function at the output neurons which equalises each neuron's chances of firing and so maximises their collective entropy. An example is given in Figure C.4 in which we show a Gaussian distribution and a sigmoid. Note that at the points of the distribution at which there are maximum values the slope of the sigmoid is greatest: the high-density parts of the probability density function

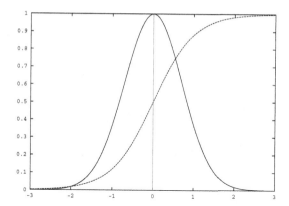

Fig. C.4. The solid line shows a Gaussian probability distribution. The sigmoid is the optimal function for evening out the output distribution so that all outputs are equally likely.

of the inputs is matched with the highly sloping parts of the sigmoid; this evens out the distribution at the outputs.

If we consider a single-output Y with a single input X joined by a single weight w, the distribution of the Y values is shown in Fig. C.5 over the weights w and inputs X where

$$y = \frac{1}{1 + \exp(-wx)} \qquad (C.9)$$

For large values of w (=1) the Gaussian nature of the input distribution is clear. For negative values of w the distribution becomes a bipolar distribution with equal probability for each of the two limiting values; but there is an intermediate value of w at which the output distribution is uniform. This last is the optimal value of w for which our learning algorithms should search.

Notice that this value depends on the actual distribution of the input data which is the sort of relationship that suits neural learning since it inherently responds to the statistics of the input distribution.

C.3.1 The Learning Algorithm

We can write the probability density function of an output y as

$$f_y(y) = \frac{f_x(x)}{\frac{\partial y}{\partial x}} \qquad (C.10)$$

where $f_y()$ is the probability density function of y and $f_x()$ is the probability density function of x. Then the entropy of the output is

Output Probability Distribution

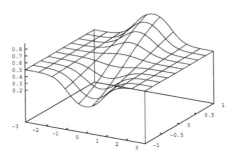

Fig. C.5. The distribution of the outputs as a function of inputs and weights. The weight which gives the most uniform distribution is about 0 in this case. Note that when the weight increases towards 1, the distribution becomes more peaked (actually Gaussian). When the weight decreases we tend to get a bimodal distribution.

$$H(y) = -E(\ln f_y(y)) = E\left(\ln\left|\frac{\partial y}{\partial x}\right|\right) - E(\ln f_x(x)) \qquad (C.11)$$

Now the second term here is unaffected by any change in the weights and therefore we can concentrate on maximising the first term with respect to w:

$$\Delta w \propto \frac{\partial H}{\partial w} = \frac{\partial}{\partial w}\left(\ln\left|\frac{\partial y}{\partial x}\right|\right) = \left(\frac{\partial y}{\partial x}\right)^{-1}\frac{\partial}{\partial w}\frac{\partial y}{\partial x} \qquad (C.12)$$

In the special case of the logistic transfer function,

$$act = wx + w_0$$
$$y = \frac{1}{1 + e^{-act}}$$

Then

$$\frac{\partial y}{\partial x} = wy(1 - y) \qquad (C.13)$$

$$\frac{\partial}{\partial w}\frac{\partial y}{\partial x} = y(1 - y)(1 + wx(1 - 2y)) \qquad (C.14)$$

Dividing (C.14) by (C.13) give the learning rule for the logistic function of

$$\Delta w \propto \left(\frac{\partial y}{\partial x}\right)^{-1}\frac{\partial}{\partial w}\frac{\partial y}{\partial x} = \frac{1}{w} + x(1 - 2y) \qquad (C.15)$$

A similar argument gives the derivation of the weight update rule for the bias term

$$\Delta w_0 \propto 1 - 2y \qquad (C.16)$$

Notice the effect of these two rules:

1. At convergence, $\Delta w_0 = 0$ and so the expected output is $\frac{1}{2}$. This in effect moves the sigmoid horizontally along its graph till the centre of the sigmoid (the steepest part) is above the peak of the input distribution $f_x()$.
2. Meanwhile the first rule is shaping the output distribution:
   - The $\frac{1}{w}$ part acts like an antidecay term and moves the weights away from one of the uninformative situations, where w is zero and y is constant regardless of the input, i.e.

$$y = \frac{1}{1 + e^{-w_0}} \qquad (C.17)$$

   - The other term is anti-Hebbian. This keeps the rule from moving to the other uninformative solution, which is that the y output is constantly 1, regardless of the input.

These forces balance out and cause the output distribution to stabilise at the maximum entropy solution.

We do not reproduce the derivation of the many input—many output rule, but merely give the results:

$$\Delta W \propto (W^T)^{-1} + (1 - 2\mathbf{y})\mathbf{x}^T$$
$$\Delta \mathbf{w_0} \propto 1 - 2\mathbf{y}$$

It is possible to use very flexible sigmoids in which, e.g. the top and bottom parts of the sigmoid are independently matched to the input signal's statistics. Bell and Sejnowski show that such a network can extract ten voices from a linear mixture of voices with great precision. Their algorithm has only failed under two conditions:

- when more than one of the sources was white Gaussian noise;
- when the mixing matrix was almost singular.

In the first case, there is no possible algorithm which can extract a single Gaussian from a mixture of Gaussians since the mixture of Gaussians is itself a Gaussian. In the second case the problem is ill-defined since we have n signals and $< n$ independent inputs to the neural network.

It has been shown that removing the correlations (the second-order statistics) from the input data greatly increases the speed of convergence of the network: Hebbian (and anti-Hebbian) learning responds mostly to the correlations in the input data; by removing these the network can concentrate on the other facets of learning.

C.4 Penalised Minimum Reconstruction Error

We discussed in Chapter 2 how Xu [185] has derived a neural network which we can describe as one which feeds activation forward to outputs and then back to inputs through the same weights. He envisages the returning signal from the outputs as trying to reconstruct the input vector. He aims to minimise the least mean square error at its inputs and shows that it is a PCA network. The error term is

$$J(W) = E(\| \mathbf{x} - \hat{\mathbf{x}} \|)$$
$$= \int p(\mathbf{x}) \| \mathbf{x} - W^T W \mathbf{x} \| \, d\mathbf{x}$$

Starting from this (which is one of the definitions of PCA), Xu derives the learning rule

$$\Delta W = \mu \mathbf{x}^T \mathbf{x} (I - W^T W) W + (\mathbf{x} - W^T W \mathbf{x})(W \mathbf{x})^T \qquad (C.18)$$

If we put this into the feedforward and feedback terms, we may use $\mathbf{y} = W\mathbf{x}$ and $\mathbf{e} = \mathbf{x} - W^T W \mathbf{x}$ and so state the rule as

$$\Delta W = \mu(\mathbf{x} \mathbf{e}^T W \mathbf{y}' + \mathbf{e} \mathbf{y}^T) \qquad (C.19)$$

C.4.1 Adding Competition

Now in the context of the discussion in the last chapter, we have a network which is wholly given over to cooperation — the neurons are taking the maximum information out of the data set. We need to add some competition to the network to find the independent sources.

One way to do this is to simply add a nonlinearity to the learning rule to get

$$\Delta W = \mu(\mathbf{x} \mathbf{e}^T f(W \mathbf{y}') + \mathbf{e} f(\mathbf{y})^T) \qquad (C.20)$$

This does have some effect in finding the individual sources (this is the same method Karhunen and Joutsensalo use) but the competition is quite weak. A better solution, developed by Xu and his colleagues, is to explicitly add a penalty term to the cost function to get

$$E(W) = J(W) + \lambda G(W) \qquad (C.21)$$

where $J()$ is the best reconstruction criterion as before and $G()$ is a competition criterion which will penalise cooperation. The Lagrange parameter λ will tradeoff competition for cooperation, so we get compromise between these two criteria.

The next section will consider the nature of the competition function $G()$.

The Nature of the Competition Function

We can develop a competition function based on information theory. Firstly let the ith activation be h_i which will then be passed through an activation function $f()$ to get the output y_i. Then, in the absence of noise on the inputs, the mutual information between inputs and outputs satisfies

$$I(\mathbf{y}, \mathbf{x}) = I(\mathbf{y}, \mathbf{h}) \tag{C.22}$$

If the postsynaptic potentials are factorial,

$$p(\mathbf{h}) = \prod_{i=1}^{M} p(h_i) \tag{C.23}$$

One possible index might be to reduce the pairwise correlation over the outputs. Thus we might have

$$G = \sum_{i=1}^{M} \sum_{j=1, j \neq i}^{M} g_{ij}$$

$$\text{where } g_{ij} = \int (y_i y_j) p(\mathbf{x}) d\mathbf{x}$$

This would certainly work for Gaussian distributions but is not sufficient for non-Gaussian distributions and so we might extend this to

$$g_{ij} = \int (y_i y_j)^k p(\mathbf{x}) d\mathbf{x} \tag{C.24}$$

Then e.g. taking $k = 2$ we would have

$$\Delta \mathbf{w}_j = \mu \left(\mathbf{x} e^T \mathbf{w}_j y_j + y_j \mathbf{e} - \lambda y_j' \mathbf{x} \sum_{l \neq j} y_i^2 \right) \tag{C.25}$$

thereby introducing higher order statistics into the learning rule.

We can imagine a number of competition functions all of which should share the property that when the output of one neuron increases the other's output must decrease, i.e. if $\sigma_i()$ is the output function of the ith neuron, then

$$\frac{\partial \sigma_i}{\partial \mathbf{y}_j} < 0, \forall j \neq i \tag{C.26}$$

C.5 FastICA

FastICA [85] is the most popular and commonly used method for performing independent component analysis. FastICA can use different measures of non-Gaussianity; i.e. different objective functions for ICA estimation. FastICA is a very efficient method for maximising the contrast functions defined in [85]. As with most ICA methods the data is assumed to have been sphered.

C.5.1 FastICA for One Unit

The one-unit version of the FastICA algorithm looks to maximise nongaussianity for one output only. This unit, or neuron, consists of a weight vector, $\mathbf{w}$ that the neuron is able to update by a learning rule. The FastICA learning rule identifies a direction; a unit vector $\mathbf{w}$ such that the projection $\mathbf{w^T x}$ maximises non-Gaussianity.

Non-Gaussianity may be measured by an approximation to negentropy [85]. Since the data is whitened, we can constrain the variance of $\mathbf{w}^T\mathbf{x}$ to unity by constraining the norm of $\mathbf{w}$ to unity.

FastICA is based on a fixed-point algorithm for finding the maximum of the non-Gaussianity of $\mathbf{w}^T\mathbf{x}$. It can also be derived as an approximative Newton method [85]. The basic form of the FastICA algorithm is:

1. Choose an initial random weight vector $\mathbf{w}$.
2. Let
$$\mathbf{w}^+ = E\left\{\mathbf{x}g(\mathbf{w}^T\mathbf{x})\right\} - E\left\{g'(\mathbf{w}^T\mathbf{x}))\right\} \tag{C.27}$$

3. Let
$$\mathbf{w} = \frac{\mathbf{w}^+}{\|\mathbf{w}^+\|} \tag{C.28}$$

4. If not converged go to step 2.

D

Previous Dual Stream Approaches

In this appendix, we review some other dual stream approaches and discuss their connections to the work presented in this book. We begin with Becker's model since we have used her artificial random dot stereogram data extensively in the book.

D.1 The I-Max Model

Becker [7] has used mutual information maximization to develop the Infomax principle. The infomax principle applies to a situation where the mutual information $I(\mathbf{y}; \mathbf{x})$ between the output vector $\mathbf{y}$ of a neural system and the input vector $\mathbf{x}$ is the objective function to be maximised.

Becker has an interesting model using error descent learning. They begin with the question "What is the best way to model the input distribution to encode interesting features in the data" and come up with the observation that they wish to constrain the learning problem by restricting the features of interest to those which are liable to be useful for later perceptual processing. In a general nonspecific environment, there are regularities ("coherence") in that any part of the environment is very likely to be predictable from other close parts of the environment e.g. any object has a finite compact surface area and so there exists a set of points all physically close to one another which share visually similar features. Similarly there exists temporal coherence in our environment. Also there is a coherence across sensory modalities — we generally see and smell and feel things at a single instant in time. This suggests that we should use coherence to extract information from the input data; one objective that might be appropriate for a network would be the extraction of redundancy (which gives rise to coherence) in raw sensory data since we do not, for example, have to use sight, smell and touch of an orange in order to identify the orange. The two output neurons are attempting to reduce the redundancy in their outputs based on their reaction to the input data which comes from the same source.

Becker [7] states that we could perform error descent on the squared error of the difference between the outputs, but one difficulty with this is that the network could simply learn to output a constant value at both neurons x_1 and x_2. So we need to force the neurons to extract as much information as possible but still ensure that they are agreeing. This suggests that the optimisation criterion should be to maximise the mutual information between the two neurons

$$I_{a,b} = H(a) + H(b) - H(a,b) \tag{D.1}$$
$$= H(a) - H(a|b) \tag{D.2}$$
$$(\text{or } = H(b) - H(b|a)) \tag{D.3}$$

Written this way we can see that by maximising the mutual information between the neurons, we are maximising the entropy (the expected information output) of each neuron while minimising the conditional entropy given the other's value (the uncertainty left about each neuron's output). So we wish each neuron to be as informative as possible while also telling us as little as possible about the other neuron's outputs.

Now if we have two binary (1/0) probabilistic units, we can estimate the mutual information by sampling their activity over a large set of input cases. This gives us an estimate of their individual and joint probabilities and so the mutual information can be calculated. Let the expected output of the jth output neuron on the Kth training example be s_j^K and define $s_j = E(y_j)$ over all training patterns. s_{ij} is defined to be the joint probability that the ith and j^{th} fire simultaneously. Becker shows that the partial derivative of the mutual information with respect to the expected output on training case K is

$$\frac{\partial I_{y_i;y_j}}{\partial s_i^K} = -P^K \left(\log \frac{s_i}{s_{\bar{i}}} - s_j^K \log \frac{s_{ij}}{s_{\bar{i}j}} - s_{\bar{j}}^K \log \frac{s_{i\bar{j}}}{s_{\bar{i}\bar{j}}} \right) \tag{D.4}$$

where P^K is the probability of training case K and $s_{\bar{i}} = E(1 - y_i)$, etc. Thus, we have a method of changing the weights (e.g. by least mean square learning) to maximise the mutual information:

$$\triangle \mathbf{w} \propto \frac{\partial I}{\partial \mathbf{w}} = \frac{\partial I}{\partial s} \cdot \frac{\partial s}{\partial \mathbf{w}} \tag{D.5}$$

It is possible to extend the IMax algorithm to both continuous variables and to multivalued spatially coherent features; the latter is interesting in that it uses a general method of ensuring that every output is a valid probability between 0 and 1. If we have a discrete random variable $A \in \{a_1, ..., a_N\}$, we can define the probability of the ith event as

$$P(A = a_i) = \frac{\exp(x_i)}{\sum_{j=1}^N \exp(x_j)} \tag{D.6}$$

which is guaranteed to give a value between 0 and 1 and to ensure that the probabilities sum to 1.

In [7], it is demonstrated that by maximizing the mutual information $I(Y_a; Y_b)$ it is possible to extract stereo disparity (depth) from random dot stereograms. We have shown in Chapter 9 that the CCA models also perform well on this data set.

D.2 Stone's Model

Stone [174] has studied a similar problem (also of depth extraction). He described a temporal learning rule based on a general assumption that inputs to sensory receptors tend to change rapidly over time. However, the physical parameters which underlie these changes vary more slowly and smoothly. For example, a film of a rigidly moving object reveals a great amount of redundant information as the image of the object appears in slightly different spatial locations on successive frames of the film. Therefore, if a neuron codes for a physical parameter, its output should also change slowly and smoothly, in spite of the rapidly fluctuating inputs. Thus, a neuron adapting to make its output vary smoothly over time can learn to code the invariances underlying its input.

Thus, we wish to minimise the short term variance of the neuron's output. But this is not enough since the neuron could simply always output one value. Therefore we wish to simultaneously maximise long-term variance.

The temporal learning rule has a single output unit. Stone jointly maximises the long-term variance of the output V of the unit and minimizes its short-term variance U by maximizing the objective function $F = \log(V/U)$ (the logarithm was used so that the derivative of F was easy to compute). The variances V and U were calculated using moving averages

$$F = \log \frac{V}{U}$$
$$= \log \frac{\sum_{t=1}^{T}(\bar{z}_t - z_t)^2}{\sum_{t=1}^{T}(\tilde{z}_t - z_t)^2} \tag{D.7}$$

where z_t was the output of the unit at a particular time, $\bar{z}_t$ was its long-term weighted average, and $\tilde{z}_t$ was its short-term weighted average (averaging over time).

The derivative of F with respect to output weight results in a learning rule that is a linear combination of Hebbian and anti-Hebbian weight update, over long and short time-scales, respectively:

$$\frac{\partial F}{\partial w_{jk}} = \frac{1}{V}\sum_{t}(\bar{z}_{kt} - z_{kt})(\bar{z}_{jt} - z_{jt}) - \frac{1}{U}\sum_{t}(\tilde{z}_{kt} - z_{kt})(\tilde{z}_{jt} - z_{jt}) \tag{D.8}$$

This rule captures the temporal smoothness constraint. Another equally valid constraint is spatial smoothness, which means that physical parameters tend to vary smoothly across space. In dealing with the temporal constraint, we need only consider a single neuron and its output over time. In the spatial case, however, we need to consider a network of neurons arranged spatially in such a way that neighbouring output neurons receive inputs from neighbouring regions of space. We can then interpret the spatial smoothness constraint as follows: if a network of neurons codes for a physical parameter its output should change slowly and smoothly across the network, despite the quite different inputs received by neighbouring neurons. Stone applied this principle to learning stereo disparity from dynamic sterograms. In the spatial model, Stone also maximises a function $F = \log(V/U)$. In this model, however, the long and short-term variances are measured as spatial variance.

D.3 Kay's Neural Models

Another information-theoretic approach to multi-stream processing was developed by Kay and Phillips [105]. Central to this work is the concept of *contextual guidance*, which refers to the guiding effect that neighbouring neurons (the context) have on a particular neuron. It is suggested that the information that the context provides can assist and guide both the learning phase as well as the processing stage of neural networks.

This idea is cast in a model in which a neural network consists of a set of local processors. Each of these local processors has a set of outputs X, receptive field inputs (RF), R, and contextual field inputs (CF), C. The receptive field inputs are connected directly to a lower layer of processing and the contextual inputs are connected to the outputs of neighbouring local processors that operate in the same context. Furthermore, there are a set of lateral connections that interconnect the outputs of a processor, which are termed the *within-processor* (WP) connections. In this way, any number of these processors can be connected to form a multilayer neural architecture; see Fig. D.1.

It is not necessary for the network to be fully connected, which we can describe by the set of indices $\delta i(r)$, $\delta i(c)$ and $\delta i(x)$, which denote the indices of the receptive field connections, the contextual field connections and the within processor connections, respectively, that are connected to the $i^t h$ output. The total integrated field inputs of all three sets of connections then become:

$$S_i(r) = \sum_{j \in \delta i(r)} w_{ij} R_j - w_{i0} \qquad (D.9)$$

$$S_i(c) = \sum_{j \in \delta i(c)} v_{ij} C_j - v_{i0} \qquad (D.10)$$

$$S_i(x) = \sum_{j \in \delta i(x)} u_{ij} X_j \qquad (D.11)$$

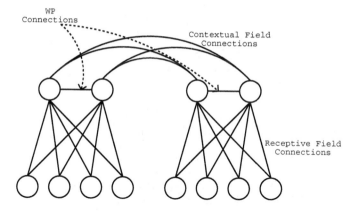

Fig. D.1. Diagram of the Contextual Guidance Network

where w_{ij} are the weights of the receptive field connections, v_{ij} are the weights of the contextual field connections and u_{ij} are the the weights of within-processor connections; w_{i0} and v_{i0} are two bias terms.

To develop learning rules for this network, an information-theoretic objective function was developed. As the processors deal with three types of information, the concept of mutual information is extended to *three-way mutual information*, which describes the information that is shared between all three variables. It is a natural extension of "normal" mutual information and can also be displayed in a Venn diagram; see Fig. D.2. From this diagram it is clear that three-way mutual information can be defined in three different ways as follows:

$$I(X,Y,Z) = I(X,Y) - I(X,Y|Z) \qquad (D.12)$$
$$= I(X,Z) - I(X,Z|Y) \qquad (D.13)$$
$$= I(Y,Z) - I(Y,Z|X) \qquad (D.14)$$

When training the network, the information from the Contextual Field, Receptive Field and the Outputs will be used. In particular we are interested in the four information-theoretic quantities that make up the entropy of the outputs,

$$H(X) = I(X; \mathbf{R}; \mathbf{C}) + I(X; \mathbf{R}|\mathbf{C}) + I(X; \mathbf{C}|\mathbf{R}) + H(X|\mathbf{R}, \mathbf{C}) \qquad (D.15)$$

All these four different terms are important when training the network. The first term relates to the information that the outputs share with the RF and CF, and the second term represents the information that the outputs share with the RF, but not with the CF. In general we want to maximise both these terms. The last two terms, denoting the information the outputs share with the CF, but not with the RF and the entropy of the outputs, given the CF and RF, should in general be minimised.

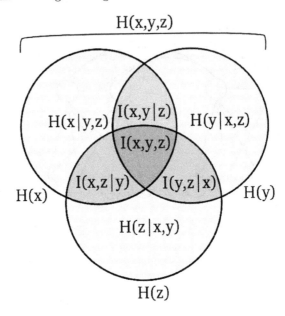

Fig. D.2. Venn-diagram of three-way mutual information

We can now write the objective function as follows:

$$J = \phi_0 I(X; \mathbf{R}; \mathbf{C}) + \phi_1 I(X; \mathbf{R}|\mathbf{C}) + \phi_2 I(X; \mathbf{C}|\mathbf{R}) + \phi_3 H(X|\mathbf{R}, \mathbf{C}) \quad \text{(D.16)}$$

This objective function consists of four different terms that we wish to maximise or minimise simultaneously and the parameters ϕ_0 to ϕ_3 indicate the relative strength that should be given to each of these objectives, and are taken between -1 and 1.

Using stochastic output units, gradient rules were derived for each of these terms, enabling the network to be trained. These rules enable a multilayer network with an arbitrary amount of processors to be built. According to experiments carried out by Kay [105], the network functions equally well for small as well as large networks.

D.4 Borga's Algorithm

Borga proposed a CCA learning algorithm in [16]. He considers the generalized eigenproblem

$$A\hat{e} = \lambda B\hat{e} \quad \text{(D.17)}$$

which is closely related to the problem of finding the extremum points of a ratio of quadratic forms:

$$r = \frac{\mathbf{w}^T A \mathbf{w}}{\mathbf{w}^T B \mathbf{w}} \quad \text{(D.18)}$$

where both A and B are symmetric, B is positive definite, and $\hat{e}$ denotes a unite vector [1]. This ratio is known as the Rayleigh quotient and its critical points, i.e. the points of zero derivatives, will correspond to the eigensystem of the generalized eigenproblem. To see this let us look at the gradient of r:

$$\frac{\partial r}{\partial \mathbf{w}} = \frac{2}{\mathbf{w}^T B \mathbf{w}}(A\mathbf{w} - rB\mathbf{w})$$

$$= \frac{2 \parallel \mathbf{w} \parallel}{\mathbf{w}^T B \mathbf{w}}(A\hat{\mathbf{w}} - rB\hat{\mathbf{w}}) = \alpha(A\hat{\mathbf{w}} - rB\hat{\mathbf{w}}) \qquad (D.19)$$

where α is a positive factor. Setting the gradient to 0 gives

$$A\hat{\mathbf{w}} = rB\hat{\mathbf{w}} \text{ or } B^{-1}A\hat{\mathbf{w}} = r\hat{\mathbf{w}} \qquad (D.20)$$

which is recognized as the generalized eigenproblem in (D.17). The solutions r_i and $\hat{\mathbf{w}}_i$ are the eigenvalues and eigenvectors, respectively, of the matrix $B^{-1}A$. If the eigenvalues r_i are distinct (i.e. $r_i \neq r_j$ for $i \neq j$) the different eigenvectors are orthogonal in the metrics A and B which means that

$$\hat{\mathbf{w}}_i^T B \hat{\mathbf{w}}_j = \begin{cases} 0, & \text{for } i \neq j \\ \beta_i > 0, & \text{for i = j} \end{cases}$$

$$\hat{\mathbf{w}}_i^T A \hat{\mathbf{w}}_j = \begin{cases} 0, & \text{for } i \neq j \\ r_i\beta_i > 0, & \text{for i = j} \end{cases}$$

This means that the $\mathbf{w}_i$'s are linearly independent. Since an n-dimensional space gives n eigenvectors which are linearly independent, $\mathbf{w}_1, ..., \mathbf{w}_n$ is a basis and any $\mathbf{w}$ can be expressed as a linear combination of the eigenvectors. The function r is bounded by the largest and smallest eigenvalue, i.e.

$$r_n \leq r \leq r_1 \qquad (D.21)$$

which means that there exists a global maximum and that this maximum is r_1. To investigate if there are any other local maxima we look at the second derivative, or the Hessian H, of r for the solutions of the eigenproblem

$$H_i = \frac{\partial^2 r}{\partial \mathbf{w}^2}\bigg|_{\mathbf{w}=\hat{\mathbf{w}}_i} = \frac{2}{\hat{\mathbf{w}}_i^T}B\hat{\mathbf{w}}_i(A - r_iB) \qquad (D.22)$$

It can be shown that the Hessians H_i have positive eigenvalues for $i > 1$, i.e. there exits vectors $\mathbf{w}$ such that

$$\mathbf{w}^T H_i \mathbf{w} > 0, \quad \forall i > 1 \qquad (D.23)$$

This means that for all solutions to the eigenproblem except for the largest root there exist a direction in which r increases. In other words, all extremum

---

[1] The hat on a vector in this section, $\hat{}$, indicates it is a unit vector.

points of the function r are saddles except for the global minimum and maximum points.

Since the only stable critical point is the global maximum it should be possible to find the largest eigenvalue and its corresponding vector by performing a stochastic gradient search on the energy function r. This can be done with an iterative algorithm:

$$\mathbf{w}(t+1) = \mathbf{w}(t) + \triangle\mathbf{w}(t) \tag{D.24}$$

where the update vector $\triangle\mathbf{w}_1$, at least on average, lies in the direction of the gradient:

$$E(\triangle\mathbf{w}) = \beta\frac{\partial r}{\partial\mathbf{w}} = \alpha(A\hat{\mathbf{w}} - rB\hat{\mathbf{w}}) = \alpha(A\hat{\mathbf{w}} - B\hat{\mathbf{w}}) \tag{D.25}$$

where α and β are positive numbers. Here we use the length of the vector to represent the corresponding eigenvalue, i.e. $\parallel \mathbf{w} \parallel = r$.

It is, of course, possible to enhance this update rule and also take second order derivatives into account. This would include estimating the inverse of the Hessian and using this matrix to modify the update direction. Such a procedures is, for the batch or off-line case, known as Gauss–Newton methods.

For finding the largest canonical correlation, with the matrices A and B and the vector w:

$$A = \begin{bmatrix} 0 & \Sigma_{12} \\ \Sigma_{21} & 0 \end{bmatrix}, B = \begin{bmatrix} \Sigma_{11} & 0 \\ 0 & \Sigma_{22} \end{bmatrix}, \mathbf{w} = \rho\begin{pmatrix} \mathbf{w}_{x_1} \\ \mathbf{w}_{x_2} \end{pmatrix} = \begin{pmatrix} \mu_{x_1}\hat{\mathbf{w}}_{x_1} \\ \mu_{x_2}\hat{\mathbf{w}}_{x_2} \end{pmatrix} \tag{D.26}$$

the update direction is:

$$E[\triangle\mathbf{w}] = \gamma\frac{\partial r}{\mathbf{w}} = \alpha\left(\begin{bmatrix} 0 & \Sigma_{12} \\ \Sigma_{21} & 0 \end{bmatrix}\hat{\mathbf{w}} - \gamma\begin{bmatrix} \Sigma_{11} & 0 \\ 0 & \Sigma_{22} \end{bmatrix}\hat{\mathbf{w}}\right) \tag{D.27}$$

This behaviour is accomplished if at each time step the vector $\mathbf{w}$ is updated with

$$\delta\mathbf{w} = \alpha\left(\begin{bmatrix} 0 & \mathbf{x}_1\mathbf{x}_2^T \\ \mathbf{x}_2\mathbf{x}_1^T & 0 \end{bmatrix}\hat{\mathbf{w}} - \begin{bmatrix} \mathbf{x}_1\mathbf{x}_1^T & 0 \\ 0 & \mathbf{x}_2\mathbf{x}_2^T \end{bmatrix}\mathbf{w}\right) \tag{D.28}$$

Since $\parallel \mathbf{w} \parallel = \gamma = \rho$ when the algorithm converages, the length of the vector represents the correlation between the variates.

For finding successive canonical correlations, let G denote the $n \times n$ matrix $B^{-1}A$. Then n equations for the n eigenvalues solving the eigenproblem,

$$GE = ED \Rightarrow G = EDE^{-1} = \Sigma\lambda_i\hat{\mathbf{e}}_i\mathbf{f}_i^T \tag{D.29}$$

where the eigenvalues and vectors constitute the matrices D and E, respectively:

$$D = \begin{bmatrix} \lambda_1 & & 0 \\ & \cdot & \\ & & \cdot \\ 0 & & \lambda_n \end{bmatrix}, E = \begin{bmatrix} | & & | \\ \hat{\mathbf{e}}_1 & \dots & \hat{\mathbf{e}}_n \\ | & & | \end{bmatrix}, E^{-1} = \begin{bmatrix} - & \mathbf{f}_1^T & - \\ & \cdot & \\ & \cdot & \\ - & \mathbf{f}_n^T & - \end{bmatrix} \tag{D.30}$$

The vectors, $\mathbf{f}_i$, in the rows of the inverse of the matrix containing the eigenvectors $\hat{\mathbf{e}}_i$, which means that

$$\mathbf{f}_i^T \hat{\mathbf{e}}_j = \delta_{ij} \tag{D.31}$$

The dual vectors $\mathbf{f}_i$ possessing the property in (D.31), can be found by choosing them according to

$$\mathbf{f}_i = \frac{B\hat{\mathbf{e}}_i}{\hat{\mathbf{e}}_i^T B \hat{\mathbf{e}}_i} \tag{D.32}$$

Now, if $\hat{\mathbf{e}}_1$ is the eigenvector corresponding to the largest eigenvalue of G, the new matrix

$$H = G - \lambda_1 \hat{\mathbf{e}}_1 \mathbf{f}_1^T \tag{D.33}$$

has the same eigenvectors and eigenvalues as G except for the eigenvalue corresponding to $\hat{\mathbf{e}}_1$, which now becomes 0. This means that the eigenvector corresponding to the largest eignvalue of H is the same as the one corresponding to the second largest eigenvalue of G.

Since the algorithm starts by finding the vector $\hat{\mathbf{w}}_1 = \hat{\mathbf{e}}_1$, it is only necessary to estimate the the dual vector $\mathbf{f}_1$ in order to subtract the correct outer product from G and remove its largest eigenvalue. To do this, the two components A and B must be modified in order to produce the desired subtraction. The two modified components, A' and B', have the following property:

$$B'^{-1}A' = B^{-1}A - \lambda_1 \hat{\mathbf{e}}_1 \mathbf{f}_1^T \tag{D.34}$$

A simple solution is obtained if only one of the matrices is modified and the other matrix is kept fixed:

$$B' = B \text{ and } A' = A - \lambda_1 B\hat{\mathbf{e}}_1 \mathbf{f}_1^T \tag{D.35}$$

For canonical correlation,

$$\begin{aligned}
G = B^{-1}A &= \begin{bmatrix} \Sigma_{11}^{-1} & 0 \\ 0 & \Sigma_{22}^{-1} \end{bmatrix} \begin{bmatrix} 0 & \Sigma_{12} \\ \Sigma_{21} & 0 \end{bmatrix} \\
&= \begin{bmatrix} 0 & \Sigma_{11}^{-1}\Sigma_{12} \\ \Sigma_{22}^{-1}\Sigma_{21} & 0 \end{bmatrix}
\end{aligned} \tag{D.36}$$

it is necessary to estimate the dual vector $\mathbf{f}_1$ corresponding to the eigenvector $\hat{\mathbf{e}}_1$, or rather the vector $\mathbf{u}_1 = \lambda_1 B\hat{\mathbf{e}}_1$:

$$\begin{aligned}
E(\Delta\mathbf{u}_1) &= \alpha(B\mathbf{w}_1 - \mathbf{u}_1) \\
&= \alpha \left(\begin{bmatrix} \Sigma_{11} & 0 \\ 0 & \Sigma_{22} \end{bmatrix} \mathbf{w}_1 - \mathbf{u}_1 \right)
\end{aligned} \tag{D.37}$$

A stochastic approximation of this rule is given by

$$\Delta\mathbf{u}_1 = \alpha \left(\begin{bmatrix} \Sigma_{11} & 0 \\ 0 & \Sigma_{22} \end{bmatrix} \mathbf{w}_1 - \mathbf{u}_1 \right) \tag{D.38}$$

with this estimate, the outer product in equation (D.35), can be used to modify the matrix A:

$$A' = A - \lambda_1 B \hat{e}_1 f_1^T = A - \frac{u_1 u_1^T}{w_1^T u_1} \tag{D.39}$$

Then the learning rule which finds the second largest canonical correlation and its corresponding directions can be written in the following form:

$$\Delta w = \alpha \left(\left(\begin{matrix} 0 & x_1 x_2^T \\ x_2 x_1^T & 0 \end{matrix} - \frac{u_1 u_1^T}{w_1^T u_1} \right) \hat{w} - \begin{bmatrix} x_1 x_1^T & 0 \\ 0 & x_2 x_2^T \end{bmatrix} w \right) \tag{D.40}$$

E

Data Sets

E.1 Artificial Data Sets

E.1.1 Gaussian Data

To illustrate PCA networks, we draw samples from zero mean Gaussian distributions. We draw each sample independently from the others. If we have an n-dimensional input sample, we draw x_1 from $N(0, \sigma_1)$, x_2 from $N(0, \sigma_2)$, etc. with $\sigma_1 > \sigma_2 > \cdots > \sigma_n$.

Thus the largest eigenvalue of the input data's covariance matrix comes from the first input, x_1, the second largest comes from x_2 and so on.

To investigate the constrained network's potential in Chapter 5, data from a distribution whose principal components are shown in Table 5.2 was used: this data set had the first principal component represent the first two inputs and the second represent the last three inputs — again, there is a clear division between the two principal components.

In Chapter 6, we changed one of these Gaussian inputs to make it (and it alone) "interesting".

In Chapter 9, we use an artificial data set where $\mathbf{x}_1$ is a four-dimensional vector, each of whose elements is drawn from the zero-mean Gaussian distribution, $N(0, 1)$; $\mathbf{x}_2$ is a three-dimensional vector, each of whose elements is also drawn from $N(0, 1)$. In order to introduce correlations between the two vectors, $\mathbf{x}_1$ and $\mathbf{x}_2$, we generate an additional sample from $N(0, 1)$ and add it to the first elements of each vector. In order to ensure that we are not simply responding to variance, the data is then normalised so that the variance in each input is identical. This data set is extended in various ways within that chapter.

A reduced rank data set was used in Chapter 13.

E.1.2 Bars Data

A standard data set which we used in Chapter 5 consists of a square grid of input values where $x_i = 1$ if the ith square is black and 0 otherwise (see Fig.

5.2). However the patterns are not random patterns: each input consists of a number of randomly chosen horizontal or vertical lines. The network must identify the existence of these lines. The important thing to note is that each line can be considered as an independent source of blackening a pixel on the grid: it may be that a particular pixel will be twice blackened by both a horizontal and a vertical line at the same time, but we need to identify both of these sources. Typically we use 64 inputs in an 8×8 grid.

Typically, on an 8×8 grid, each of the 16 possible lines are drawn with a fixed probability of $\frac{1}{8}$ independently from each of the others. The data set then is highly redundant in that there exists 2^{64} possible patterns and we are only using at most 2^{16} of these.

E.1.3 Sinusoids

Five mixtures of sine waves was input data to the network of Chapter 5 so that

$$x_0 = \sin(t) + \sin(2t)$$
$$x_1 = \sin\left(t + \frac{\pi}{8}\right) + \sin\left(2t + \frac{\pi}{4}\right)$$
$$x_2 = \sin\left(3t + \frac{3\pi}{7}\right)$$
$$x_3 = \sin\left(4t + \frac{4\pi}{3}\right) + \sin(5t)$$
$$x_4 = 2\sin(5t)$$

The first two mixtures, x_0 and x_1, are identical but slightly out of phase, the third is a totally independent sine wave and the last two contain the same sine wave, however, one has another sine wave mixed with it. Therefore the relationship between the outputs of the sources is straightforward in the case of x_3 and x_4 but time-varying in the case of x_0 and x_1 where the underlying source is emitting different phase signals.

In Chapter 11, we embedded two sine waves on the surface of a sphere and on the plane. We generate artificial data according to the prescription (see Fig. 11.3):

$$x_{11} = \sin(s_1) * \cos(s_2)$$
$$x_{12} = \sin(s_1) * \sin(s_2)$$
$$x_{13} = \cos(s_1)$$
$$x_{21} = s_1$$
$$x_{22} = s_2$$

where e.g. $s_1(t) = \sin(t/8) * \pi, s_2(t) = \sin\{(t/5 + 3) * \pi\}$ for each of $t = 1, 2, ..., 500$. The results from the nonlinear mixture, $\mathbf{x_1}$, and $\mathbf{x_2}$ are shown in Fig. 11.4 .

Other varying mixtures of sinusoids are also used in this chapter.

E.1.4 Random Dot Stereograms

In Chapter 9, we have a data set in which each input vector consists of a one-dimensional random strip which corresponds to the left image and a shifted version of this which corresponds to the right image. The left image has components drawn with equal probability from the set $\{-1, 1\}$ and the right image is generated by choosing a randomly chosen global shift — either one pixel left or one pixel right — and applying it to the left image.

A slightly more complex data set was developed by James Stone [174]; a description of Stone's method is found in Appendix B. This data simulates a moving surface with a slowly varying depth. The disparity between the left and right images of a stereo pair was generated by convolving a circular array of 1000 uniformly distributed random numbers with a Gaussian function, and then normalising these values to lie between ± 1. For example, consider an array, $\mathbf{r}$ of 1000 random numbers, and a 1000 $\times$ 1000 matrix $\mathbf{G}$ with values given by:

$$G_{a,b} = \frac{1}{\sigma} e^{\frac{-(\min(|a-b|, 1000-|a-b|))^2}{2\sigma^2}} \tag{E.1}$$

The min( ) function in (E.1) ensures that the Gaussian wraps around the $\mathbf{r}$ array, which effectively makes it circular. Then, the unnormalised array of disparity values, $\mathbf{d}$, is given by the product in (11.46). Fig. 11.17 shows an example of an array of 1000 disparity values generated using this method.

E.1.5 Nonlinear Manifolds

In Chapter 11, we generate data according to the prescription:

$$x_{11} = 1 - \sin\theta + \mu_1$$
$$x_{12} = \cos\theta + \mu_2$$
$$x_{21} = \theta - \pi + \mu_3$$
$$x_{22} = \theta - \pi + \mu_4$$

where θ is drawn from a uniform distribution in $[0, 2\pi]$ and $\mu_i, i = 1, ..., 4$ are drawn from the zero mean Gaussian distribution $N(0, 0.1)$. Equations (11.1) and (11.2) define a circular manifold in the two-dimensional input space while (11.3) and (11.4) define a linear manifold within the input space, where each manifold is only approximate due to the presence of noise ($\mu_i, i = 1, ..., 4$). The subtraction of π in the linear manifold equations is merely to centre the data.

In Chapter 14, we create two sets of two-dimensional artificial data which are known to have a correlation using $x_1(t) = \sin(t) + \mu_1, y_1(t) = \cos(t) + \mu_2, x_2(t) = t + \mu_3, y_2(t) = \frac{t}{3} + \sin(t) + \mu_4$, where t is drawn from a uniform distribution in $[0, 2\pi]$ and $\mu_i \sim N(0, 0.2)$ is Gaussian noise. Examples of this data are shown in the top row of Fig. 14.1.

A similar data set is shown in the top line of Fig. 14.3: it comprises two sets of two-dimensional artificial data from $x_1(t) = \sin(t) + \mu_1, y_1(t) = \cos(t) + \mu_2, x_2(t) = t + \mu_3$, if $t \in [0, 2\pi]$ else $x_2(t) = (4\pi - t) + \mu_3, y_2(t) = \frac{t}{3} + \sin(t) + \mu_4$ if $t \in [0, 2\pi]$ else $y_2(t) = \frac{4\pi - t}{3} + \sin(4\pi - t) + \mu_4$ where t is drawn from a uniform distribution in $[0, 4\pi]$ and $\mu_i \sim N(0, 0.2)$ is Gaussian noise.

E.2 Real Data Sets

E.2.1 Wine Data

This is a data set which may be retrieved from the UCI Repository of machine learning databases and is based on wine: it is a 13-dimensional data set of 178 samples from three different classes of wine. Extensive experimentation (Lina Petrakieva, personal communication) has shown that it is almost one-dimensional (the first principal component contains the overwhelming majority of the variance).

E.2.2 Astronomical Data

This is a remote sensing data set, the 65-colour spectra of 115 asteroids used by [130]. The data set is composed of a mixture of the 52-colour survey by Bell et al. [11] together with the 8-colour survey conducted by Zellner et al. [187] providing a set of asteroid spectra spanning 0.3–2.5μm. A more detailed description of the data set is given in [130].

E.2.3 The Cetin Data Set

This is an artificial data set constructed by Cetin [23] which consists of 16 sets of spectra in which each set contains 32×32 samples arranged in a chequerboard grid. It was used for comparative study of a variety of algorithms. The spectra themselves are samples taken from a spectral library.

E.2.4 Exam Data

In Chapter 9, we use a data set reported in [125], page 290; it comprises 88 students' marks on five module exams. The exam results can be partitioned into two data sets: two exams were given as closed-book exams (C) while the other three were open-book exams (O). The exams were on the subjects of Mechanics(C), Vectors(C), Algebra(O), Analysis(O), and Statistics(O). We thus split the five variables (exam marks) into two sets—the closed-book exams (x_{11}, x_{12}) and the open-book exams (x_{21}, x_{22}, x_{23}).

E.2.5 Children's Gait Data

The children's gait data has been used in [116] and was collected by the Motion Analysis Laboratory at the Children's Hospital, San Diego, California, (full details in [146]). The data set consist of the angular rotations in the sagittal plane of the hip and knee of 39 normal five-year-old children. The observations are taken over a gait cycle consisting of one double step taken by each child, and time is measured in terms of the cycle which has been discretized to a regular grid of 20 points; the data are illustrated in Fig. 13.1.

E.2.6 Speech Signals

In Chapters 3, 8 and 12, we use three speech signals (each speaker said "perhaps the most frequent use of ICA is in the extraction of independent causes") and mix them linearly. In Chapter 3, we used an ill-conditioned mixing matrix $A = \begin{pmatrix} 0.8 & 0.4 \\ 0.9 & 0.4 \end{pmatrix}$ whose determinant is -0.04. Its eigenvalues are 1.2325 and -0.0325. In Chapter 8, we linearly mixed using the mixing matrix A:

$$A = \begin{pmatrix} -0.3 & 0.4 & 0.2 \\ 0.6 & -0.9 & 0.4 \\ -0.5 & 0.5 & -0.3 \end{pmatrix} \tag{E.2}$$

The kurtosis of the individual signals is 6.8444, 7.7582 and 3.6833 respectively.

In Chapter 6, five samples of five seconds of natural speech was recorded using the standard telecom sampling rate of 8 kHz. Two adult male and female voices were used along with that of a female child. The speakers each spoke their name and a six-digit number. The samples were then linearly mixed using the 5×5 mixing matrix shown in Table 6.2, which is well conditioned with a determinant value of 1.42. The fourth-order statistics of the original signals are shown in Table 6.3.

In Chapter 7, we used raw voice data sampled at 8 kHz and subjected to no preprocessing. We isolated single vowel sections and trained the network only on these vowel sections. Also in this chapter, we used artificial sound data containing a range of frequencies between 200 Hz and 10 000 Hz. This data was high-pass filtered by taking the differences between successive samples. The training vector consisted of 64 consecutive samples of sound recorded at 8 kHz giving an input window duration of 8.75 ms. The one-dimensional map consisted of 24 modules, each of which had a subspace dimensionality of two. Eight sets of inputs that were adjacent in time formed a single episode. Representative winners were selected using the maximum energy criterion.

Each input window was multiplied by a Gaussian weighting function with a full-width half-maximum, FWHM(t), of eight samples and this was increased to 20 during training. The initial learning rate was 0.008 and was reduced by 0.0005 after every 20 episodes.

E.2.7 Bank Database

In Chapter 6, we use a small database of bank customers, containing 1000 records, with 12 fields. A few records are shown in Fig. 6.7. Information stored in the database includes a unique identifier, age, sex, salary, type of area in which they live, whether married or not, number of children and then several fields of financial information such as type of bank account, whether they own a Personal Equity Plan[1], etc.

E.2.8 World Bank Data

The world bank data set consists of entries for 124 countries, with fields for the countries' name, gross national product (GNP), percentage growth, GNP per capita plus the percentage growth and finally GNP PPP (productivity per person) and percentage growth of GNP PPP. We first used the EPP network of Chapter 6 on the data set.

E.2.9 Exchange Rate Data

We use the U.S. dollar against the British Pound and split the data set into two sets: 1706 samples were used as the training data and 1706 for the test data. Each training input comprised a particular day's exchange rate plus the previous n days' exchange rates where values of n ranged from 5 to 25.

E.2.10 Power Load Data

In Chapter 11, we use a data set from the Taiwan Power Company; for the supervised networks two years (1992 and 1993) were used to train the neural network, the 1994 data was used for validating, and the 1995 data has been used for the practical test. The results are shown in Table 11.3. Table 11.2 shows the 12 values which are used as input to the neural networks.

E.2.11 Image Data

In Chapter 11, we use the Columbia Object Image Library (COIL-20) which is a database of grayscale image of 20 objects (see Fig. 11.13). The objects were placed on a motorized turntable (see Fig. 11.14) against a black background. The turntable was rotated through 360 degrees to vary object pose with respect to a fixed camera. Images of the objects were taken at pose intervals of 5 degrees. This corresponds to 72 images per object. The images used have been size normalized. The object is clipped out from the black background using

---

[1] Personal Equity Plans were a savings scheme devised by the British government to encourage savers to invest. They are no longer available to purchase but continue to exist as historical savings vehicles.

a rectangular bounding box. The bounding box is resized to 128×128 using interpolation decimation filters to minimize aliasing [136]. When resizing, the aspect ratio is preserved.

In Chapter 12, natural images depicting landscapes, trees, people, etc. are used. In order to reduce complexity, the images are typically divided into patches taken from random positions, and PCA is performed on these patches. Fig. 12.6 shows the results of this experiment, using 12×12 pixels image-patches taken from nine grayscale images. In these experiments, the visual data is first mean centred and then preprocessed, using a filter with frequency response $R(f) = f \exp(-(f/f_0)^4)$. This is a widely used whitening/low-pass filter that ensures the Fourier amplitude spectrum of the images is flattened and which also decreases the effect of noise by eliminating the highest frequencies. We randomly sampled the preprocessed images by taking 12×12 pixel patches, on which FastICA with 100 outputs was used.

When we create artificial stereo images, we also use the above images but we also use real stereo images for a two stream sparse coding experiment. We selected 14 natural stereo images (downloadable from http://www.undersea3d.com/), which were preprocessed as above. An example of an unpreprocessed stereo image pair used is displayed in Fig. 12.11. The images were sampled by randomly taking 5000 12×12 patches from each image.

References

1. J. Ambrose-Ingerson, R. Granger and G. Lynch. Simulation of paleocortex performs hierarchical clustering. *Science*, 247:1344–1348, 1990.
2. J.A. Anderson. A memory model using spatial correlation functions. *Kybernetik*, 5:113–119, 1968.
3. H. Attias. Independent factor analysis. *Neural Computation*, 11:803–851, 1999.
4. P. Baldi and K. Hornik. Neural networks and principal component analysis: Learning from examples without local minima. *Neural Networks*, 2:53–58, 1988.
5. H.B. Barlow. Unsupervised learning. *Neural Computation*, 1:295–311, 1989.
6. E. Bates and J. Elman. Connectionism and the study of change. Technical Report 9202, Centre for Research in Language, University of California, San Diego, CA 92093-0526, 1992.
7. S. Becker. *An Information-theoretic Unsupervised Learning Algorithm for Neural Networks*. PhD thesis, University of Toronto, 1992.
8. S. Becker. Mutual information maximization: Models of cortical self-orgainization. *Network: Computation in Neural Systems*, 7:7–31, 1996.
9. A.J. Bell and T.J. Sejnowski. Fast blind separation based on information theory. In *Proceedings of 1995 International Symposium on Nonlinear Theory and Applications, NOLTA-95*, volume 1, pages 43–47, 1995.
10. A.J. Bell and T.J. Sejnowski. An information maximization approach to blind separation and blind deconvolution. *Neural Computation*, 7:1129–1159, 1995.
11. J.F. Bell, P.D. Owensby, B.R. Hawke and M.J. Gaffey. The 52 colour asteroid survey: Final results and interpretation. *Lunar Planet Sci. Conf, XiX*, 57, 1988.
12. E.L. Bienenstock, L.N. Cooper, and P.W. Munro. Theory for the development of neuron selectivity: Orientation specificity and binocular interaction in the visual cortex. *The Journal of Neuroscience*, 2(1):32–48, Jan 1982.
13. C. Bishop. *Neural Networks for Pattern Recognition*. Oxford:Clarendon Press, 1995.
14. C. Bishop, M. Svensen, and C.K.I. Williams. GTM: A principled alternative to the self-organizing map. In *Proceedings of the 1996 International Conference on Artificial Neural Networks, ICANN96*, pages 165–170. Springer-Verlag, 1996.
15. R.J. Bolton, D.J. Hand, and A.R. Webb. Projection techniques for nonlinear principal component analysis. *Statistics and Computing*, 13:267–276, 2003.

16. M. Borga. *Reinforcement Learning Using Local Adaptive Models*. PhD thesis, Linkoping University, Sweden, Thesis 507, ISBN 91-7871-590-3, 1998.

17. R.W. Brause. A symmetrical lateral inhibition network for pca and feature decorrelation. In S. Gielen and B. Kappen, eds., *ICANN93*, pages 486–490. Springer-Verlag, 1993.

18. R.W. Brause. Transform coding by lateral inhibited neural nets. In *Proceedings of IEEE Tools with Artificial Intelligence*, 1993.

19. L. Breimen. Bagging predictors. *Machine Learning*, (24):123–140, 1996.

20. T.H. Brown and V. Chatterji. Hebbian Synaptic Plasticity *in The Handbook of Brain Theory and Neural Networks*, MIT Press, 1995.

21. A. Carlson. Anti-Hebbian learning in a nonlinear neural network. *Biological Cybernetics*, 64:171–176, 1990.

22. G.A. Carpenter. Neural network models for pattern recognition and associative memory. *Neural Networks*, 2:243–257, 1989.

23. H. Cetin and D.W. Levandowski. Interactive classification and mapping of multi-dimensional remotely sensed data using n-dimensional probability density functions. *Photogrammetric Engineering and Remote Sensing*, 57(12):1579–1587, 1991.

24. D. Charles. *Unsupervised Artificial Neural Networks for the Identification of Multiple Causes in Data*. PhD thesis, University of Paisley, 1999.

25. D. Charles and C. Fyfe. Modelling multiple cause structure using rectification constraints. *Network: Computation in Neural Systems*, 9:167–182, May 1998.

26. D. Charles, C. Fyfe, D. MacDonald and J. Koetsier. Unsupervised neural networks for the identification of minimum overcomplete basis in visual data. *Neurocomputing*, 47:119–143, August 2002.

27. S.J. Chuang, C. Fyfe and R.C. Hwang. Power load forecasting by neural network with weights-limited training algorithm. In *European Simulation Symposium and Exhibition, ESS98*, 1998.

28. M.A. Cohen, S. Grossberg and D.G. Stork. chapter Speech Perception and Production by a Self-organizing Neural Network, *in Evolution, Learning, Cognition, and Advanced Architectures*, pages 217–231. World Scientific Publishing, 1988.

29. E. Corchado. *Early Vision and Artificial Neural Networks*. PhD thesis, Universidad de Salamanca, 2002.

30. E. Corchado and C. Fyfe. Maximum likelihood hebbian learning. In *Tenth European Symposium on Artificial Neural Networks, ESANN2002*, 2002.

31. T.M. Cover and J.A. Thomas. *Elements of Information Theory*. Wiley-Interscience, 1991.

32. D. DeMers and G. Cottrell. Nonlinear dimensionality reduction. Neuroprose ftp site, Feb 1993.

33. P. Dayan and G.E. Hinton. Varieties of Helmholz machine. *Neural Networks*, 9:1385–1403, 1996.

34. P. Dayan, G.E. Hinton, R.M. Neal, and R.S. Zemel. The Helmholz machine. *Neural Computation*, 7:889–904, 1995.

35. P. Dayan and R.S. Zemel. Competition and multiple cause models. *Neural Computation*, 7:565–579, 1995.

36. P. Delicado. Another look at principal curves and surfaces. *Journal of Multivariate Analysis*, 77:84–116, 2001.

37. S. Dennis and J. Wiles. Integrating learning into models of human memory: the Hebbian recurrent network. Technical report, University of Queensland, 1993.

38. S. Dennis, J. Wiles, and M. Humphries. What does the environment look like? Setting the scene for interactive models of human memory. Technical Report 249, University of Queensland, 1992.

39. P. Diaconis and D. Freedman. Asymptotics of graphical projections. *The Annals of Statistics*, 12(3):793–815, 1984.

40. K. Diamantaras and S.Y. Kung. Multilayer neural networks for reduced rank approximation. *IEEE Transactions on Neural Networks*, 5(5):684–697, 1994.

41. K.I. Diamantaras. *Principal Component Learning Networks and Applications*. PhD thesis, Princeton University, 1992.

42. R.O. Duda, P.E. Hart, and D.G. Stork. *Pattern Classification*. Wiley, (second edition), 2001.

43. J. Elman. Incremental learning, or the importance of starting small. Technical Report 9101, University of California, San Diego, CA, March 1991.

44. J. Elman. Distributed representations, simple recurrent networks, and grammatical structure. *Machine Learning*, 1992.

45. D.J. Field. What is the goal of sensory coding. *Neural Computation*, 6(4):559–601, 1994.

46. P. Földiák. *Models of Sensory Coding*. PhD thesis, University of Cambridge, 1992.

47. Y. Fregnac. chapter Hebbian Synaptic Plasticity: Comparative and Developmental Aspects, *in The Handbook of Brain Theory and Neural Networks*, MIT Press, 1995.

48. B.J. Frey. *Graphical Models for Machine Learning and Digital Communication*. MIT Press, 1998.

49. J. Friedman, T. Hastie and R. Tibshirani. Additive logistic regression: a statistical view of boosting, Technical report, Statistics Dept., Stanford University, 1998.

50. J.H. Friedman. Exploratory projection pursuit. *Journal of the American Statistical Association*, 82(397):249–266, March 1987.

51. C. Fyfe. Introducing asymmetry into interneuron learning. *Neural Computation*, 7(6):1167–1181, 1995.

52. C. Fyfe. *Negative Feedback as an Organising Principle for Artificial Neural Networks*. PhD thesis, University of Strathclyde, 1995.

53. C. Fyfe. A comparative study of two neural methods of exploratory projection pursuit. *Neural Networks*, 10(2):257–262, 1997.

54. C. Fyfe and R. Baddeley. Nonlinear data structure extraction using simple Hebbian networks. *Biological Cybernetics*, 72(6):533–541, 1995.

55. C. Fyfe and D. MacDonald. Epsilon-insensitive Hebbian learning. *Neurocomputing*, 47:35–57, 2002.

56. C. Fyfe, D. MacDonald, P.L. Lai, R. Rosipal and D. Charles. chapter Unsupervised Learning using Radial Kernels, *in Recent Advances in Radial Basis Networks*,pages 193–218. Physica Verlag, 2000.

57. P. Geladi and B. Kowalski. Partial least squares regression: a tutorial. *Anal. Chim. Acta*, 185:1–58, 1986.

58. M. Girolami. *Self-organising Neural Networks for Signal Separation*. PhD thesis, University of Paisley, 1998.

59. M. Girolami. *Self-organising Neural Networks: Independent Component Analysis and Blind Signal Separation*. Springer-Verlag, 1999.

60. Z.K. Gou. *Canonical Correlation Analysis and Artificial Neural Networks*. PhD thesis, University of Paisley, 2003.

61. S. Grossberg. Some nonlinear networks capable of learning a spatial pattern of arbitrary complexity. *Proceedings of the National Academy of Sciences (USA)*, 59:368–372, 1968.

62. S. Grossberg. Unitization, automaticity, temporal order, and word recogntion. *Cognition and Brain Theory*, 7:263–283, 1984.

63. S. Grossberg. Nonlinear neural networks: principles, mechanisms and architectures. *Neural Networks*, 1:17–61, 1988.

64. S. Grossberg and N.A. Schmajuk. Neural dynamics of adaptive timing and temporal discrimination during associative learning. *Neural Networks*, (2):79–102, 1989.

65. P. Hall. On polynomial-based projection indices for exploratory projection pursuit. *The Annals of Statistics*, 17(2):589–605, 1989.

66. Y. Han. Filter formation using artificial neural networks. Master's thesis, University of Paisley, 2001.

67. Y. Han. *Analysing Times Series using Artificial Neural Networks*. PhD thesis, University of Paisley, 2004.

68. Y. Han, E. Corchado and C. Fyfe. Forecasting using twinned principal curves and twinned self-organising maps. *Neurocomputing*, (57):37–47, 2004.

69. Y. Han and C. Fyfe. Finding underlying factors in time series. *Cybernetics and Systems: An International Journal*, 33:297–323, March 2002.

70. T. Hastie and W. Stuetzle. Principal curves. *Journal of the American Statistical Association*, 84(406), June 1989.

71. S. Haykin. *Neural Networks – A Comprehensive Foundation*. Macmillan, 1994.

72. D.O. Hebb. *The Organisation of Behaviour*. Wiley, 1949.

73. J. Hertz, A. Krogh and R.G. Palmer. *Introduction to the Theory of Neural Computation*. Addison-Wesley Publishing, 1992.

74. G.E. Hinton, P. Dayan, B.J. Frey and R.M. Neal. The "wake-sleep" algorithm for unsupervised neural networks. *Science*, 268:1158–1161, 1995.

75. G.E. Hinton, P. Dayan and M. Revow. Modelling the manifolds of images of handwritten digits. *IEEE Transactions on Neural Networks*, 8(1):65–74, Jan 1997.

76. G.E. Hinton and Z. Ghahramani. Generative models for discovering sparse distributed representations. *Philosophical Transactions of the Royal Society, B*, 352:1177–1190, 1997.

77. G.E. Hinton and S.J. Nowlan. The bootstrap Widrow-Hoff rule as a cluster-formation algorithm. *Neural Computation*, 2(3):355–362, Fall 1990.

78. G.E. Hinton and T. Shallice. Lesioning an attractor network: Investigations of acquired dyslexia. *Psychological Review*, 98(1):74–95, 1991.

79. J. Hopfield. Neural networks and physical systems with collective computational abilities. *Proceedings of the National Academy of Sciences, USA*, 79:2554–2558, 1982.

80. R.L. Horswell and S.W. Looney. A comparison of tests for multivariate normality that are based on measures of multivariate skewness and kurtosis. *Journal of Statistical Computing Simulations*, 42:21–38, 1992.

81. P.O. Hoyer and A. Hyvarinen. Independent component analysis applied to feature extraction from color and stereo images. *Network: Computation in Neural Systems*, 11(3):191–210, 2000.

82. S.C. Huang and Y.F. Huang. Principal component vector quantization. *Journal of Visual Communication and Image Representation*, 4(2):112–120, June 1993.

83. D.H. Hubel and T.N. Wiesel. Receptive fields, binocular interaction and functional architecture in the cats visual cortex. *Journal of Physiology (London)*, 160:106–154, 1962.

84. P.J. Huber. Projection pursuit. *Annals of Statistics*, 13:435–475, 1985.

85. A. Hyvarinen, J. Karhunen, and E. Oja. *Independent Component Analysis*. Wiley, 2001.

86. N. Intrator. A neural network for feature extraction. In *NIPS 2*, pages 719–726. Morgan Kaufman, 1990.

87. N. Intrator. Feature extraction using an unsupervised neural network. *Neural Computation*, 4(1):98–107, January 1992.

88. N. Intrator. Combining exploratory projection pursuit and projection pursuit regression with application to neural networks. *Neural Computation*, 5:443–455, 1993.

89. N. Intrator. On the use of projection pursuit constraints for training neural networks. In B. Spatz, ed., *NIPS 5*, pages 3–10, Morgan Kaufmann, 1993.

90. N. Intrator and L.N. Cooper. Objective function formulation of the BCM theory of visual cortical plasticity: Statistical connections, stability conditions. *Neural Networks*, 5:3–17, 1992.

91. N. Intrator and J.I. Gold. Three-dimensional object recognition using an unsupervised BCM network: The usefulness of distinguishing features. *Neural Computation*, 5:61–74, 1993.

92. R.A. Jacobs, M.I. Jordan, S.J. Nowlan and G.E. Hinton. Adaptive mixtures of local experts. *Neural Computation*, 3:79–87, 1991.

93. I.T. Jolliffe. *Principal Component Analysis*. Springer-Verlag, 1986.

94. M.C. Jones and R. Sibson. What is projection pursuit? *Journal of The Royal Statistical Society*, pages 1–37, 1987.

95. H. Jonker. *Information Processing and Self-Organisation in Neural Networks with Inhibitory Feedback*. PhD Thesis, 1992.

96. M.I. Jordan and R.A. Jacobs. Hierarchical mixtures of experts and the EM algorithm. *Neural Computation*, 6:181–214, 1994.

97. C. Jutten and J. Herault. Blind separation of sources, part 1: An adaptive algorithm based on neuromimetic architecture. *Signal Processing*, 24:1–10, 1991.

98. N. Kambhatla and T. Leen. Dimension reduction by local PCA. *Neural Computation*, 9, 1997.

99. J. Karhunen and J. Joutsensalo. Nonlinear Hebbian algorithms for sinusoidal frequency estimation. In I. Aleksander and J. Taylor, eds., *Artificial Neural Networks, 2*, pages 1099–1103. North-Holland, 1992.

100. J. Karhunen and J. Joutsensalo. Nonlinear generalizations of principal component learning algorithms. In *IJCNN'93*, pages 2599–2602, 1993.

101. J. Karhunen and J. Joutsensalo. Representation and separation of signals using nonlinear PCA type learning. *Neural Networks*, 7(1):113–127, 1994.

102. J. Karhunen and J. Joutsensalo. Generalizations of principal component analysis, optimisation problems, and neural networks. *Neural Networks*, 1995.

103. J. Karhunen and J. Joutsensalo. Learning of robust principal component subspace. In *1993 International Joint Conference on Neural Networks*, pages 2409–2412, 1993.

104. J. Kay and W.A. Phillips. Activation functions, computational goals and learning rules for local processors with contextual guidance. Technical Report

CCCN-15, Centre for Cognitive and Computational Neuroscience, University of Stirling, April 1994.

105. J. Kay and W.A. Phillips. Activation functions, computational goals and learning rules for local processors with contextual guiding. *Neural Computation*, 9(4), 1997.

106. B. Kegl, A. Krzyzak, T. Linder and K. Zeger. Learning and design of principal curves. *IEEE Transactions on Pattern Analysis and Machine Intelligence*, 22(3):281–297, 2000.

107. A. Kehagias. Stochastic recurrent networks: Prediction and classification of time series. Neuroprose ftp site, March 1991.

108. J. Koetsier. *Context Assisted Learning in Artificial Neural Networks*. PhD thesis, University of Paisley, 2003.

109. T. Kohonen. An adaptive associative memory principle. *IEEE Transactions on Computers*, C-23:444–445, 1974.

110. T. Kohonen. *Self-Organization and Associative Memory*. Springer-Verlag, 1984.

111. T. Kohonen. *Self-Organising Maps*. Springer, 1995.

112. B. Kosko. chapter Adaptive Bidirectional Associative Memories, *in Pattern Recognition by Self-Organising Neural Nets*,pages 207–236. 1991.

113. P.L. Lai. *Neural Implementations of Canonical Correlation Analysis*. PhD thesis, University of Paisley, 2000.

114. P.L. Lai and C. Fyfe. A neural network implementation of canonical correlation analysis. *Neural Networks*, 12(10):1391–1397, 1999.

115. P.L. Lai and C. Fyfe. Kernel and nonlinear canonical correlation analysis. *International Journal of Neural Systems*, 10(5):365–377, 2001.

116. S.E. Leurgans, R.A. Moyeed and B.W. Silverman. Canonical correlation analysis when the data are curves. *Journal of the Royal Statistical Society, B*, 55(2), 1993.

117. A. Levin. Optimal dimensionality reduction using Hebbian learning. In *Connectionist Summer School*, pages 141–144. 1990.

118. E. Linsker. Self-organization in a perceptual network. *IEE Computer*, pages 105–117, March 1988.

119. R. Linsker. From basic network principles to neural architecture. In *Proceedings of National Academy of Sciences*, 1986.

120. S. Lipschutz. *Theory and Problems of Linear Algebra*. McGraw-Hill, 1968.

121. D. MacDonald. *Unsupervised Neural Networks for Visualisation of Data*. PhD thesis, University of Paisley, 2001.

122. D. MacDonald and C. Fyfe. Data mining using unsupervised neural networks. In *The Third International Conference on Soft Computing, SOCO99*, 1999.

123. D. MacDonald, S. McGlinchey, J. Kawala and C. Fyfe. Comparison of Kohonen, scale-invariant and gtm self-organising maps for interpretation of spectral data. In *Seventh International Symposium on Artificial Neural Networks, ESANN99*, pages 117–122, 1999.

124. C. von der Malsburg. Self-organization of orientation sensitive cells in the striate cortex. *Kybernetik*, 14:85–100, 1973.

125. K.V. Mardia, J.T. Kent and J.M. Bibby. *Multivariate Analysis*. Academic Press, 1979.

126. J. McClelland, D.E. Rumelhart and The PDP Research Group. *Parallel Distributed Processing*, volume 1 and 2. MIT Press, 1986.

127. S. McGlinchey. *Transformation-Invariant Toplogy Preserving Maps.* PhD thesis, University of Paisley, 2000.
128. S. McGlinchey, D. Charles, P.L. Lai and C. Fyfe. Unsupervised extraction of structural information from high dimensional visual data. *Applied Intelligence,* 12(1/2):63–74, 2000.
129. J.M. Mendel. *A Prelude to Neural Networks: Adaptive and Learning Systems.* Prentice Hall, 1994. (Ed).
130. E. Merenyi. Self-organising anns for planetary surface composition research. *Journal of Geophysical Research,* 99(E5):10847–10865, 1994.
131. S. Mika, B. Scholkopf, A. Smola, K.-R. Muller, M. Scholz and G. Ratsch. Kernel PCA and de-noising in feature spaces. In *Advances in Neural Processing Systems, 11,* pages 536–542, 1999.
132. K. Miller and D. MacKay. The role of constraints in Hebbian learning. *Neural Computation,* 1992.
133. R. Möller. A self-stabilizing learning rule for minor component analysis. *International Journal of Neural Systems,* 14(1):1–8, 2004.
134. F. Mulier and V. Cherkassky. Self-organisation as an iterative kernel smoothing process. *Neural Computation,* 6(6):1165–1177, 1995.
135. P.C. Murphy and A.M. Sillito. Corticofugal feedback influences the generation of length tuning in the visual pathway. *Nature,* 329:727–729, 1987.
136. S. Nene, S. Nayer and H. Murase. Columbia object image library (coil-20). Technical Report CUCS-006-96, Columbia University, 1996.
137. E. Oja. A simplified neuron model as a principal component analyser. *Journal of Mathematical Biology,* 16:267–273, 1982.
138. E. Oja. Neural networks, principal components and subspaces. *International Journal of Neural Systems,* 1:61–68, 1989.
139. E. Oja and J. Karhunen. On stochastic approximation of the eigenvectors and eigenvalues of the expectation of a random matrix. *Journal of Mathematical Analysis and Applications,* 106:69–84, 1985.
140. E. Oja and J. Karhunen. Nonlinear PCA: Algorithms and applications. Technical Report A18, Helsinki University of Technology, 1993.
141. E. Oja, H. Ogawa and J. Wangviwattana. PCA in fully parallel neural networks. In I. Aleksander & J. Taylor, eds., *Artificial Neural Networks,2,* 1992.
142. E. Oja, H. Ogawa and J. Wangviwattana. Principal component analysis by homogeneous neural networks, part 1: The weighted subspace criterion. *IEICE Trans. Inf. & Syst.,* E75-D:366–375, May 1992.
143. E. Oja, H. Ogawa and J. Wangviwattana. Principal component analysis by homogeneous neural networks, part 2: Analysis and extensions of the learning algorithms. *IEICE Trans. Inf. & Syst.,* E75-D(3):375–381, May 1992.
144. E. Oja, J. Ogawa and J. Wangviwattana. Learning in nonlinear constrained Hebbian networks. In T. Kohonen, K. Makisara, O. Simula, and J. Kangas, eds., *Artificial Neural Networks,* pages 385–390. Elsevier Science Publishers, 1991.
145. B. Olshausen and D. Field. Sparse coding with an overcomplete basis set: A strategy employed by v1? *Vision Research,* 37:3311–3325, 1997.
146. R.A. Olshen, E.N. Biden, M.P. Wyatt and D.H. Sutherland. Gait analysis and the bootstrap. *Annals of Statistics,* 17:1419–1440, 1989.
147. C.G. Osorio and C Fyfe. Initialising exploratory projection pursuit networks. In *3rd Annual IASTED International Conference on Visualization, Imaging, and Image Processing (VIIP 2003),* 2003. (accepted for publication).

148. S.E. Palmer. *Vision Science, Photons to Phenomenology*. MIT Press, 1999.

149. F. Palmieri. Linear self-association for universal memory and approximation. In *World Congress on Neural Networks*, pages 2–339– 2–343. Lawrence Erblaum Associates, 1993.

150. F. Palmieri, J. Zhu, and C. Chang. Anti-Hebbian learning in topologically constrained linear networks: A tutorial. *IEEE Transactions on Neural Networks*, 4(5):748 – 761, Sept 1993.

151. A.E.C. Pece. Redundancy reduction of a Gabor representation: a possible computational role for feedback from primary visual cortex to lateral geniculate nucleus. In *ICANN92*, pages 865–868. North Holland, 1992.

152. L. Petrakieva and C. Fyfe. Bagging and bumping self-organising maps. In B. Gabrys and A. Nuernberger, eds., *European Symposium on Intelligent Technologies, Hybrid Systems and their implementation on Smart Adaptive Systems, EUNITE2003*, 2003.

153. P. Plumbley. On information theory and unsupervised neural networks. Technical Report CUED/F-INFENG/TR. 78, University of Cambridge, Aug 1991.

154. K.R. Popper. *The Logic of Scientific Discovery*. Harper Torchbooks, 1968.

155. J.O. Ramsay and B.W. Silverman. *Functional Data Analysis*. Springer, 1997.

156. D.A. Robinson. chapter : Why Vusuomotor Systems Don't Like Negative Feedback and How They Avoid It, *in Vision, Brain, and Cooperative Computation*, pages 89–107. MIT Press, 1987.

157. S. Romdhani, S. Gong and A. Psarrou. A multi-view nonlinear active shape model using kernel PCA. In *BMVC99*, 1999.

158. J. Rubner and K. Schulten. Development of feature detectors and self-organisation. *Biological Cybernetics*, 1990.

159. J. Rubner and P. Tavan. A self-organising network for principal component analysis. *Europhysics Letters*, 10(7):693–698, Dec 1989.

160. T.D. Sanger. Analysis of the two-dimensional receptive fields learned by the generalized Hebbian algorithm in response to random input. *Biological Cybernetics*, 63:221–228, 1990.

161. E. Saund. A multiple cause mixture model for unsupervised learning. *Neural Computation*, 7:51–71, 1995.

162. J. Schmidhuber and D Prelinger. Discovering predictable classifications. Technical Report CU-CS-626-92, University of Colorado, 1992.

163. B. Scholkopf, S. Mika, C. Burges, P. Knirsch, K.-R. Muller, G. Ratsch and A. J. Smola. Input space vs feature space in kernel-based methods. *IEEE Transactions on Neural Networks*, 10:1000–1017, 1999.

164. B. Scholkopf, S. Mika, A. Smola, G. Ratsch and K.-R. Muller. Kernel PCA pattern reconstruction via approximate pre-images. In L. Niklasson, M. Boden, and R. Ziemke, editors, *Proceedings of 8th International Conference on Artificial Neural Networks*, pages 147–152. Springer Verlag, 1998.

165. B. Scholkopf, A. Smola and K.-R. Muller. Nonlinear component analysis as a kernel eigenvalue problem. Technical Report 44, Max Planck Institut fur biologische Kybernetik, Dec 1996.

166. B. Scholkopf, A. Smola and K.-R. Muller. Nonlinear component analysis as a kernel eigenvalue problem. *Neural Computation*, 10:1299–1319, 1998.

167. B. Scholkopf, A. Smola and K.-R. Muller. chapter Kernel Principal Component Analysis, *in Support Vector Machines*, pages 327–370. 1999.

168. C. Shannon. A mathematical theory of communication. *Bell System Technical Journal*, 1948.

169. J.L. Shapiro and A. Prugel-Bennett. Unsupervised Hebbian learning and the shape of the neuron activation function. In I. Aleksander and J. Taylor, eds., *Artificial Neural Networks, 2*, pages 179–183. North-Holland, 1992.

170. A.J. Smola, O.L. Mangasarian and B. Scholkopf. Sparse kernel feature analysis. Technical Report 99-04, University of Wiscosin Madison, 1999.

171. A. J. Smola, S. Mika, B. Scholkopf and R.C. Williamson. Regularized principal manifolds. *Machine Learning*, pages 1–28, 2000.

172. A. J. Smola and B. Scholkopf. A tutorial on support vector regression. Technical Report NC2-TR-1998-030, NeuroCOLT2 Technical Report Series, Oct. 1998.

173. K. Steinbuch. Die lernmatrix. *Kybernetik*, (1):36–45, 1961.

174. J.V. Stone. Learning perceptually salient visual parameters using spatiotemporal smoothness constraints. *Neural Computation*, 8(7):1463–1492, 1996.

175. J. Sun. Some practical aspects of exploratory projection pursuit. *SIAM Journal of Scientific Computing*, 14(1):68–79, January 1993.

176. D.J. Tholen. *Asteroid Taxonomy from Cluster Analysis of Photometry*. PhD thesis, University of Arizona, 1984.

177. M. Tipping. The relevance vector machine. In S.A. Solla, T.K. Leen and K.-R. Muller, eds., *Advances in Neural Information Processing Systems, 12*. MIT Press, 2000.

178. V. Vapnik. *The nature of statistical learning theory*. Springer Verlag, New York, 1995.

179. H.D. Vinod. Canonical ridge and econometrics of joint production. *J. Econometrics*, 4:147–166, 1976.

180. L. Wang and J. Karhunen. A unified neural bigradient algorithm for robust PCA and MCA. *International Journal of Neural Systems*, 1995.

181. R.H. White. Competitive Hebbian learning: Algorithm and demonstration. *Neural Networks*, 5:261–275, 1992.

182. R.W. White. Competitive Hebbian learning 2:an introduction. In *World Congress on Neural Nets*, pages 557–560, 1993.

183. R.J. Williams. Feature discovery through error-correcting learning. Technical Report 8501, Institute for Cognitive Science, University of California, San Diego, 1985.

184. D.J. Willshaw, O.P. Buneman and H.C. Longuet-Higgins. Non-holographic associative memory. *Nature*, 222:960–962, 1969.

185. L. Xu. Least mean square error reconstruction principle for self-organizing neural-nets. *Neural Networks*, 6(5):627 – 648, 1993.

186. L. Xu, E. Oja and C.Y. Suen. Modified Hebbian learning for curve and surface fitting. *Neural Networks*, 5:441–457, 1992.

187. B. Zellner, D.J. Tholen and E.F. Tedesco. The eight colour asteroid survey: Results from 589 minor planets. *Icarus*, pages 355–416, 1985.

188. Q. Zhang and Y.-W. Leung. A class of learning algorithms for principal component analysis and minor component analysis. *IEEE Transactions on Neural Networks*, 11(1):200–204, Jan 2000.

189. Y. Zhao. *On Projection Pursuit Learning*. PhD thesis, MIT, 1992.

Index